SCIENCE AND SCIENTISM
IN NINETEENTH-CENTURY EUROPE

Science and Scientism in Nineteenth-Century Europe

RICHARD G. OLSON

UNIVERSITY OF ILLINOIS PRESS
URBANA AND CHICAGO

Manufactured in the United States of America
1 2 3 4 5 C P 5 4 3 2 1

This book is printed on acid-free paper.

Library of Congress Cataloging-in-Publication Data
Olson, Richard, 1940–
Science and scientism in nineteenth-century Europe /
Richard G. Olson.
p. cm.
Includes bibliographical references and index.
ISBN-13 978-0-252-03188-5 (cloth : alk. paper)
ISBN-10 0-252-03188-1 (cloth : alk. paper)
ISBN-13 978-0-252-07433-2 (pbk. : alk. paper)
ISBN-10 0-252-07433-5 (pbk. : alk. paper)
1. Science—Social aspects—Europe—History—
19th century. 2. Europe—Intellectual life—19th century.
3. Science and civilization. I. Title.
Q175.52.E85047 2008
303.48'309409034—dc22 2007005146

To my incredibly supportive wife,

Kathy Collins Olson

CONTENTS

ACKNOWLEDGMENTS

The specific topics discussed in this work have emerged out of a series of undergraduate seminars on Science in Nineteenth-Century European Culture given at Harvey Mudd College over nearly a decade and I owe a great debt to the students in those seminars for focusing my attention. Several colleagues and former students have read all or portions of this material in earlier and usually considerably more expanded versions. Among those, I owe special thanks to Cathy Corder, Marianne de Laet, Charles Salas, Lisa Sullivan, and Andre Wakefield for providing both encouragement and sound advice, some of which I accepted. I have received additional support and excellent advice from Kevin Downing at Greenwood Press, whom I have worked with on other projects, and from Kerry Callahan at the University of Illinois Press, as well as two anonymous reviewers who pushed me to cut the material included to a tolerable length. Kristen Ehrenberger of the University of Illinois Press offered particularly useful comments. Finally, I owe my wife, Kathy Collins Olson, big time, for her unflagging support, unending patience, and ability to make word-processing programs do wonderful things that I could not even imagine.

I have benefited from sabbatical support from Harvey Mudd College and from a fellowship granted by the John Randolph Haynes and Dora Haynes Foundation during the period in which I was working on this project. The University of California Press has graciously granted permission to use figure 7.1, which is reproduced from Peter J. Bowler, *Evolution: The History of an Idea*, rev. ed. (Berkeley: University of California Press, 1986), p. 85.

SCIENCE AND SCIENTISM
IN NINETEENTH-CENTURY EUROPE

Introduction

> To tell everything is to tell too much; one must not tire the mind. . . . The historian heads straight for the general idea through the facts which prove it, stops only to explain better by means of impressive details and to show the end of his trip on the horizon. One feels with him that one is going somewhere and progressing; the narration becomes interesting because the facts are chosen.
>
> —Hippolyte Taine, *Essai sur Tite-Live* (1856)

When I took my first tenure-track teaching job at the opening of Crown College, the first science-oriented college at the University of California at Santa Cruz, one of my initial tasks was to join in planning a required course for all students that dealt with the place of the sciences in Western civilization. That experience has shaped virtually everything I have done since. Today the traditional celebrations of Western civilization have rightly given way to more critical stances. Nevertheless, it is sill important to consider the roles that the natural sciences have played in shaping our values and our understandings of our selves and our social institutions; for these are often hidden or misunderstood.

In three earlier works[1] I have sought to explore the many ways in which ideas, practices, methods, and attitudes that were initially developed in connection with the natural sciences shaped the development of self-consciously scientific approaches to human individuals and societies—from the time of Ancient Greek civilization to the period of the French Revolution. Moreover, I extended that exploration to the shaping of less-formal but pervasive social and political ideologies. I have persistently thought of those early thinkers who sought to bring insights, especially methodological insights, from the natural sciences into the human social domain as engaging in scientism—a term that I intended to indicate the transfer of ideas, practices, attitudes, and methodologies from the context of the study of the natural world (which was assumed to be independent of human needs and expectations) into the study of humans and their social institutions, without imposing any judgment on the legitimacy of such an appropriation. I was, however, consistently warned by knowledgeable colleagues that "scientism" could not have existed prior to the nineteenth century for at least two reasons. First, the term "science" did not achieve its modern meaning until then, so nothing approaching the notion of scientism could have either. Second, even if there had been something that could reasonably be called science before the nineteenth century, it was not until after the French Revolution that science had been invested with substantial cultural authority and power as a consequence of its application to socially important technological innovation. Furthermore, no one would have expected an audience to accept their views because they claimed to be scientific until science was acknowledged to be an important source of authority.

For reasons that will become clear, I have not found either of these claims to be compelling; but in what follows, I finally get to openly and unabashedly explore the phenomenon of scientism as it is more widely construed; for by the beginning of the nineteenth century, first in France, then in Germany, and finally in Britain, the natural sciences were invested with substantial authority by ruling elites and those exercising economic power. Social theorists and those literary and artistic figures who molded the larger public culture continued through the nineteenth century to borrow heavily from developments in the natural sciences in formulating their understandings of humans and their societies. And they probably did so with greater persuasive effect on a much broader audience.

For the purposes of this work, I take science to be: "A cultural institution characterized for each particular time and place by a set of activities and habits of mind aimed at contributing to an organized, universally valid, and testable body of knowledge about phenomena. These characteristics are usually embodied in systems of concepts, rules of procedure, theories, or model investigations that are generally accepted by groups of practitioners—the scientific specialists."[2] Note that the claim that scientists generally *seek* to produce universally valid and testable knowledge implies nothing about whether such knowledge can be achieved—any more than the Christian admonition to love our neighbors as ourselves implies that humans can achieve that state. In both cases, whether the goal is attainable or not, the practices that it has engendered have been of tremendous importance. In addition, note that this definition acknowledges that precisely what counts as scientific may vary over time, so that modern notions of science may differ with respect to allowable practices and attitudes from ancient or medieval science. Yet there are also continuities over time and space that, for example, make it reasonable to think of Aristotle's speculations about local motion as having such a close relationship to modern physics that they can be considered part of a single continuous enterprise.

No definition of a term of general use, like science, is likely to please everyone; for every choice incorporates and/or promotes certain values while it undermines others. The present choice may seem too broad for some, but it allows me to deal in what I hope to be illuminating ways with many topics. Ultimately, its justification depends upon the insights it allows us to develop. One thing should be clear from my definition of science and my long-standing notion of scientism: it is possible, but not necessary, for any attempt to extend natural scientific ideas, practices, and/or attitudes to social phenomena to be scientific as well as scientistic.

One other issue with regard to my use of the term "scientism" is important. For the most part, those who have used this term have used it to condemn certain behaviors either explicitly or implicitly. One of the most famous definitions, for example, comes from to the twentieth-century economist and anti-communist social theorist, F. A. Hayek (1882–1992), who was specifically concerned with nineteenth century socialists' claims to be scientific. Identifying their doctrines as the source of modern scientism, he wrote: "Wherever we are concerned, not with the general spirit of disinterested inquiry but with slavish imitation of the method and language of Science, we speak of *scientism* or *the scientistic* prejudice. . . . It should be noted that, in the sense that we shall use these terms, they describe, of course, an attitude which is decidedly unscientific in the true sense of the word, since it involves a mechanical and uncritical application of habits of thought to fields different from those in which they have been formed. The scientistic as distinguished from the scientific view is not an unprejudiced but a very prejudiced approach which, before it has considered its subject, claims to know what is the most appropriate way of investigating it."[3]

In chapter 2 I will suggest that early socialist considerations regarding the applicability of scientific methods to the study of society were anything but "slavish imitations" or "mechanical and uncritical" applications; but that does not mean that I am a special advocate of early socialist perspectives either. My goal is neither to applaud nor to condemn the attempts of nineteenth-century thinkers to bring methods, concepts, practices, and attitudes from the investigation of the natural world to bear on human activities and institutions. No doubt I have occasionally slipped and allowed my irritation with the arrogance and smugness of some authors or my admiration for the insights of others to show through, but, in the words of the seventeenth-century Dutch philosopher, Baruch Spinoza, "I Have made a[n] . . . effort not to ridicule, not to bewail, not to scorn human actions, but to understand them." I chose in earlier discussions to explore certain features of a variety of natural scientific traditions in order to understand how different earlier social thinkers drew vastly different consequences from different natural scientific traditions. Similarly, in this work I offer some detailed discussions of several nineteenth-century traditions of natural science—associated with a revival of Hippocratic approaches to medicine and with the term "Positivism" in France, with the term *Naturphilosophie* in Germany, and with the investigation of biological descent with modification in Britain—in order to understand something of the variety of natural scientific resources from which nineteenth-century social theorists and opinion makers had to draw. Furthermore, I explore some of the mechanisms by which natural scientific ideas and attitudes became common intellectual currency within different segments of the population in order to understand to whom scientistic arguments could be expected to appeal and why some sciences were more appealing to certain groups than others.

Why Focus on Scientisms?

Everyone who stops to think about it knows that our lives at the beginning of the twenty-first century are dominated, for better or for worse, by technological systems that have been developed through the exploitation of scientific knowledge. It was not always so. At least until the mid seventeenth century, and probably up until the early nineteenth century, technological developments were largely divorced from any search for natural knowledge that was not directed to immediate use. Premodern agricultural techniques, methods of transportation and communication, architecture, mechanical devices, and crafts of all kinds, changed slowly and almost exclusively as a consequence of chance innovations by craft practitioners. But the relationship between science and technological innovation in the modern world has become quite different. Virtually all major technological innovations since the mid nineteenth century can be directly traced to new knowledge in physics, chemistry, biology, or the earth sciences.

Some intellectuals argue that science impacts the lives of ordinary citizens *only* through the technologies in which it is implicated. Thus, for example, the distinguished German social theorist, Jürgen Habermas, argues that the social life-world in which we live is the realm of literary expression rather than of scientific discourse. Citing Aldous Huxley, he defines that world as, "the world in which human beings are born and live and finally die; the world in which they love and hate, in which they experience triumph and humiliation, hope and despair; the world of sufferings and enjoyments, of madness and common sense, of silliness, cunning and wisdom; the world of social pressures and individual impulses, of reason against passion, of instincts and conventions, of shared language and unsharable feelings and sensations."[4] Then he insists that the sciences penetrate the social life-world only "through the technological exploitation of their information." He continues, "Such information is not on the same level as the action-orienting self understanding of social groups. Hence, without mediation, the information content of the sciences cannot be relevant to that part of practical knowledge which gains expression in literature. It can only attain significance through the detour marked by the practical results of technological progress."[5]

Others argue that when we acknowledge that we live in a scientific culture today we mean that both the information and the methods, attitudes, and practices developed in connection with the natural sciences are every bit as important in shaping our mental lives as they are in shaping our material circumstances. The feminist philosopher, Sandra Harding, for example, argues that "scientific rationality has permeated not only the modes of thinking and acting of our public institutions but even the ways we think about the most intimate details of our private lives. . . . If [projects cannot be justified] by appeal to objective, dispassionate, impartial, rational knowledge seeking, then in our culture they cannot be legitimated at all. Neither God nor tradition is privileged with the same credibility as scientific rationality in modern cultures."[6]

It seems to me that Harding is correct in suggesting that our mental habits in the contemporary world are as surely shaped by the sciences as those of people in the medieval world were shaped by religion. That is, according to my definition, they are scientistic. Where I disagree with Harding is in arguing that there is no monolithic "scientific rationality" that shapes our views about a myriad of topics. Instead, there are many different sciences and many different scientific practices, or rationalities, each of which functions in our culture to affect the views of different groups of people for different purposes. Thus, for example, I will argue that communism and modern American conservative ideology, which is grounded in a particular variant of Social Darwinism, are equally scientistic, though they are derived from different nineteenth-century scientific traditions.

It is also the case that the information content of scientific knowledge claims and theories may be as important as the methods and practices of science in shaping our understandings of social issues. The methods, practices, and attitudes that informed Charles Darwin's (1809–82) *Origin of Species* and those that informed *The Descent of Man* were, for example, identical, but the first emphasized individual competition in the struggle for survival, while the second emphasized the roles of in-group cooperation and sexual selection. As a consequence, as we shall see, those who drew primarily from the first in their social thought focused on competition for wealth and power within societies, while those who drew primarily from the second focused on cooperation and on competition for approval.

If we want to be able to make conscious and informed choices among the mental tools and concepts that are available to us with which to think about ourselves and our societies at the beginning of the twenty-first century, we must understand the variety of scientisms through which we think, and acknowledge that many of those scientisms came into existence during the nineteenth century.

Major Themes

In the first chapter I turn to Paris to explore a group of self-styled social scientists who came to be called *Idéologues.* This group rejected the egalitarian social science of their intellectual predecessors. They did so both because of their negative experiences of egalitarianism during the French Revolution and because their new leader, the physician Pierre Cabanis (1757–1808), rejected the analytic approach to science that focused on causal explanations. This causal and analytic approach had worked well in mechanics and had dominated pre-revolutionary French social theory. But during the late eighteenth century, French medical education, especially at Montpellier, had moved away from the seventeenth-century practice of theorizing about the body as a machine in favor of a return to Hippocratic practices and to careful and nonspeculative physiological description, which seemed better able to account for the complex circumstances that govern human diseases. Cabanis brought this new (or revived) way of thinking about illness and disease to Paris and sought to apply it to moral as well as physical phenomena, giving rise to a new scientistic approach to humans and their society, *Idéologie*. Turning against the simplistic egalitarianism and crude utilitarianism (i.e., the contention that humans are incapable of acting except out of direct material self-interest) of their immediate predecessors, the *Idéologues* distinguished between what they considered to be "natural" inequalities based on temperament, age, sex, and so on, on the one hand, and artificial inequalities of opportunity imposed by ill-designed social institutions on the other. The first group was unavoidable and had to be acknowledged and accommodated, whereas the second might be eliminated by reforming social institutions. In addition, the *Idéologues* rejected the purely reactive model of human behavior that seemed implicit in earlier theory in favor of an insistence on the human will to self-motion and self-expression.

The first chapter explores Cabanis' central arguments, then it follows the work of J. B. Say (1767–1832) in working out many of the economic implications of *Idéologie* and the transformation of one branch of *Idéologie* into an apology for extreme individualism at the hands of Destutt de Tracy (1754–1836), the French nobleman and friend of Thomas Jefferson, whose *Eléments d'idéologie* (1801–15) gave its name to the entire movement.

Chapter 2 takes up the story of Henri Saint-Simon (1760–1825) and his immediate followers, who variously called themselves Saint-Simonians or Socialists. An eccentric hanger-on at the *École polytechnique,* which was the leading science and engineering school in the world at the beginning of the nineteenth century, Saint-Simon passed through an *Idéologue* phase before developing a more historically grounded scientism, which he used to become a chief critic of De Tracy's individualism and of the competitive nature of capitalist economics. Beginning around 1812, Saint-Simon was able to gather around himself a group of extremely able young *polytechniciens.* The chapter explores why Saint-Simonianism appealed to able young engineering and medical students and how this group of Saint-Simonians developed a series of doctrines that were foundational to all subsequent European socialisms as well as to the growth of a Continental banking system that emphasized the monopolization of industries and the creation of powerful cartels.

Chapter 3 explores the early growth of an extremely widespread and long-lasting scientistic movement called Positivism at the hands of Auguste Comte (1798–1857). The chapter considers Comte's initial establishment of what he called the fundamental Positivist Law of Three Stages, which claimed that the understanding of every class of phenomena goes through a progression that begins with religions perspectives, moves on to philosophical analyses that are inevitably shaped by metaphysical assumptions, and ends in a "positive," or, scientific, stage in which metaphysics as well as religion are abandoned and what we now call "objective" knowledge can be attained. Chapter 3 goes on to explore Comte's opposition to psychology and his attempt to construct a comprehensive social physics, or "sociology," that would not depend upon an understanding of the individual members of society. This social physics would form the core of most academic social theory in France and the United States through much of the century and would have a significant impact in Britain and Germany. Finally, the chapter discusses the early development of Comte's Religion of Humanity, which was designed to replace earlier, outmoded religions, and which drew a small but extremely influential following, especially in Britain, in colonial India, and in South and Central America.

A form of natural science that was dramatically different from that characterized by Comte and his followers emerged in early nineteenth-century Germany. The growth of this "Romantic" science, which was created out of a complex amalgam of elements that came together in the philosophy of F. W. J. Schelling (1798–1854), is the central topic of the first portions of chapter 4. I explore the spread of Schelling's speculative *Naturphilosophie* into academic work in both physics and physiology in order to display those characteristics that differentiated it from the scientific developments associated with Positivism. I then investigate the close linkages between Romantic natural science and early-nineteenth-century German cultural and political nationalism, which drew from the patterns of organic development posited among *Naturphilosophen* to understand patterns of cultural and national development.

Chapter 5 begins with a discussion of the appropriation of Romantic culture by religiously and politically conservative forces in Germany to reassert the power of the traditional nobility in Germany after the defeat of Napoleon. It explores the consequent turn of young German intellectuals against traditional religion and *Naturphilosophie* and toward either a very liberal anthropology of religion coupled with liberal politics, or an open antagonism toward religion, sometimes coupled with increasingly radical politics. We follow the rise of a formally materialist but uniquely German philosophy with Ludwig Feuerbach (1804–72) and the growth of a massively appealing and aggressively secular scientific materialism linked with liberal politics (materialist philosophies claim that the most fundamental reality is constituted by material, sensible objects, and that ideas are built out of our experiences of these objects). Chapter 5 continues with a brief discussion of a less aggressively comprehensive and politicized version of materialism associated with more pragmatic scientists, and it concludes with a longer discussion of early Marxism, which I view as a deeply scientistic project that rejected scientific materialism in favor of an equally atheistic "dialectical materialism" that harkened back to concepts and practices associated with *Naturphilosophie.* Both scientific materialism and Karl Marx (1818–83) and Friedrich Engels's (1820–95) dialectical materialism were simultaneously products of scientistic trends and immensely important to the growth of both middle-class liberalism and working-class communism on the Continent.

The complex circumstances that shaped the way in which science and scientism entered into British culture are the foci of chapter 6. Unlike the situation in France or Germany, there was almost no governmental support for science or scientific education in Britain at the beginning of the nineteenth century. It thus fell to voluntary organizations and private institutions to promote scientific literacy and activity. The chapter begins with a discussion of the origins of the Mechanics Institute movement and the Society for the Diffusion of Useful Knowledge, both of which promoted scientific knowledge of a certain kind within the labor elite, as well as the establishment of the *Edinburgh Review,* which was the first of a series of widely circulated intellectual journals to spread scientific intelligence within the middle classes.

On the Continent, after the initial Romantic phase, science was often closely identified with challenges to Christian religion. In Britain, on the other hand, while there were some connections between science and religious heterodoxy, at least during the first half of the nineteenth century much of the scientific elite of the nation were orthodox clergy and the general feeling was that scientific activity promoted traditional Christianity rather than challenged it. I explore these connections in chapter 6 to help provide a context for the shock felt by many during the second half of the century as scientific naturalism—often linked to evolutionary scientism—seemed to undermine the old positive relationships between religion and science.

Chapter 7 begins by arguing that by mid century, increasing numbers of British liberal and conservative intellectuals alike had become dissatisfied with the guidance that classical and neoclassical political economy, utilitarian political theory, or the Anglican Church offered for humane public policies. It suggests that many influential thinkers turned to the historically oriented elements of both Comte's Positivism and German thought and began to view the patterns of progressive change seen in theories of Earth history and biological evolution as models for thinking about society. It explores the major developments in geology and evolutionary biology from the time of the French natural historians, George Louis Leclerc, Comte de Buffon (1707–88) and J. B. Lamarck (1744–1829), to those reflected in the anonymously published *Vestiges of the Natural History of the Creation,* which first appeared in London in 1844. Then it concludes by considering Herbert Spencer's (1820–1903) widely admired scientistic attempt to fashion a compelling and comprehensive social theory out of elements of

Positivism, Lamarckian evolutionary doctrines, and the newly emerging science of Energetics.

Charles Darwin's recasting of evolutionary biology in terms of natural selection in *The Origin of Species* of 1859 and in terms of sexual selection and group selection in *The Descent of Man and Selection in Relation to Sex* of 1873 provide the initial subjects of chapter 8. The chapter continues by suggesting that just as Isaac Newton's two major works, the *Principia* and the *Opticks,* led to significantly different patterns of social theorizing, the two major works of Darwin supported dramatically different scientistic approaches to social phenomena and grounded important new patterns for discussing issues connected with race, sex, and mental phenomena.

After exploring the variety of meanings associated with the term "Social Darwinism" in the nineteenth century at the beginning of chapter 9, I consider a selection of evolutionary social theories associated with a wide range of political outlooks, from the most reactionary to the most openly communistic. I explore how each adopted and adapted Lamarckist or Darwinian ideas, shaping them into what became the almost universal language of both domestic and imperialist politics throughout Europe by the last decade of the century.

Finally, in chapter 10, I explore the theme of "degeneration" that arose in part as improving methods of collecting social statistics seemed to indicate that crime, drunkenness, and insanity were increasing dramatically during the second half of the nineteenth century. Coupled with the Lamarckist evolutionary notion that acquired characteristics might be transmitted to future generations and with the pessimistic implications of the second law of thermodynamics, which projected an ultimate "heat death" of the universe as all matter approached thermal equilibrium, the evidence of social collapse generated a near obsession with degeneration. The popular novelist Émile Zola made a career out of a series of novels tracing the gradual degeneration of a French family, and the fear of degeneration stimulated the beginnings of several new social movements, including the temperance movement and the race-hygiene movement, whose common goal was to stave off human degeneration for as long as possible.

The brief conclusion looks forward into the twentieth century to suggest that several of the scientistic movements initiated during the nineteenth century were the direct predecessors of movements that continued into the twentieth. The degenerationist concerns and race-hygiene movement morphed into the powerful eugenics movement of the 1920s and 1930s, which sought to improve the character of the human race. Moreover, certain forms of Social Darwinism remain alive even into the twenty-first century, both within the academic community, especially among evolutionary psychologists, and in popular politics, among conservatives in Europe and America. But the main thrust of the conclusion is to explain why many of the scientistic patterns of thought that dominated

European intellectual life receded in importance during the last decades of the century.

The Temporal Scope of This Work

In spite of the existence of continuities across the beginning of the twentieth century, I have chosen to end the central arguments of this book during the early 1890s rather than to follow the pattern of exploring the "long nineteenth century"—which is commonly held to end with the beginning of the First World War—for three basic reasons. First, as H. Stuart Hughes convincingly argued almost half a century ago in his *Consciousness and Society: The Reorientation of European Social Thought, 1890–1930,*[7] the generation of intellectuals and social theorists who came into prominence in the late 1880s and 1890s, especially Friedrich Nietzsche (1844–1900), Sigmund Freud (1856–1939), Max Weber (1864–1920), and Benedetto Croce (1866–1952), turned away from the relatively simple patterns of nineteenth-century scientism, rejecting Comte's version of Positivism and Marx's version of dialectical materialism, and turned Darwin's transformationist notions inward to explore the unconscious and irrational elements of human behavior. Above all, this new generation of thinkers sought to evade the fatalistic and pessimistic implications of the predominantly materialistic scientisms of the mid nineteenth century. Nietzsche called for a return to a tragic culture in which wisdom and artistic creativity would replace scientific knowledge and in which the indifference of the scientist would be replaced by a passionate engagement "which seeks to grasp, with sympathetic feelings of love, the eternal suffering as its own."[8] And Freud's goal was to bring our hidden personal and cultural traumas into the conscious mind so that intentional choice might replace unconscious reactions in guiding our behaviors.

Second, the 1890s and the first decade of the twentieth century saw a series of revolutions within philosophy and the natural sciences that were ultimately to challenge many of the underlying principles of the dominant nineteenth-century scientisms. Ernst Mach's (1838–1916) phenomenalist philosophy of science, which rejected the use of any variables presumed to underlie the direct sensory experiences of scientists, challenged the traditional claim that scientists seek to discover "truths" about the natural world in favor of the notion that scientific knowledge is always constructed rather than discovered and that its claim to our attention is grounded in its utility rather than its certainty or objectivity. The discovery of x-rays, of natural radioactivity, and of subatomic structure completely transformed the focus of modern physics. The subsequent development of quantum mechanics and relativity undermined many of the foundational concepts of the traditional sciences—including the notions of determinism, space, and time—and once again challenged the confidence that many classical scientists and humanists had in the ability of the natural sciences to comprehend an independently existing reality.[9] In the biological sciences, the discovery of chromatin and subsequently of the genetic basis for inheritance undermined the Lamarckists' notions of progressive organic evolution and challenged those evolutionary social theories that depended on teleological assumptions. Dorothy Ross has summarized the consequent situation at the end of the century particularly clearly: "Through most of the nineteenth century, thinkers in Europe and the United States believed that scientific and historical knowledge would provide synthetic and normative foundations for modern life of the kind that philosophy and religion had traditionally provided. By the end of the century that hope could no longer be sustained. The implications of evolutionary theory and the critique of knowledge mounted by philosophers . . . were destroying the belief that science yielded unequivocal knowledge of nature. . . . In the human sciences, as in every cultural domain, this new uncertainty led to efforts to reconstruct knowledge, value, and representation."[10]

Finally, as western Europe saw increasing challenges to scientistic social thought, the center of scientism shifted to the United States and Latin America. As European intellectuals were turning many of their critical efforts against Positivism, communism, and Social Darwinisms, positivist-based social theories were gaining ground within American academic and popular culture. In spite of Ross's claim that scientism was on its way out in the United States as well as in Europe by the end of the nineteenth century, chairs in Sociology held by self-proclaimed Positivists proliferated in American universities during the last decades of the nineteenth century and the first three decades of the twentieth century.[11] Moreover, when the Columbia geneticist, Herman Muller, wrote in 1910 that, "Science, in the form, especially of psychology and sociology, will discover what qualities are desirable for the most efficient cooperation and for the best enjoyment of life; and science, in the form especially of physiology and genetics . . . will discover what the elementary bases of these qualities are and how to procure them for man,"[12] he was expressing ideas that were widely shared by those in the nascent eugenics, conservation, and scientific management movements, which shaped progressive American politics for at least the first three decades of the twentieth century. Thus, the exploration of scientistic trends after the early 1890s must shift foci from Europe to America.

Dominant Forms of Social Theory at the Beginning of the French Revolution

Early nineteenth-century scientisms were principally reactions against scientifically grounded social theories that had been developed during the seventeenth and eighteenth centuries and that seemed to be implicated in the origins of the French Revolution. For that reason, we set the stage in this section of the introduction by quickly characterizing the social theories that the central characters of the early chapters of this work were reacting against.

When Marie-Jean-Antoine-Nicolas Caritat de Condorcet (1743–94) submitted his *Report on the General Organization of Public Instruction* to the French National Assembly in 1792, he proposed that three professors of the social sciences be attached to each lyceum level *Institut.* One would deal with the analysis of sensations and ideas, particularly the generation of moral ideas, and then move to an analysis of the implication of these ideas for the creation of suitable political structures. This instructor would present the core of what was then called associationist or sensationalist psychology and its extension to utilitarian political theory. A second professor would explore the elements of commerce and political economy and their implications for legislation. The third would deal with what the Scots were calling "philosophical history" or what some Germans were beginning to call anthropology, a subject that emphasized the ways in which different communities produced special customs and patterns of organization in response to the particular environments in which they grew. Condorcet justified this organization on the grounds that it reflected the organization "which has spontaneously developed in the last half century, during which all branches of human knowledge have made rapid progress."[13] While it may be the case that to claim three totally independent traditions of self-consciously scientific social thought in the eighteenth century is to oversimplify the situation because many authors and works combined insights from more than one perspective, Condorcet's categories do seem to offer a useful schema for thinking about Enlightenment social thought, one that contemporaries would have acknowledged.

Of the three traditions discussed by Condorcet, the most important for present purposes—because it was that which had provided the greatest support for revolutionary activity and it was that against which many early-nineteenth-century attempts at developing a new scientific knowledge of society were rebelling—was the first, which began with the analysis of sensations and ideas. The chief spokespersons for eighteenth-century associationist psychology—David Hume (1711–76) and David Hartley (1705–57) in Britain and Etienne Condillac (1714–80) in France—as well as those who tried to derive a preferred political and legal structure from the characteristics of human nature implied by associationism—Claude-Adrienne-Helvetius (1715–71) and Condorcet in France, Jeremy Bentham (1743–1832) and Mary Wollstonecraft (1759–97) in England, and Cesare Beccaria (1738–94) in Italy—all saw themselves as sharing an intellectual heritage. This heritage depended heavily on an openly acknowledged dependence on John Locke's (1632–1704) *Essay Concerning Human Understanding* of 1692, as well as a typically unacknowledged debt to Thomas Hobbes's (1588–1679) *Leviathan,* from which eighteenth-century associationists drew the notion that humans seek to repeat those actions that bring pleasure and to avoid those that bring pain. To the foundational works of Locke and Hobbes were added the two major works of Isaac Newton (1643–1727), his *Principia* of 1687 and his *Opticks,* first published in 1704.

From Locke came the basic claim that because all normal healthy persons possess the same mental apparatus and because all of our ideas come from sensation and our reflection on our sensations, all persons are intrinsically equal. Each person's mind is a blank slate upon which experience writes; what makes people different is nothing but the set of experiences, the education, to which they are exposed. No person or group of persons deserves to be treated differently from any other by virtue of some special innate right of birth. Locke was convinced that all legitimate ways of thinking about those ideas that come from sensation involve a reasoned analysis of the relationship between ideas; but he acknowledged both that different levels of certainty are available for different kinds of knowledge and that some ideas become "associated" with one another unintentionally when they chance to appear together frequently.

While Locke insisted that one should constantly be on guard against the influence of such accidental associations because they are likely to mislead us in our actions, most of the subsequent developers of sensationalist or associationist psychology followed the lead of David Hume and David Hartley in focusing attention on the positive role of associations. Returning to the Hobbesian notion that private happiness is the proper and ultimate end of all our actions and that happiness, or pleasure, and unhappiness, or pain, are linked to certain objects or actions solely through nonrational associations, these thinkers turned associations into the foundation of all morality and turned to an investigation of how and under what circumstances various kinds of association operated to inform human actions, for better or for worse. When they did so, they openly modeled their investigations on the patterns set by Newtonian natural philosophy. Indeed, Hume announced his desire to be the Newton of the social sciences.

Some, led by David Hartley, drew from the openly speculative physical model of nervous and brain function that Newton had offered in Queries 12, 23, and 24 to the fourth edition of his *Opticks.* According to this model—which Hartley adopted as a heuristic device to suggest relationships in the theory of associations—the vibrations of a subtle medullary substance of the brain correspond to the presence of ideas in the mind, so that when these vibrations are caused to change in frequency, direction, place, or intensity by vibrations transmitted through the nerves, our ideas change in some related way. Associations occur because two experiences that occur together frequently will leave vibrations in the brain that are strongly linked to one another. Moreover, simple ideas can combine into complex ones easily because the vibrations of two or more simple ideas can be superimposed upon one another to interfere constructively or destructively in the same way that waves in water do. (If the peaks of two waves occur in the same place at the same time, the resultant wave has an amplitude which is the sum of the two. If, on the other hand, the trough of one wave corresponds to the peak of the other, the wave height at that point and time will be the difference between the two amplitudes.) By positing that greater amplitudes of vibration produce greater pleasure up to the point at which there is a tearing of the medullary substance, which produces pain, Hartley was able to explain—to his own and to a number of important followers' satisfaction—a huge range of mental and moral phenomena. He purported, for example, to be able to explain why intellectual pleasures are likely to be more pleasurable and less painful than sensory ones (the vibrations are more complex and widespread so they have large amplitudes but very few sharp edges that might cause pain), and why the pursuit of "gross" self-interest is likely to be self-defeating (because it deprives us of the vibrations associated with the pleasures linked to being admired by others). One interesting consequence of the vibratory theory was that because vibrations can be summed by superposition, complex ideas also came from a linear combination of simpler ones.

Most associationists, however, including Hume and Condillac, found their inspiration in the more cautious experimental and nonhypothetical emphases of Newton's *Principia.* Drawing especially from the Rules of Right Reasoning in Philosophy at the beginning of Book 3 of the *Principia* and from the final Scholium in which he argued that even though he could not offer a cause for gravity, "I frame no hypothesis; for whatever is not deduced from the phenomena is to be called an hypothesis; and hypotheses, whether metaphysical or physical, whether of occult qualities or mechanical, have no place in experimental philosophy. In this philosophy particular propositions are inferred from the phenomena, and afterwards rendered general by induction. . . . And to us it is enough that gravity really does exist and act according to the laws that we have explained, and abundantly serves to account for all the motions of our celestial bodies and of our sea."[14]

By the same token, Hume argued that although we cannot explain why the mind produces "a secret tie or union among particular ideas," it is nonetheless the case that such ties are empirically established and that "these are the only links that bind . . . or connect us with any person or object exterior to ourselves."[15]

From Newton's first Rule of Right Reasoning, which insisted that, "We are to admit no more causes of natural things than such as are both true and sufficient to explain their appearances . . . for nature is pleased with simplicity, and affects not the pomp of superfluous causes,"[16] virtually all of the sensationalist/associationists drew the notion that the explanations for natural phenomena should be reduced to the greatest possible simplicity. Hume sought "those few simple principles" upon which all human nature is founded,[17] and Condillac was so obsessed with the notion of simplicity that he argued that one should seek to account for human nature with a single principle. He criticized Locke for positing two sources of knowledge—sensation *and* reflection, and directed the whole of his *Treatise of Sensations* to showing that all ideas can be traced to a single source in sensations.

In connection with their notion of simplicity, all sensationalist/associationist psychologists made a fundamental and very conscious assumption about the way in which complex ideas or phenomena were constituted out of simple causal relations. This assumption was most fully articulated in Condillac's *Logic, or the First Development of the Art of Thinking,* but it was implicit or explicit for all, and it involved a recasting of an older mechanical and Cartesian explanatory strategy. According to Condillac, in order to understand any complex idea or phenomenon we must abstract from the complex whole the most salient simple features. "This operation," he says, "is absolutely necessary for limited minds who can consider only a few ideas at a time."[18] Once one has decomposed (analyzed) a complex situation into its component simple parts, then one can re-compose the complex situation by combining the now-understood parts, much as a mechanic can take apart and reassemble a clock or as a seamstress can understand and recreate a dress by tearing apart the seams and then putting together the set of patterned pieces. Generally speaking, advocates of this kind of analysis insisted that simple causes did not interact with one another, so that the consequence of multiple causes acting on some object or person could be discovered by considering the consequences of each cause acting independently and then simply summing up the various effects. When such an assumption was demonstrably untrue—as in the case of calculating the velocity of a body falling through a viscous medium in which the resistance to motion is a function of the velocity—it was assumed that one could consider first the most salient feature of the phenomenon (in this case, the velocity attained in free fall through a vacuum), then add in the complicating factor, assuming as an initial approximation that the velocity was that of free fall, in order to solve the problem in a few iterations.

When social theorists such as Helvetius, Beccaria, and Bentham explored the social and political implications of associationist psychology, they argued that one should design social institutions to maximize the happiness and minimize the unhappiness of the greatest number of members of society. This greatest happiness of the whole would simply be the sum of the individual calculations of happiness and unhappiness for each separate member. Furthermore, the total happiness of each member could be established by adding up those experiences and actions that independently produced pleasure and subtracting those that produced pain, adjusting for a variety of circumstances including duration, frequency, anticipation, immediacy, and so on. Those who sought to imagine how such calculations could be made and used to direct legislation and social institutions came to be called Utilitarians.

While not all liberal to radical social theorists in the period leading up to the French Revolution were full-fledged Utilitarians, virtually all did seek to increase the general welfare, virtually all accepted the intrinsic equality of all persons, and virtually all assumed the necessity of the kind of analytic approach to human nature that Condillac had promoted.

According to many historians, beginning with Edmund Burke (1729–97), whose *Reflections on the Revolution in France* appeared in 1791, the claim of intrinsic equality among humans and the call for reducing the great gaps in wealth between the nobility and the working classes that pervaded political writings grounded in associationist psychology, were among the most important causes of the French Revolution. This revolution, which began with attempts to establish a constitutional monarchy, eventuated in the bloody Terror of 1794, frightening and disturbing even many of those who had favored the early and more moderate stages of the revolution. In reaction, liberal, or moderate, intellectuals sought an understanding of humans and society that could justify a political and social structure that would be more just than the old regime, but that was not founded on assumptions of radical equality among humans. In order to accomplish that goal they had to abandon the causal, simplicity-driven, and analytic model of science drawn from seventeenth- and eighteenth-century mechanics and begin anew using a different scientific model. That new model was promoted by the *Idéologues,* to whose ideas we turn in chapter 1.

PART I

Scientisms in the Aftermath of the French Revolution

CHAPTER 1

Ideology

During the French Revolution, extremists of both the right and left attacked science, scientists, and scientific institutions. Elsewhere I have discussed the largely religious rationale for conservative opposition in the writings of Joseph DeMaistre (1753–1821),[1] but many secular revolutionaries also opposed scientists and scientific institutions, largely because claims to scientific expertise smacked of the elitism and anti-egalitarianism of the old regime.[2] Indeed, in August 1793, the Jacobin-dominated Convention abolished the *Académie des science* along with the other learned academies of France. Moreover, a number of prominent scientists, including the chemist Antoine Lavoisier (1743–94), were attacked for their antidemocratic activities, and Lavoisier was eventually guillotined at least in part because of the enmity of the radical revolutionary, Jean Paul Marat (1743–93), whose work on the chemistry of fire Lavoisier had unfavorably refereed as an academician.[3]

In spite of such attacks, after the Terror and through the Napoleonic aftermath of the revolution, it seems clear that the status and authority of the technically trained intellectual elite became higher than it had ever been in France and that scientific learning became a dominant force in shaping the French intellectual and political leadership. In large part this change in status developed because political power had passed from the hands of a commercially uninterested landholding aristocracy into the hands of a commercially active bourgeoisie. Furthermore, military leadership had passed into the hands of technically trained personnel and out of the hands of aristocrats who provided their own troops. The military successes of the revolutionary armies had been heavily dependent on the work of scientists in establishing reliable munitions supplies and in providing leadership.

Key scientists such as the chemists Jean-Antoine Chaptal (1756–1832) and Claude Berthollet (1748–1822), and the mathematicians Gaspard Monge (1746–1818) and Antoine Fourcroy, as well as key industrialists such as the manufacturers Louis Ternaux (1763–1833) and Jacques-Etienne Montgolfier (1745–99), believed that France could not compete commercially with Britain without a technically trained work force and without science-based technological innovation, and these were the figures who came to power in the Napoleonic aftermath of the revolution. Monge, Berthollet, and Fourcroy were all on the commission established to plan the *École polytechnique,* which was to become the greatest engineering school in the world during the early nineteenth century. They were all also involved in founding the voluntarist *Société d'Encouragement pour l'Industrie Nationale.*[4] Under Chaptal's influence as Napoleonic Minister of the Interior, new educational systems emphasizing applied science were initiated throughout the empire, especially in Belgium, Holland, and parts of Austria, and France instituted policies promoting technological innovation.[5]

By 1812, though Napoleon had become impatient with the arrogance, atheism, and uncompromising attitudes of many within the scientific elite, Henri Saint-Simon (1760–1825) was probably not too far wrong when he boasted, "Such is the difference in this respect between the state of . . . even thirty years ago and that of today that while in those not distant days, if one wanted to know whether a person had received a distinguished education, one asked: 'Does he know his Greek and Latin authors well?' Today one asks: 'Is he good at mathematics? Is he familiar with the achievements of physics, of chemistry, of natural history, in short, of the positive sciences and those of observation?' "[6]

At one level, the domination of French education and intellectual life by the *École polytechnique,* the *École de santé,* the *Institut de France* (the reconstituted *Académie des science*), the Museum of Natural History, and the more problematic *École normale supérieur,* during the first third of the nineteenth century represented the fruition of the dreams of Etienne Condorcet and the eighteenth-century liberal educational reformers associated with him.[7] But at a more fundamental level, the institutions of the early nineteenth century perverted that vision because they failed to build upon the base of extensive and broad-ranging elementary and secondary education that men like Condorcet had envisioned.[8] This situation emerged in part because, though different parties during the revolution and its aftermath were able to agree upon the need for and shape of institutions for higher education in the sciences, medicine, and engineering, they diverged bitterly over broader issues relating to basic education for citizenship.[9] As a consequence, no system of education through the middle level received enough support to be adequately staffed and funded until the Third Republic during the last third of the century.

Under Napoleon, the *École polytechnique* became a training ground for men headed for more advanced work in applied science and engineering at schools like the *École de l'Artillerie et du Genie,* the *École des Ponts and Chaussées,* or the *École vétérinaire,* which produced respectively military engineers, civil engineers, and veterinary surgeons. Even applicants for the *École normale supérieure* after 1808 were often graduates of the *École polytechnique.* To a slightly lesser extent, the *École de santé* (the school of health), which offered a training as narrow in its own domain as the *École polytechnique,* offered the only major alternative to those with modest means who were seeking a high-status career. In effect, basic training in science and engineering had become the gateway to higher careers in the military, in education, in medicine, and in public life; and almost no common background in humanistic learning, philosophy, history, languages, and so on, could be expected of *techniciens* or physicians.

Several French social theorists sought to appropriate both the analytic strengths and the prestige associated with scientific training in order to build broad-ranging visions for society based on scientific knowledge and methods during the first three decades of the nineteenth century. Few of these theorists were themselves graduates of the *École polytechnique,* though one, Auguste Comte (1798–1857), had briefly been a student and later, a part-time examiner in mathematics, there. And none had studied at the Faculty of Medicine in Paris.[10] But their audiences and followers in France were often provincial working-class or lower-middle-class *polytechniciens* or graduates of the *Ecole de santé.* It was the mentality of the engineering student and medical student, largely ignorant of traditional learning, but hungering for more than mere technical facility and open to rational and radical social experimentation, that allowed the scientistic and idiosyncratic visions of men like Henri Saint-Simon and Auguste Comte to generate significant and enthusiastic followings.

The *Idéologues*

Before turning to the socialist and positivist thinkers who fired the imagination and commanded the commitment of many young intellectuals during the early nineteenth century, in this chapter we explore key views of the *Idéologues,* a loosely affiliated group of self-styled social scientists who provided both the foundations for nineteenth-century French liberalism and a transition between the views of pre-revolutionary associationist or sensationalist-psychology-based social theorists and the social prophets of the early nineteenth century. Almost all of those prophets and their followers passed through an *Idéologue* stage before they moved on.

Ideology was a relatively short-lived movement centered in Paris, but it had very substantial impact on subsequent French intellectual life and upon American ideas. In fact, two of the major works of Antoine-Louis-Claude Destutt de Tracy (1754–1836), one of the most prolific of the *Idéologues* and the man who coined the term "*Idéologie,*" were first published in America in translations done or supervised by Thomas Jefferson, and the *Treatise on Political Economy* of the most distinguished *Idéologue* economist, Jean-Baptiste Say (1767–1832), also translated under Jefferson's supervision, dominated the American textbook market in political economy from 1814 to the period of the Civil War.

The nucleus of the group seems to have been the physician, Pierre-Jean-Georges

Cabanis (1757–1808); the army officer, Destutt de Tracy; the political economist, Jean-Baptiste Say; the educational administrator, Dominique-Joseph Garat (1749–1833); and the physician and reformer of mental hospitals, Philippe Pinel (1745–1826). These men were linked together initially by their attendance at the Auteuil Salon of Mme Helvetius, the widow of one of the most prominent early proponents of both associationist psychology and the early phases of the revolution. Most also became members of the Moral and Political Sciences Class of the *Institute de France* when it was established in 1794, and they used the class sessions as platforms for expressing their ideas until Napoleon suppressed the class in 1803.

Cabanis was the first of the group to become fascinated by the social sciences. Moreover, it was his treatment of the physiological grounds of morality that provided methodological and conceptual starting points for the work of Say and De Tracy, the two best-known members of the group, as well as for a broader medico-moralist tradition that spread among Paris medical reformers in the early nineteenth century.[11] Born in 1757 near Brive, Cabanis moved to Paris in 1771. After a brief stint in Poland as secretary to the prince-bishop of Vilna, he returned to Paris in 1775. There he began to study medicine. He was introduced to the salon of Mme Helvetius in 1778 by a family friend, Anne-Robert Turgot (1727–81). There he soon became committed to the great goal of associationist psychology, which was to ground a complete theory of morality in a theory of human psychology different from the psychology of Helvetius.[12]

Like almost all of his politically active friends, Cabanis had strong liberal leanings that emphasized individual rights and laissez-faire economic ideals. He entered the revolutionary period favoring a constitutional monarchy. Through the period of radical Jacobin leadership he sympathized with the moderate Girondins rather than with the *sans-culottes;* and by 1799, he had become so concerned with the need for political stability that he supported Napoleon's coup d'etat, though he broke away from Napoleon before his death in 1808. Unlike many of his liberal and laissez-faire colleagues, however, Cabanis shared Helvetius's belief that because the poor were being artificially suppressed, they deserved a large amount of public assistance, especially in connection with health care and education. Indeed, Cabanis personally provided free medical care for the indigent in the region around Auteuil during the revolution, and his reputation as a friend of the people kept him from being arrested during the Terror, when virtually all of his intellectual associates were imprisoned.

In spite of his sympathy for the poor, Cabanis agreed with most of his contemporary medical colleagues that there was an ineradicable *natural inequality* among men and women that the theories of Condillac, Helvetius, and Condorcet had failed to admit. Cabanis, moreover, could not accept the relatively crude utilitarianism of Helvetius, who had insisted that humans were incapable of acting except out of direct material self-interest. In fact, Cabanis was generally opposed to dogmatic claims that a single cause could account for any phenomenon or class of phenomena, and he was inclined to believe that more complex entities are governed by more complex laws.[13] Since humans are undeniably among the most complex of all beings, they are likely to be governed by the most complex set of laws.

Cabanis and the Hippocratic Model for Studying Humans and Society

The complexity of human beings and their thoughts and emotions brings us to the most important of Cabanis's deviations from his near predecessors among the French associationists. Cabanis was convinced that *different* scientific methods were appropriate to different subject matters. In particular, the form of analysis used in mechanics and promoted as appropriate for the investigation of the genesis of human ideas by Helvetius and Condorcet—a form in which the effects of single causes were isolated and analyzed independent of their context before the effects of complicating factors were considered—seemed to Cabanis totally inappropriate for considering human ideas and behaviors.[14]

According to Cabanis, it was the adoption of such a method that led to the absurd claims of universal human equality.[15] Condillac, Helvetius, and Condorcet had all gone wrong in their attempts to understand the origins of our ideas because they lacked knowledge of physiology and of the centrality of the interaction of different elements in the formation of human ideas and emotions. Cabanis could correct their errors by bringing to bear a method of analysis proper to medicine, or physiology, on the origins of human ideas and feelings and thence, on the foundations of morality.

Whether Cabanis's need to correct and improve the associationist tradition was more powerfully motivated by his need to find a justification for the anti-egalitarian elements of moderate politics or by his primarily professional and intellectual commitment to the methods of Montpellier medicine, as opposed to those of mechanics and mathematics, is virtually impossible to say. It is clear that as he presented his arguments—principally in a series of twelve memoirs presented to the Class of Moral and Political Sciences of the *Institute de France* beginning in 1796 and published as *Rapports du Physique et du Moral de l'Homme* (*On the Relations between the Physical and Moral Aspects of Man*) in 1802—he presumed that moral and political consequences followed from the more fundamental application of appropriate method to the analysis of the physical structure of human beings as modified by key features of their environment.

Cabanis, however, very clearly distinguished between the "path which the mind must follow in its research" on the one hand and "the order it must follow in exposing its works," on the other,[16] so the fact that he presented his ideas as if moral and political imperatives followed logically from medical methods applied to human interactions says little about how he arrived at his moral and political views. When readers approached his works, they had access only to the presentation, which gave priority to the physical and methodological elements in constructing moral and political ideas, so they were virtually forced to respond as if human morality and politics flowed from our physical existence rather than to think of our medical knowledge as an elaborate construct developed to support our moral and political presuppositions or biases.[17] Ironically, as "*Idéologie*" was developed by Cabanis and his followers, especially by Destutt de Tracy, it provided the intellectual tools that suggest that our ideas are often shaped by our biases and that apparently objective and disinterested arguments can produce a "false and blind conscience" that masks rather than reveals the true state of affairs.[18]

Cabanis explored his commitment to a particular methodological approach to medicine in three different places: in *Du degré de certitude de la medicine*, written by 1788, published in 1798, and translated into English as *An Essay on the Certainty of Medicine* in Philadelphia in 1823; in *Coup d'oeil sur les revolutions et sur la reforme de la medicine*, written about 1795, published in 1804, and translated into English as *A Sketch of the Revolutions of Medical Science, and Views Relating to Its Reform* in London in 1806; and in the first of his memoirs to the Institute de France in 1796. Most of his views on method closely followed ideas developed during the eighteenth century at Montpellier,[19] so Cabanis's major achievement was in bringing them to a broad audience interested in moral and political issues and in connecting them with the sensationalist/associationist psychology derived from Locke and Condillac.

Rejecting the reductionist claims of both the followers of Paracelsus (1493–1541), who sought to offer chemical causes for diseases, and the followers of William Harvey (1578–1657), René Descartes (1596–1650), and Giovanni Borelli (1608–79), who sought mechanical causes, Cabanis and the Montpellier school insisted that the ultimate causes of diseases must inevitably remain unknowable. Drawing guidance especially from the descriptive works of Hippocrates (ca. 470 BC–410 BC) and the philosophy of Sir Francis Bacon (1561–1626), they promoted an autonomous and purely phenomenological approach to medicine based on extensive observations of the courses of events associated with illness.

It was not the case that Cabanis rejected the findings of chemistry and mechanics. Indeed, he felt that both had much to offer the physician. What he insisted on was that medicine borrow from the collateral sciences nothing but "what can be transported into it *without hypothesis*."[20] It was clear that some bodily functions could be illuminated using mechanical and chemical knowledge, but what could not be *known* was whether mechanical and/or chemical processes were adequate to account for *all* human phenomena. In fact, Cabanis was very clear in arguing that available evidence suggested that many phenomena, including the critically important phenomena of sensibility (the capacity of nerves to transmit impressions to the mind when touched) and irritability (the capacity of muscles to contract when touched), emerge as characteristics of complex organic structures and are inexplicable by any previously known principles.

The idea that sensibility, irritability, and even thought, might emerge as properties of highly organized matter in a way that is ultimately inaccessible to human knowledge was, of course, not new. It had been explicitly suggested in Locke's *Essay Concerning Human Understanding* and in Julian LaMettrie's (1709–51) *L'Homme machine.*[21] But it was Cabanis's formulation of this notion, along with his extended discussion of why we are driven to accept the claim that organized entities may manifest a variety of ultimately inexplicable properties, that formed the explicit foundation for a series of biological and social applications of the principle. It was to Cabanis, for example, that Xavier Bichat (1771–1802) appealed in his explication of the role of tissues in physiology, that J. B. Lamarck (1744–1829) appealed in formulating his theory of the historical development and complexification of living beings, and that Henri Saint-Simon appealed in arguing for the crucial role of social organization.

Cabanis insisted that the ultimate causes of irritability and sensibility are literally "beyond the reach of the knife and the microscope," that is, they are beyond the capacity of our sensory experiences to determine.[22] Since, for Cabanis, any "cause" that is external to ourselves can only be inferred from its sensory impact on us, any presumed cause that is not accessible to our experiences, "must be excluded from the objects of our research."[23] To fail to exclude all but directly experienced entities, or facts, from scientific discussion was, for Cabanis, to allow "metaphysics" into science, and in a move that was to be extremely important for Auguste Comte and his Positivist followers, Cabanis proudly insisted that in his or in any other truly scientific work, "the reader will not find . . . what was in the past called metaphysics."[24]

With one notable exception to be discussed later, Cabanis insisted that all essentialist notions of causality be abandoned in favor of the correlational notion that he believed had been adopted by Hippocratic physicians, by Bacon, and by David Hume. Indeed it was the works of ancient physicians, especially Hippocrates' *Epidemics* and *Airs, Waters, and Places* that provided the best model for the descriptive approach to science that was alone appropriate to studies of human physiology and behavior:

> A series of phenomena takes place in each ill person. These phenomena are all that is evident and sensible in the illness. Hippocrates applies himself to describing them in striking, ineradicable brush strokes. . . . Each history forms a particular picture: the sex, age, temperament, diet, and profession of the patient are carefully noted. The location of the site, its exposure, the nature of its production, the work of its inhabitants, its temperature, the season of the year, the changes that the air has undergone during the preceding seasons—such are the circumstances that he gathers around his pictures. From them, simple

> rules are born, according to which the illnesses can be divided into general and particular, and the effect of these various circumstances for *causing* the illnesses, determined by easy comparisons and combinations, is expressed through direct and immediate generalizations.[25]

The point of reproducing this passage is to emphasize the huge disparity between the method that Cabanis proposed to follow in dealing with humans and that which Helvetius had insisted upon and that was modeled on Galilean mechanics. Helvetius had pointed out that students of mechanics succeed in solving complex problems by making simplifying assumptions: by assuming the existence of resistanceless fluids and frictionless pulleys, for example. By the same token, he urged those concerned with human laws to "pay no regard to the resistance of prejudices, or the friction of contrary and personal interests, or to manners, laws, and customs already established."[26] In direct contrast to that approach was the Hippocratic approach to illness advocated by Cabanis as a strategy for studying the origin of sensations and passions. The whole point for Hippocrates was to identify and establish the effects of the many circumstances presumed to influence the body and its responses to sickness.

By discovering empirically how both individual circumstances and combinations of two or more circumstances produced differential responses, one could learn to control or manipulate responses to illness in order to bring humans back to health even without knowing the basic causes of sickness. By the same token, according to Cabanis, our ideas and emotions are influenced by a wide variety of conditions. We may not understand the ultimate cause through which sensibility produces our ideas and "ethical affections," but we can demonstrate that and how a wide range of factors influence the process. In particular, we can demonstrate and investigate the influences of age, sex, temperament, disease, diet, climate, instincts, sympathy, sleep, delirium, and many other factors that affect the physical condition of any parts of the body.[27] Once these influences are understood, they can be used to control and manipulate circumstances so as to bring the ideas and passions of any individual "back to the natural order."

"It is certain," wrote Cabanis, "that knowledge of the human organization and of the modifications that temperament, age, sex, climate, and illness can bring to the physical states, throw a singular light on the formation of ideas; that without this knowledge it is impossible to have accurate notions of the way in which the instruments of thought act to produce it, the way in which the passions and the will develop; finally that such a knowledge is sufficient, in this respect to dissipate a large number of both ridiculous and dangerous prejudices."[28]

According to Cabanis, "It is from this point of view that the physical study of man is chiefly interesting. It is on these points that the philosopher, the moralist, the legislator, must concentrate their attention, and it is from the study of these points that they can find both new lights on human nature and fundamental views on its perfecting."[29]

Given the wide range of factors, both internal and external, influencing the development of any individual's ideas and emotions, it is clear to Cabanis that the mental world of each individual is likely to be unique and that any claim of de facto equality is false. At the same time, it was also obvious to Cabanis that some forms of inequality, based on temperament, age, sex, physical condition, and so forth, were "natural" and unavoidable, while others—especially those based on inequalities of opportunity—were artificial and imposed on persons not by nature or even by their own choices, but by ill-designed social institutions. The corrupt old regime, with its discriminatory inheritance laws, its sales of monopolies, and its inherited privileges, propagated artificial inequalities and made it virtually impossible to get back to the natural order that was presumed to optimize the general welfare. Instead, society should be organized to allow for natural inequalities, but to promote "the free exercise of the faculties of each and the peaceful enjoyment of the goods these faculties produce."[30]

Cabanis on the Centrality of Self-motion, or the Will

Mention of the free exercise of one's faculties raises a second major issue on which Cabanis departed significantly from most of his predecessors. One of the most important of Cabanis's moves, in the long run, emerged out of the unique place that he granted *motion* as an essential rather than as a purely descriptive "cause." Cabanis begins by arguing that

> life is a sequence of movements that are made by virtue of the impressions received by the different organs; that the operations of the soul or mind also result from the movements executed by the cerebral organ; and that these movements result from impressions either received and transmitted by the sensitive extremities of the nerves in the different parts, or activated in that organ by means that appear to act immediately upon it.
>
> Without sensibility [the ability of our nerves to convert the motions impressed by outside bodies into impressions] we would not become aware of the presence of external objects; we would not even have means of perceiving our own existence. But from the moment we feel, we exist.[31]

Up to this point Cabanis seems to be following the well-trodden path of Locke, Condillac, Helvetius et al., which saw the human body as an essentially *passive* receptor of impressions, responding in a deterministic fashion to its environment. But then Cabanis reformulated his statement to put a new twist on it, a twist that paralleled the slightly earlier formulation of Immanuel Kant (1724–1804). In his *Metaphysical Foundations of the Natural Sciences* of 1786, Kant had argued that it is more appropriate to think of our perception of external bodies as grounded in the resistance that we feel to our own self-generated motion than in the passive reception of some externally generated motion. That is, we initially learn that an external body exists if and only if the voluntary motion of some part of our body is impeded. Similarly, Cabanis now argued that an object produces sensations only "by the resistances which such an object opposes to our will."[32]

The key issue here is that from this point of view human *voluntary activity* is the precondition for experience, and awareness of voluntary self-motions precedes and serves as a suggestive model for the derivative understanding of the determinate motions of nonliving objects: "It is through progressive and *voluntary* movement that man particularly distinguishes his own life from that of other animals: movement is for him the true sign of vitality. When he sees a body move, his imagination animates it. Before he has any idea of the laws that make rivers flow, seas rise, clouds move in the air, he gives to these different objects a soul. But as his knowledge broadens, he perceives that many movements are executed like those of his arm when a foreign force displaces it without his own participation, or even against his will. He does not need to reflect very much to perceive that these latter movements bear no relation to those determined by his will, and soon he no longer attaches the idea of life to any but voluntary movement."[33]

In this brief passage and the commentary surrounding it, Cabanis makes a series of important points. First, he argues that we can discover that our own actions are a mix of voluntary and involuntary ones, or conscious and unconscious ones, but it is as conscious beings that we are uniquely human. Furthermore, like other nonmechanical materialists, Cabanis argued that the source of consciousness is not something spiritual and nonmaterial; it is simply a consequence of our complex physical structure.

Most important for our present purposes, Cabanis argued that since self-motion, or *willed* motion, is the source of all awareness: "the first need of all animals is that of exercising their faculties, of developing them, of extending them, in a sense of ensuring consciousness through them."[34] This claim was transformed by his friend Destutt de Tracy into an argument that self-motion produces our most important pleasures and that our ability to exercise our own powers through labor provides our most primitive notion of wealth.[35]

From this perspective, which was later adapted by the utopian social theorist, Charles Fourier (1772–1837), and by Karl Marx (1818–83), labor is *naturally* at least as important as a form of self-expression as it is as a means for the production of commodities intended for exchange. To treat labor solely as a source of commodities is to pervert its meaning. To be able to freely exercise one's faculties is every bit as important to humans as to be free to enjoy the goods that are also fruits of our actions.

Jean-Baptiste Say and Political Economy

Even if the *Idéologues* had possessed a coherent political agenda, the open materialism of Cabanis's views would have alienated both Napoleon and all of the major political leaders of the restoration, thus limiting their political influence. But a second reason for the limited direct influence of the *Idéologues* on French politics arose out of the fact that virtually all of the *Idéologues* viewed politics as a secondary activity and as relatively unimportant. As a consequence, they were willing to support almost any government that promised to bring stability to society, as long as it claimed to favor some guarantees of economic liberty and freedom of the press. Indeed, they included supporters of every change of government that occurred between 1798 and 1830, as successive regimes failed to improve the economy or generated so many enemies that social instability was a certainty. The general opinion of the group was well expressed by Pierre Danou in an 1818 article in the *Censeur européenne*: "We shall repeat it a thousand times. Man's concern is not with government [which is] no more than a very secondary affair—we might say almost a minor thing. His goal is industry, labor, and the production of everything needed for his happiness. . . . The height of perfection would be attained if all the world worked and no one governed."[36]

Because of their focus on the importance of economic issues relative to governmental ones, Jean-Baptiste Say, though one of the youngest of the *Idéologues*, was one of the most influential as the chief and most able political economist of the group. Say's economic doctrines began from those of Adam Smith (1723–90), whom he had studied during a stay in England in the 1780s. But they diverged substantially from those of Smith and even more from those of David Ricardo (1772–1823), the most influential English developer of political economy in the early nineteenth century. First, there were a group of methodological differences connected with Say's desire to make political economy more scientific by grounding it in the procedures advocated by Cabanis. Second, and growing out of their methodological differences, was Say's rejection of the Smith-Ricardo labor theory of value (to be discussed below) in favor of a theory that grounded value in utility. Third, and of particular importance for the theme of this work, was Say's emphasis on the role of technological innovation relative to the division of labor in productivity increases and his concomitant emphasis on the economic significance and institutional place of science in modern economies.

In keeping with the general Idéological devaluation of the political relative to the economic, Say took an even more aggressive stance against taxation and governmental expansion than had Adam Smith. This was a feature of his work adopted by other *Idéologues*, and one that undoubtedly helped to account for his tremendous popularity in Jacksonian America.

Surprisingly, Say distrusted quantification and claims of mathematical precision in political economy because of the sheer complexity of the issues it addresses. Consider a problem as simple as determining the demand for wine at a particular place and time, Say suggests. One must know the price at which it will be sold, the former stock on hand, the numbers, tastes, and means of the consumers, the prosperity of the local economy, the prices of possible substitute liquors such as beer and cider, as well as "an infinite number of less important considerations." Given this situation, Say says that no one knowledgeable about applied mathematics would attempt to calculate demand, "not only on account of the numerous data, but in consequence of the difficulty of characterizing [many] of them with anything like precision, and of combining their separate influences." Those who have attempted to estimate the answers to such problems have inevitably failed, according to Say, because they "have not been able to enunciate these questions in analytical language without divesting them of their natural complication by means of simplifications and arbitrary suppressions, of which the consequences . . . always essentially change the condition of the problem and pervert all its results."[37]

To support this view, Say drew directly from Cabanis, whose views on the science of medicine he saw as offering a perfect parallel with his own views on the proper way of investigating social phenomena in general and those addressed by political economy in particular. In Cabanis's words, the phenomena "depend upon so many unknown springs, held together under such various circumstances, which observation vainly attempts to appreciate, that these problems, from not being stated with all their conditions, absolutely defy calculation. Hence whenever writers on mechanics have endeavored to subject the laws of life to their method, they have furnished the scientific world with a remarkable spectacle. . . . The terms they have employed were correct, the process of reasoning strictly logical, and nevertheless, all the results were completely erroneous. . . . It is by application of this method of investigation to subjects to which it is altogether inapplicable, that systems the most whimsical, fallacious, and contradictory have been maintained."[38]

Say was particularly opposed to the postulational approach to economics that had been promoted by the Physiocrats in the eighteenth century but that was espoused in the early nineteenth century by the highly respected David Ricardo. Though Say was perfectly willing to admit the *prima facie* plausibility of Ricardo's assumptions and the legitimacy of his reasoning, he likened Ricardo to a mechanical theorist who purported to be able to prove from the law of the lever that ballet dancers could not execute the leaps that they did. "The reasoning proceeds in a straight line, but a vital force, often unperceived, and always inappreciable, makes the facts differ very far from our calculations. From that instant, nothing in the author's work is represented as it really occurs in nature."[39] The science of political economy, he wrote, "should not teach what *must necessarily* take place, even if deduced by legitimate reasoning, and from undoubted premises; it must show, in what manner that *which in reality does* take place, is the consequence of other facts equally certain."[40]

Once again, the opposition between necessary inference and contingent observation of events and the opposition between foundational premises, or assumptions, and facts, are taken directly from Cabanis, who constantly insisted that the sciences are competent to deal only with observable relationships and not with traditional causality, or necessary connection, which smacks of metaphysics. All of these views, including the language in which they were expressed, were adopted and widely spread by Auguste Comte through his *Cours de philosophie positive.*

Utility as the Source of Value

Because of his emphasis on empiricism, Say insisted that in political economy, claims about rights, which could purportedly be established through reason alone, but which were largely matters of opinion, be replaced with statements about factual relations that could be demonstrated by observations.[41] For this reason, Say argued that Smith and Ricardo had gone horribly wrong in their theory of value, according to which a person was presumed to acquire a right in something by investing effort in acquiring it. The value of some entity then became what the owner of some entity could demand by virtue of the labor expended. On the contrary, Say argued, nothing has a value until someone is willing to exchange something else for it, and the reason they do so is that the object or service is capable of satisfying some want of the purchaser. The capacity that something has to satisfy a human want, Say calls "utility." Thus, "the value that mankind attach to objects originates in the use it can make of them." Moreover, the production of wealth is the creation, not of material objects, but of utility, and, "when one man sells any product to another, he sells him the utility vested in that product, [and] the buyer buys it only for the sake of the use he can make of it."[42]

Though value "originates" in utility, it is clear and critically important to Say that political economy is incapable of determining whether value is directly proportional to the "actual utility" that a good has because there is no way of estimating "actual utility." Political economy must take buyers' judgments of utility as measured by their willingness to pay for goods as raw data without understanding how the judgments of utility come about. It is clear that wants and their intensity vary according to a huge range of biological, social, and cultural circumstances, including "the physical and moral constitution of man, the climate he lives in, [and] the laws, customs, and manners of the particular society in which he may happen to be enrolled." But these factors are of little or no concern to the political economist, whose business is not to ask how human wants originate. He "must take them as existing *data,* and reason upon them accordingly."[43]

It is certainly true for Say that there is a close relationship between the investment of labor in a product and the long-term market price of that product; for producers would not continue to sell something for which they were not appropriately compensated. But Say turned Ricardo's views directly on their head by insisting that "the value of products is not founded upon that of productive agency, as some authors have erroneously affirmed; . . . it is [rather] the ability to create utility . . . that gives value to productive agency."[44] All value is thus grounded in exchange value for Say, and the value of any object or service (Say uses the term "immaterial object") must fluctuate according to the relationship between supply and demand, without regard for how much effort must go into making it.

Say on Science as a Factor of Production

One major consequence of Say's rejection of Smith and Ricardo's labor theory of value was his reassessment of the various factors that contribute to production and his special emphasis on the use we make of the powers of nature, which in turn depends upon the extent of our scientific knowledge. Say argued that Smith's obsession with labor as the ultimate source of value led him to vastly overestimate the role of the division of labor relative to machinery and technological innovation as a source of productivity increases, and hence of increases of wealth.[45] "What makes machines and other technical devices and processes productive?" asks Say. Tools and machines are merely "expedients, more or less ingenious, for turning natural powers to account." Consider the steam engine, for example, "[It] is but a complicated method of taking advantage of the alteration of the elasticity of water reduced to vapor, and of the weight of the atmosphere. So that, in point of fact, a steam engine employs . . . a variety of natural agents, whose gratuitous aid may perhaps infinitely exceed in value the interest of the capital invested in the machine."[46] Thus, according to Say, increases in productivity and wealth—especially during the preceding century—came primarily from "the increased command acquired by human intelligence over the productive powers and agents presented gratuitously by nature." Moreover, the most important source of that "command acquired by intelligence" was natural scientists. Indeed, argued Say, natural science had led to vast improvements in productivity, and the only thing that was preventing people from reaping the benefits was the hitherto slow progress of the moral and political sciences.[47]

Many intellectuals since the time of Bacon had been claiming that science offered great benefits to humankind, but Say was the first to incorporate scientific knowledge into political economy as a major factor—perhaps *the* major factor by the early nineteenth century—in production. However, Say argued, scientific knowledge has characteristics that differentiate it radically from other factors of production. Labor, land, capital—even entrepreneurial ability, which Say distinguished from the first three—were all forms of private property. Scientific knowledge, on the other hand, he argued, was public knowledge. As such, it was largely available to all, free of charge. In particular, it crossed international boundaries without difficulty, so that any given nation could acquire it without nurturing it: "The knowledge of the man of science, indispensable as it is to the development of industry, circulates with ease and rapidity from one nation to all of the rest.

And men of science have themselves an interest in its diffusion; for upon that diffusion they rest their hopes of fortune, and, what is more prized by them, their reputation too. For this reason, a nation in which science is but little cultivated, may, nevertheless carry its industry to a very great length, by taking advantage of the information derivable from abroad."[48]

What was particularly necessary for the exploitation of scientific knowledge was a class of entrepreneur-industrialists interested in and able to apply that knowledge to productive purposes. It was the existence of such a class in Britain—one that included Josiah Wedgewood (1730–95), Matthew Boulton (1728–1809), and James Watt (1736–1819) et al.—that explained why, even though France was equal or superior to England in the cultivation of scientific knowledge, England had nonetheless become vastly more wealthy during the previous 150 years. And this was why French education, not just for the intellectual elite, but also for the classes of entrepreneurs and skilled workers, should focus on scientific knowledge.

The major problem raised for a liberal political economist such as Say by the generally public nature of scientific knowledge was that, since public knowledge was free for the taking, no individual had an economic incentive to produce it, for it could confer little competitive advantage. Here was a case where a clear economic benefit to society—increased aggregate wealth—conferred little or no direct economic benefit to its producer. For this reason, Say argued, the production of most scientific knowledge must be left up to "rich and philanthropic" individuals, whose "whims and caprices" allocate some of their revenues to useful ends, or to governments. The encouragement of science was one of the very few non-security-related issues to which Say felt tax monies were legitimately applied, for, as he wrote, "The portion of them so spent is scarcely felt at all, because the burden is divided among innumerable contributors," and, "the advantages resulting from success being a common benefit to all, it is by no means inequitable that the sacrifices, by which they are obtained, should fall on the community at large."[49]

Of course, Say recognized that the nearer scientific knowledge got to immediate productive application, the more likely it could provide at least some "period of exclusive advantage" for the initial producer. Moreover, Say analyzed which industries were most likely to repay the capital risked in experiments aimed at applications. Thus, in modern terms, Say distinguished between more or less "basic" science, which demanded—or at least warranted—government support, and "applied" science, which should be supported for economic reasons by entrepreneurs and which could be encouraged at least in some cases by patent protection.[50] All in all, Say was able to offer a compelling and detailed economic analysis of the foundations of the expanding authority of science as well as a set of arguments that are still used throughout the world in explaining why governments should support science and scientific education.

A New Attack on Luxury

One of the critical features of nineteenth-century liberalism was its need to distance itself from the attitudes of the landed aristocracy. But given its emphases on the inviolability of property and the freedom of all persons to dispose of their property as they saw fit, the question was how to condemn the activities of the idle rich, who simply seemed to be meeting their own perceived needs through the use of their own resources. This problem was particularly acute because liberal theorists from Bernard Mandeville (1670–1733) through Adam Smith had argued that modern economic growth had, in fact, been initiated by the vanity-inspired conspicuous consumption of the landed aristocracy, though it was certainly agreed that the aristocrats had not intended to benefit others.

Drawing from and amplifying an argument that had been introduced by the earlier political economist, Richard Cantillon (1680–1734), Say launched an all-out attack on "luxury," or what we now call conspicuous consumption. Say argued that consumption could take two and only two basic forms. Either it was unproductive, or it was productive. If, for example, fuel is burned in a fireplace merely to keep people warmer than they need to be for subsistence, value is consumed in meeting a human want without in any way being replaced. This is unproductive consumption. On the other hand, if fuel is burned to heat dyeing ingredients to make cloth, then the value of the fuel has been used to produce new value.[51] The fuel serves like capital for a new round of production and we say that it has been consumed productively. In general, the luxury items sought by the very rich, items such as perfumes, jewelry, fine lace, pyrotechnic entertainments, and so on, are consumed unproductively, effectively depriving the country of capital to invest. Moreover, they use up a relatively large amount of wealth to provide a marginal amount of utility. Consider the following example: "The wants of the rich man occasion the production and consumption of an exquisite perfume, perhaps those of the poor man, the production and consumption of a good warm winter cloak; supposing the value to be equal, the diminution of the general wealth is the same in both cases; but the resulting gratification will, in the one case, be trifling, transient, and scarcely perceptible; in the other, solid, ample, and of long duration."[52] In general, then, where private wealth is most unequally distributed, there is a huge unproductive consumption that provides very little utility. At the same time, there is a huge population of poor who can provide no effective demand to stimulate the production of wealth. Concluding his section on private consumption, Say argues, "It is time for the rich to abandon the puerile apprehension of losing the objects of their sensuality, if the poor man's comforts be promoted. On the contrary, reason and experience concur in teaching, that the greatest variety, abundance, and refinement of enjoyment are to be found in those countries, where wealth abounds most, and is most widely diffused."[53]

Say, of course, did not expect all of the wealthy to be open to the positive argument that the investment of wealth in productive activities ultimately magnifies the wealth of the rich and poor alike. For those who could not see the compulsion of this argument, Say pointed out that the French Revolution demonstrated that if the misery that inevitably accompanies luxury becomes too intense, it will eventually lead to open rebellion. So the rich are well advised to seek a more equitable distribution of wealth purely for their own protection.[54]

Most of Say's arguments for a redistribution of wealth unrealistically hoped that enlightened self-interest might change the behaviors of the wealthy, and his acceptance of the basic insights of Malthus (population will increase until it overshoots the maximum carrying capacity of its environment) left him very doubtful that a situation could ever be achieved in which some part of the population doesn't die annually from "mere want."[55] Yet he did offer one interesting insight as to what particular conditions made the laboring poor most subject to exploitation, and this insight was later incorporated into Marxian analyses as well as into American Jeffersonian Republicanism.

Say begins from Adam Smith's analysis of the advantage that masters have over workers in salary negotiations. Since masters generally have some resources that will allow them to subsist for a time idle, while workers do not, masters can hold out for longer and workers must take what they are willing to offer. Say focuses on what allows the masters to withhold the opportunity to work. It is at least in part the fact that they control the tools, or means of production. Whenever laborers own their own means of production they are not subject to this form of coercion: "The workman that carries about with him the whole implements of his trade, can change his locality at pleasure, and earn his subsistence wherever he pleases: in the other case, he is a mere adjective, without individual capacity, independence, or substantive importance, when separated from his fellow laborers, and obliged to accept whatever terms his employer thinks fit to impose."[56]

Destutt de Tracy and the Hardening of Liberalism into *Individualisme*

Though older than either Cabanis or Say, Destutt de Tracy (1754–1836) came to the salon of Mme Helvetius and into the domain of the younger *Idéologues* only in 1792 after a career as an army officer, minor diplomat, land speculator, and member of the liberal nobility in the National Assembly in 1789. De Tracy's imprisonment from November 2, 1793, through December 1, 1794, changed his intellectual path in two critical ways. In the first place it led him to an almost pathological fear of the "barbaric anarchy" of the *sans-culottes* and to an obsession with the restoration of morality and order. This obsession with order in turn led him to support Napoleon and later, the restoration of the Bourbon Monarchy. Second, it offered him time to study both Locke and Condillac closely and to conclude that the scientific analysis of sensations and ideas must be the explicit starting point of all well-founded human knowledge and activity. Thus, when de Tracy was invited to join the Section of the Analysis of Sensations and Ideas of the Class of Moral and Political Sciences of the *Institut de France*, in his first *memoire* he announced his intention of establishing a new acausal science of ideas, or "*Idéologie*," rejecting the common term "psychology," because it seemed to presume the existence of a "psyche," or soul, independent of the brain—something that de Tracy was unwilling to do. The ultimate goal of *Idéologie* would be to provide the foundation for "regulating society in such a way that man finds there the most help and the least possible annoyance from his own kind."[57] The traditional religious source of morality would be ignored because either it "comes from God, in which case it is above human reason, or else [it comes] from human dreaming, in which case it is below reason."[58]

To see how de Tracy's approach turned the ideas of Cabanis and Say in new and often interesting directions, we will focus on the fourth and fifth parts of his *Elements of Ideology*, published originally in 1815 as *Traité de la volonte et des effets* (*Treatise on the Will and Its Effects*), translated under Jefferson's supervision as *A Treatise on Political Economy* in 1818, and reedited as *Traité d'economie politique* in 1823. The alternative titles of this work are symptomatic of the change in orientation that de Tracy brought to political economy.

The basic orientation of all political economy prior to 1815 had been toward what we now call macroeconomics, that is, toward the wealth and well-being of national aggregates. While political economists from William Petty (1623–87) through Smith had assumed that individuals made economic choices in order to serve their own interests or increase their own pleasures, their emphasis was on how these individual economic choices served, even if unintentionally, to maximize the wealth of the nation. Given this orientation, a thinker such as Pierre Boisguilbert could warn his readers that two-party exchanges freely entered into by both parties guaranteed a net gain by each party and, *only when there were no unaccounted for third-party effects,* a gain to the economy as a whole.[59] Similarly, Say could admit that since national autonomy was a legitimate goal of the state, a free market in grain might wisely be suspended in order to develop agriculture within national boundaries, even in the absence of a productive advantage in grain cultivation.

De Tracy turned this macroeconomic emphasis completely on its head, focusing attention on the needs and wants of individuals. Basing economics on the human will, de Tracy viewed economic exchanges solely as the means by which private individuals satisfied their desires. Given this emphasis, third-party effects could be completely ignored, for no one had any obligation to care whether someone else was injured as a consequence of the attempt to satisfy his or her own wants. It was one's sole duty to attend to one's own needs. De Tracy characterizes this situation by stating that "our rights are always without limits, or at least equal to our wants; and our duties *are never but the general duty of satisfying our wants.*"[60] Only where social conventions explicitly protect the property and persons of others does one incur an obligation to respect them; so unless a particular transaction is explicitly prohibited, there is no duty to consider any but its direct consequences in satisfying the wants of the two direct participants.[61]

While it is the case that in exercising their free choices to enter into exchanges individuals do generally increase the aggregate wealth of the society, that increase in wealth is important, not for its own sake, but because a wealthier society offers greater opportunities for its members to satisfy their individual private desires. Indeed, societies exist solely because they offer three conditions under which productivity, or the creation of greater utility (the satisfaction of more wants), can be increased. First, combined labor is often more productive. For example, two or more individuals can lift objects that were too heavy for a single person. Second, social organization allows for the increase and dissemination of knowledge through language, and for de Tracy as for Say, knowledge is a critical factor in production. Finally, social organization allows for the division of labor, so that the talents of each person can be most efficiently used.[62]

De Tracy on the Cooperation between Scientists, Entrepreneurs, and Workers

De Tracy's understanding of the productive advantages offered by society was linked to a new way of understanding the nature of production that built on Say's reflections on the role of science in increasing productivity. Ironically, this new understanding of production also provided the foundation for Saint-Simonian socialist theories of cooperation, against which the individualistic liberalism of de Tracy and many of his followers struggled during the first half of the nineteenth century. According to de Tracy, Say had identified three critical factors involved in production and three different kinds of laborer associated with them that offered a particularly illuminating way of analyzing the productive process. Everything contains theory, application, and execution; and the three related species of laborers are scientists, entrepreneurs, and workmen or hirelings. The scientist discovers the properties of materials and the laws of nature that allow materials to be combined and fabricated into useful objects. The entrepreneur, or industrialist, figures out how to utilize the scientist's knowledge to produce something that will increase utility and provide the capital advances needed to initiate the process. In this way, the entrepreneur gives "a useful direction" to the efforts of the third group, those who are usually called the laborers, or workers, who execute the entrepreneur's plans.

Several consequences follow from viewing production, or "industry," in this way. First, the entrepreneur, industrialist, or "active" capitalist becomes the lynchpin of society. These "undertakers of industry" are, for de Tracy and for the Saint-Simonians, "the heart of the body politic, and their capitals, its blood," for it is quite literally their advances that feed the hirelings while the productive process gets under way.[63] Second, and partially as a consequence of the central leadership role of the entrepreneur, the most effective society is anything but democratic and egalitarian. It is the nature of both humans and the industrial process that some must lead and others must do as they are directed.

Since it is through the cooperation of scientists, industrialists, and workers that social organization increases the wealth of all, earlier associationists, such as Helvetius, the Jacobins, and the subsequent radical socialists—all of whom viewed the fundamental relationship between workers and the wealthy as one of unending class conflict—were absolutely wrong. One of the central functions of *Idéologie* was to educate members of each class to recognize both their common interests and their different roles. On the one hand, workers had to recognize the need for concentration of wealth in the hands of the industrialists and the need for industrialists to direct their actions. On the other hand, the entrepreneurs had to learn that it was not in their best interests to keep wages as low as possible and that historical evidence demonstrates that, "in all times and places, whenever the lowest class is too wretched, its extreme misery, and its abjectness—which is a consequence of it—is the death of industry, and the principle of infinite evils, even to its oppressors."[64]

The Divergence of Socialist Cooperation and Liberal Individualism

Both Saint-Simonian socialist doctrines and subsequent Continental liberalism incorporated de Tracy's understanding of the hierarchical structure of the industrial world, the importance of class cooperation for the benefit of all, and the need for all classes to recognize their common interests in a stable and efficient social order. But the socialists and liberals diverged radically on how the details of cooperation were to be worked out and how the wealth of the industrial order was to be distributed. In the next chapter we will see how the Saint-Simonians worked these issues out. Here we focus on de Tracy's liberal resolution of the issues.

In order to understand de Tracy's approach to the distribution of wealth, we need briefly to return to his understanding of the nature and inviolability of property. For de Tracy, the very process of individuation—of the separation of self from the rest of the world—involved recognition of the ability to control the motions of our bodies in order to satisfy our desires. Moreover, the recognition that we have an absolute right to use our persons to satisfy our desires provides the foundation of our notion of property, which thus involves the notion of a sole and inviolable right of use.[65] Without this "natural" property in our persons, we could not have developed the artificial and conventional notion of property in goods, which simply extends the notion of exclusive use to satisfy needs to objects beyond our selves.[66]

One consequence of this notion of property is that it undermines the Rousseauist idea that property, embodied in the notions of thine and mine, is the unfortunate artifact of civilized society. There is, for de Tracy, no point in asking whether property is good or bad or in discussing the possibility of rejecting the institution of private property. Indeed, property is a necessary consequence of individuation, and it has no meaning except in terms of an individual's capacity to satisfy desires. There can be no legitimate way in which any individual or group can acquire the right to appropriate the property of another. Persons may voluntarily enter into conventions by which their rights to use their property is limited, but they will only do so if they have reason to believe that this will lead to a net increase in their capacity to satisfy their wants.

Since individuals are differently endowed with intelligence and other abilities, it is inevitable that property must also be unequally divided within society. Any attempt to eradicate this inequality must be in vain, "for nothing which has its existence in nature can be destroyed by art."[67] Like Cabanis and Say, de Tracy was inclined to see inequalities in the distribution of wealth as the most important source of evil in society. But he was less inclined than his colleagues to see this evil as avoidable: "The inequality of means are, then, conditions of our nature, as are sufferings and death. I do not conceive that there can be men sufficiently barbarous to say that it is a good; nor can I any more conceive that there can be men sufficiently blind, to believe that it is an evitable evil. I think this evil a necessary one, and that we must submit to it."[68]

Given his commitment to the inviolable character of private property, de Tracy insisted that the poor, who possess only their own labor, have an even greater interest in the protection of property than the rich, for with their superior resources, the wealthy have more opportunities to deprive the poor of their property than vice versa.[69]

It would be both futile and a violation of the most fundamental principles of political economy for the state to coerce the wealthy into sharing their property with hirelings, so all projects to limit or redistribute wealth are simultaneously doomed and abhorrent. Whereas Say and all of the socialists were inclined to see the entrepreneurial class as most responsible for poverty by virtue of its greed, de Tracy and many of his middle-class liberal followers were more inclined to place the blame primarily with the working class because of the workers' willful choice to procreate so fast that they create a perpetual labor surplus in most places. Taking his cue from Thomas Malthus, de Tracy wrote, "the poor, who everywhere constitute the great number . . . always multiply imprudently and without foresight, and *plunge themselves into inevitable misery*. . . . The interest of men under every consideration, then, is to diminish their fecundity."[70]

Thus, de Tracy's arguments were symptomatic of and contributed to the increasing nineteenth-century divergence between Continental liberalism and more radical working-class movements. Unlike British liberalism, which had long been associated with wealthy merchants and landowners and with a notion of utility that focused on the general or aggregate well-being, Continental liberalism had intensely egalitarian roots and a tradition of acknowledging the existence of class conflict. But at the hands of the *Idéologues,* and especially of de Tracy, liberalism on the Continent took on a distinctly individualist and upper-middle-class character, sending workers alternatively into their own radical working-class political movements or back into support for reactionary monarchist politics.

CHAPTER 2

Saint-Simon, Saint-Simonianism, and the Birth of Socialism

During the period between 1810 and 1830, Paris saw the emergence of two scientifically based socio-religious cults that have played important long-term roles in European and American social and intellectual life. Each was founded on the ideas of a brilliant visionary thinker who at least began his intellectual odyssey hoping to bring order out of the chaos of post-revolutionary French society by establishing a social science—a social physiology, social physics, or sociology. Each thinker drew a small band of disciples, drawn primarily from engineering students, mathematicians, medical students, and journalists, and sought to lead men and women into a more perfect industrial world that would be managed rather than governed by an elite of talent and knowledge. The senior and more complex visionary of the two was Count Claude Henri de Rouvroy de Saint-Simon, whose work provided an important foundation for those forms of socialism that appealed to the intellectual elite of nineteenth-century Europe. Saint-Simon will play the central role in this chapter. The junior and more broadly influential Auguste Comte, who was centrally in the nineteenth-century liberal camp, will provide the focal point of the next.

Henri Saint-Simon

After a brief military education, Saint-Simon accepted a commission at age seventeen, serving in the French artillery that fought at Yorktown in support of the American revolutionaries against the British. Captured during a Caribbean naval battle in 1782, Saint-Simon used his time incarcerated in Jamaica to work out a proposal for the first of many transportation projects that were to become a hallmark of the Saint-Simonians. He unsuccessfully promoted a canal through Lake Nicaragua to link the Atlantic and Pacific Oceans. Later Saint-Simonians would initiate the building of both the Panama and Suez Canals and would be the major architects and financiers of the French and Belgian railroad systems.

Back in France during the revolutionary period, Saint-Simon formally adopted a peasant name, Bonhomme, and made a public issue of supporting republican and peasant projects while he simultaneously speculated heavily in land confiscated by the revolutionary governments. As a consequence of his carefully orchestrated work with local *sans-culottes,* though he was arrested along with many of his financial-speculator colleagues during the Terror, he escaped the guillotine. But Saint-Simon's experiences during the Terror shaped his subsequent views, so that throughout the remainder of his life he saw the anarchy associated with violent revolution as a condition to be avoided at all costs.[1]

In the aftermath of the revolution, temporarily wealthy, Saint-Simon established a salon and began a period of personal private study of mathematics, the physical sciences, the life sciences, and, most importantly for our interests, the social doctrines of the *Idéologues.* Then, in 1802, after a brief trip through England and Switzerland, Saint-Simon wrote the first of a huge number of works, *Lettres d'un habitant de Geneve.*

Saint-Simon's Scientistic Phase

Though virtually unread in its initial privately printed editions of 1802 and 1803, the *Lettres* raised in a preliminary form almost all of the themes that Saint-Simon developed in much greater detail and with varying emphases and modifications over the next twenty-three years.[2] Like many of his works, the *Lettres* is presented in the form of a proposal for a reform of society. Moreover, like all of them, it purports to offer a scientific analysis of social conditions grounded in the understanding that "our social relations are considered as physiological phenomena."[3]

The centerpiece of *Lettres d'un habitant de Geneve* is a project to establish a subscription in Newton's memory to support three mathematicians, three physicists, three chemists, three physiologists, three authors, three painters, and three musicians in a way that would make them completely independent of all factional pressures (including ones brought to bear by their academic peers), so that they could focus on research of importance to all of mankind rather than to issues of interest to patrons. In supporting such a project, Saint-Simon argues, subscribers "will permanently entrust to mankind's twenty-one most enlightened men, the two great instruments of power: prestige and wealth."[4] In the *Lettres,* Saint-Simon does not yet explain just how the power placed in the hands of this intellectual elite will be exercised on behalf of all mankind, but he offers an intriguing argument about why it is critical for the future of mankind that such a plan be followed. All humans have, to some extent, the desire to dominate others. Furthermore, up until recently, the wealthy, who constitute less than one percent of the population, have been able to exercise their power to serve their own interests. Only the scientists are capable of recognizing the difference between "things which benefit a part of mankind at the expense of the rest, and those which increase the happiness of the whole of mankind." Thus only the scientists are capable of exercising power for the benefit of all.[5]

In subsequent works Saint-Simon argued initially that scientists share in the universal desire for power over others, they just exercise that power through their knowledge of nature and on behalf of all of humanity. By the early 1820s, however, he had come to argue that science has radically changed the character of the human desire for domination, turning it away from the domination of humans to the domination of nature.[6]

In the *Lettres,* Saint-Simon argued that all of mankind was now divided into three basic classes. In the first were scientists, artists, and the knowledgeable, active "proprietors" whom Say had included in his class of "industriels." In the second were those proprietors who simply collect income from inherited wealth and/or involve themselves in traditional state functions, whom Say had identified as social parasites. In the third were the rest, dominated by wage laborers, whose slogan during the revolution had been "equality."

Though sympathetic to the plight of the working poor, Saint-Simon was absolutely convinced that their call for equality was completely misplaced, because it sought to avoid the hierarchical arrangement essential to the structure of all complex organisms. The first class alone, he insisted, was sufficiently enlightened to provide guidance to society. This class was further divided into three basic groups to correspond to the three basic human physiological types identified by Xavier Bichat, that is, those associated with intellectual activity, emotions, and motor activity, respectively.[7] Every organism is composed of tissues with three different basic functions and every person tends to be dominated by one of the three basic abilities. Those with the highest rational abilities are the scientists of the world. Those with the greatest emotive capacities are the artists and preachers. And those with the greatest motor capacities are the entrepreneurs and engineers of society, while the less talented and most numerous group of motor-ability-dominated persons constitutes the numerous third class—the workers of society.

Focusing on the features of positive science as defined by Cabanis, Saint-Simon wrote, "A scientist, my friends, is a man who foresees; it is because science provides the means to predict that it is useful, *and the scientists are superior to all other men.*"[8] Ultimately, it was because the scientists have developed a procedure for verifying their predictions that they have gradually "disentangled themselves from the facts created by their imagination."[9] Scientists did not pretend to know things about which they were ignorant. As a consequence, astronomy gradually drove out astrology, chemistry gradually drove out alchemy, and in general, positive scientific knowledge must drive out philosophical and metaphysical claims as a foundation for human action.[10]

What Saint-Simon added to the views of the *Idéologues* on this issue was a sense of an ongoing and inevitable dynamic process by which topics of increasing complexity, beginning with astronomy and moving on to physics then chemistry came under the control of the scientists.

Because of the increasing predictive ability of science and the confidence that it engendered within the public, scientists had, in fact, largely replaced the clergy of Europe as leaders of public opinion, though many were unaware of their new role. Saint-Simon pleaded with the scientists on behalf of everyone to embrace the power that was now available to them.[11]

Anticipating a theme that he would develop much more extensively later, Saint-Simon concluded the first of the *Lettres* with his vision of the power and authority relations of the ideal society toward which he insisted we are moving: "I think that all classes of society would be happy in the following situation: spiritual power in the hands of the scientists; temporal power in those of the [active] proprietors; power to nominate those called to carry out the functions of the great leaders of mankind in the hands of everyone; the reward for those who govern to be—esteem."[12]

In the last of the *Lettres,* Saint-Simon even suggested the formal establishment of a "Religion of Newton" in which the scientists functioned as priests, in which the chief commandment was to work in the service of all humanity, and in which redemption came through science rather than through Christ.[13]

Saint-Simon on the Role of Scientists in the French Revolution

The French Revolution, Saint-Simon argued, "was secretly fomented by scientists and artists," and was "aimed at the destruction of all the institutions which had wounded their self-esteem." The scientists and artists ultimately succeeded in destroying what they wanted to, but in order to do so, because they were resisted by the traditional landowning aristocracy, they were forced to "inflame the passions of the ignorant and to burst all bounds of subordination which, until then, had contained the rash passions of those without property."[14]

By forcing the scientists and artists to ally themselves with the propertyless masses, the property owners of the old regime created a situation in which the scientists lost control of the revolution they had initiated, and it fell to Napoleon to restore order. In spite of their mistakes, however, Saint-Simon argued, the scientists' post-revolutionary prestige continued to grow, so the kind of crises that France saw were bound to occur elsewhere throughout Europe. The only hope for traditional property owners throughout the rest of Europe was to accept the reforming leadership of the scientists and artists. While it is true that the primary villains of Saint-Simon's story were the propertied classes of the old regime, he also insisted that scientists were capable of caving in to practical pressures and of using their analytic skills in destructive as well as constructive ways, creating a problem that Saint-Simon and his followers were to explore at great length throughout their works.

Shortly after he produced the *Lettres,* Saint-Simon's economic and intellectual fortunes plummeted. Yet he continued to churn out works promoting his vision of scientific progress and the scientific direction of society: *Essai sur l'organisation sociale* in 1804, *Introduction aux Travaux Scientifiques du XIXe siecle* in 1807–8, and

Esquisse d'une nouvelle encyclopedie in 1809–10. In all of these works, he sought to present a unified system or general theory that would incorporate all phenomena, from those of the astronomical world through those of the social world.

Between 1807 and 1810, Saint-Simon refined his notions of scientific advance and extended the pattern that he found in that development to human social history as well. The history of science, and consequently all human history, he argued, proceeds through a sequence of periods dominated alternatively by analytic and synthetic activities or by a posteriori and a priori theories, combining the languages of Condillac and Immanuel Kant. During periods dominated by analysis, later termed critical periods, emphasis is on undermining prevailing scientific theories or social structures and on the unintegrated pursuit of specialist "positive" sciences or factional interests in society. During periods dominated by synthesis, on the other hand—later termed organic periods—emphasis is on constructing comprehensive scientific theories and on unifying social institutions under comprehensive guiding principles.

In 1812, Saint-Simon's fortunes took a positive turn. His family settled a modest life income on him so that his financial condition would not continue to embarrass them, and though he did not gain the recognition he sought from the scientific elite—Pierre Laplace, the powerful mathematician and natural philosopher, thought he was a crackpot—he did begin to attract a series of young and brilliant colleagues and disciples. Beginning with the mathematician Denis Poisson (1781–1840), the first of his "adoptive sons," there were Augustin Thierry (1795–1856), who became his secretary in 1813; Auguste Comte, who replaced Thierry in 1817; the mathematician Olinde Rodrigues (1794–1851), who became his closest confidant during the final years of his life; Hippolyte Carnot (1801–88), son of the famous mathematician, Lazare, and brother to Sadi Carnot, the founder of thermodynamics, who was the principle author of *Doctrine de Saint-Simon* through which most persons learned of Saint-Simon; and Prosper Enfantin (1796–1864), the Polytechnicien who became the leader of the Saint-Simonian movement after the master's death.

One of the intriguing questions is why Saint-Simon's doctrines appealed to so many bright young scientifically trained Parisians in the early decades of the nineteenth century. A substantial part of the answer is surely identical with that given by Mary Midgely for the tremendous appeal that Marxism had for a comparable group of young British scientific intellectuals, including J. D. Bernal, Joseph Needham, J. B. S. Haldane, and Herbert Crowther, just over a century later:

> It seemed like a means of extending the reliability of science over the whole area of practical thinking—a way of spreading it that would be free from doubtful value judgments, since the theory was impartial, nonsectarian, essentially scientific. The modesty of science was to be combined with the constructive achievement of a new and central moral insight.
>
> This hope appealed to the architectonic intelligence in many bright scientists.

It satisfied that urge toward a general, comprehensive understanding which had brought them into science in the first place. It balanced the fragmentation of their specialized studies, allowing them to relate scientific aims to a wider humanitarian idealism.

[Finally, it] was not only a theory, but a faith. . . . once accepted [it] was emotionally very sustaining. It provided, in a most striking form, the promise of a better future.[15]

Saint-Simon's Turn to Industrialism

In *De la Reorganisation de la Societe Européenne* of 1814, Saint-Simon continued to emphasize the claim that only the application of scientific methods to political issues could optimize human well-being. Moreover, he argued that beginning from the principle that "one can only be truly happy in seeking for happiness in the happiness of others," which was as well grounded in experience as Newton's second law, one was led to the conclusion that Europe would be better off unified than divided into separate independent states.[16] But when he came to discuss who should lead both the subordinate national parliamentary bodies and the central coordinating all-European parliament, he now placed businessmen, administrators, and judges on a par with scientists.

The trend toward maintaining a central role for the scientific investigation of political and social issues while diminishing the centrality of the role of scientists within the proposed new order continued in a series of writings that brought Saint-Simon's doctrines to their mature state with the full incorporation of Jean-Baptiste Say's political economy into Saint-Simon's historicist framework. Up to this point it had been implicit in his work that traditional political issues were becoming less critical than social and economic issues. Beginning with *L'Industrie* in 1817, however, society becomes for Saint-Simon, nothing but "the whole unified body of men who are engaged in useful work." Politics becomes nothing but "*the science of production*—that is to say, that science whose object is the creation of an order of things most favorable to every kind of production."[17]

Saint-Simon's initial acceptance of Say's insistence on the ability of free-market exchanges to optimize production gradually gave way to a conviction that lack of knowledge among consumers and producers alike makes the market an extremely inefficient allocator of economic resources and activities. Ultimately, he argued that the best decisions "can only be the result of scientific demonstrations, absolutely independent of any human will."[18]

Soon after publication of the *Reorganisation,* Saint-Simon initiated a personal project in which he cast himself as the paid spokesperson for the growing commercial middle class. He solicited subscriptions from bankers, businessmen, manufacturers, and scientists, including the astronomer François Arago (1786–1853), the zoologist Georges Cuvier (1769–1832), and his friend J. B. Say, to establish a new journal, *L'Industrie.* To attract his new financial supporters he wrote, "I undertake to free you from the supremacy exercised over you by the courtiers, by the idle, by the nobles, by the phrase-makers. I promise to employ only legal, loyal, and inoffensive methods. I promise, moreover to obtain for you in a short period of time the greatest degree of public esteem and a dominant influence over the direction of public affairs. . . . As soon as industrial opinion is crystallized, nothing will be able to resist it."[19]

Unfortunately, in volume 4 of *L'Industrie,* Saint-Simon went well beyond the bounds of his promise to remain loyal and inoffensive in presenting his ideas on religion. Though he insisted that the existing religious institutions should be maintained, he argued for such a radical change in doctrine that all but a handful of his most loyal backers abandoned him. Attacking the very foundations of morality as seen by the cautious middle class, Saint-Simon wrote: "It is sufficient to note that in fact, ideas of the supernatural are almost everywhere destroyed, that they will every day continue to lose their hold and that the hope of paradise and the fear of hell can no longer serve as the basis for the conduct of men. The human mind has advanced since the establishment of Christian morality and as a result of this progress the facts are that the epoch of theology is past beyond recall, and it would be folly to try to base morals on prejudices whose absurdity is becoming more obvious every day."[20]

At this point, Saint-Simon and his current *protégé,* Auguste Comte, began a gradual falling out. Comte argued that Saint-Simon's views were so threatening to most interests that he should stop trying to promote political and social action in the present. Instead, Comte urged that he and Saint-Simon should concentrate on the development of the pure theory of their scientific system in the conviction that social forces would eventually lead others to acknowledge its virtues. Saint-Simon, on the contrary, sought to recoup his lost support, in part by recasting his religious views in a way that linked them to Christianity and that focused on the positive historical role played by medieval Catholicism.

While he continued to see the spiritual leadership of the new society as coming from the scientists, he identified such leadership with a new form of Christianity and with New Testament morality: "I believe that the new spiritual power will in the beginning be composed of all the academies of science in Europe and of all persons who deserve to be admitted to those scientific corporations. I believe that once this nucleus is formed, those who compose it will organize themselves. . . . I believe the pure morality of the New Testament will serve as the foundation for a new system of public education, and that this education will be pushed as far as possible in the direction of positive knowledge. . . . Finally I believe that the new spiritual power will have as [its] principle mission to inflame [its] spiritual charges with a passion for the public weal."[21]

The strategy of looking for positive roles for Christianity led Saint-Simon into a new and quite remarkably fruitful historical synthesis. He now understood history to oscillate between organic periods, dominated by the unification of all aspects of culture, and critical, analytic, periods, dominated by the undermining of old patterns of natural knowledge, social institutions, and spiritual life. Overlaid on this recurring pattern of organic and critical periods was a progressive pattern in which natural knowledge moved increasingly away from anthropomorphizing and metaphysical foundations to positive scientific ones. Simultaneously, the military organization of society gradually gave way to an industrial organization as human "association" expanded its scope from immediate families to nation states to all of human kind. Finally, spiritual progress paralleled that of knowledge and society as religion moved from fetishism through polytheism to an increasingly inclusive monotheism.

Initially, Saint-Simon focused his historical attention on three eras, the first of which was the medieval period in Europe, during which culture was unified under social feudalism, religious catholicism, and rationalist metaphysics. These systems interlocked in such a way that they supported one another in creating an organic structure that Saint-Simon identified as the *feudal-ecclesiastical* regime. Within this regime, a kind of stability based on military might, metaphysical speculation, and the ignorance of the masses was achieved:

> The combination of physical force (possessed preeminently by those who bear arms) and the wily and foxy methods invented by the priests had invested the leaders of the clergy and the nobility with sovereign powers and had enabled them to subjugate all the rest of the population.
>
> No better system could have been established at that time; for, on the one hand, all the knowledge that we then possessed was still superficial and vague, and general metaphysics contained the only principles which could serve as a guide to our medieval ancestors and, consequently, the general metaphysicians had to guide society in scientific matters.[22]

There was, however, a kind of immorality and "monstrousness" about this system in the way that the nobility and clergy exploited the masses for their own gain.

The second period that Saint-Simon focused on was that of the coming industrial order, during which specialized positive sciences will have been replaced by a single unified general theory; society will be organized under the leadership of the scientists, artisans, and artists (who are also the leaders of industry) for optimal productivity; and religion will preach universal love for all humankind. Once again the world will see an organic period, but now without the negative features of the feudal-ecclesiastical order with its exploitative character.

The third period, that which received Saint-Simon's greatest attention, was the transitional period that had lasted basically from the beginning of the Protestant Reformation to the end of the Enlightenment. During this period the great projects of humankind were to break down the conjectural foundations of the feudal-ecclesiastical regime and to carve out room for the growth of the positive foundations of industrialism. This necessary act of destruction could only be accomplished by those with special analytic and critical skills—that is, by lawyers and metaphysical moralists. Thus, argued Saint-Simon, "I cannot see how the old system could have been modified and the new developed, without the intervention of the jurists and metaphysicians."[23]

Saint-Simon's Critique of Liberalism

If their functions were absolutely essential at one point, however, the jurists and metaphysicians had now outlived their usefulness: "By the very fact of the end it had in view, the political influence of the jurists and the metaphysicians was limited to an ephemeral existence; because it was of necessity modificatory and transitional. It had fulfilled its whole natural function at the moment when the old system had lost the greater part of its power. . . . whereas it has now, in fact become positively harmful because it has outlived its natural limit."[24]

In some respects this emphasis on the limitations of critical rationalism is completely consistent with all of Saint-Simon's earlier works, but it also signals an important reversal with respect to issues associated with his prior position regarding economic freedoms and voting rights. Saint-Simon argued that the lawyers and metaphysicians of the Enlightenment were constantly harping on the maintenance of individual freedom because the notion of freedom to act in the economic realm was a powerful destructive tool when used against the feudal order. Now, however, such a conception of freedom must function to hinder industrial progress: "It would act against the development of civilization and the organization of a well regulated system, which demands that the parts be firmly linked with the whole and dependent on it."[25] Industrial efficiency must henceforth be seen increasingly as something that demands administration by those particularly competent rather than something that follows from the uncoordinated free economic decisions of producers and consumers.

One of the reasons that the traditional conception of freedom was so powerful, of course, was that it fit so well into the "metaphysical" doctrine of natural law and natural rights that dominated early modern jurisprudence and the social thought of the Enlightenment. On this issue, Saint-Simon provides a fascinating bridge between the ideas of J. B. Say and those of Karl Marx and Friedrich Engels (1820–95). Say had rejected the nonempirical notion of rights in connection with economic value to replace it with the positive notion of measurable preferences derived from physical needs and wants. In this way he had justified replacing the dominant labor theory of value with a utility-based theory of value. Saint-Simon now argued that the great advantage of industrialism over liberalism in general for the new age was that, while the latter was entirely grounded in the old appeals to unverifiable rights, the former is grounded solely in observable "interests."[26]

For most men, who are oriented toward motor skills, political activity is not related to any need or interest. Thus Saint-Simon now argued, contra his earlier argument in *Lettres d'un habitant de Geneve,* that the separation of voting rights from a special "requirement of competence" made no sense, because the notion of political rights or liberties was grounded in a metaphysical notion of rights rather than in an understanding of the material interests of the masses. Those interests will be much better served if the most intelligent and knowledgeable persons make political decisions. Accordingly, in the fully realized industrial society, "the cultivation of politics will be exclusively trusted to a special class of scientists."[27] The notion that the political participation of the proletariat was not just unimportant, but that it was also potentially dangerous, was reemphasized by the later Saint-Simonians and by important elements within twentieth-century Marxism-Leninism.

If there was no real political need on the part of most men, according to Saint-Simon, there *was* a crucial need for self-expression through productive labor. Drawing from the discussions of self-motion that had been emphasized by Cabanis and Destutt de Tracy, Saint-Simon claimed that there was indeed one important liberty demanded by every human: "the Liberty to use his hands [and] his industry."[28] In his last words to his disciple, Olinde Rodrigues, Saint-Simon said that the highest goal of his entire doctrine was "to afford all members of society the greatest possible development of their faculties."[29] This meant that he sought to create a system that would give each man and woman the opportunity to labor in a job that made maximum use of their talents.

Recast by Marx in the language of dialectical materialist logic, and with capitalists incorporated into the important role of Saint-Simon's lawyers and clergy, the Saint-Simonian historical vision of a feudal order that contained the seeds of its own destruction giving way to a transitional period that must eventuate in the rise of an apolitical industrial world managed for the sake of the proletariat had a long-time hold on the beliefs of many communists throughout the world.

In Saint-Simon's industrial world, however, the traditional class conflicts that had thus far been the key engine of historical change would not disappear with the triumph of a single class. Instead, they would give way to a fully cooperative relationship between the managing class and the working class. Because access to positions of high prestige and power would be completely open to those with appropriate talent, those from the working class would have no reason to believe themselves inappropriately excluded from the managerial class by any accident of birth, and thus would not feel resentful. At the Hands of Vilfredo Pareto (1848–1923) and Antonio Gramsci (1891–1937) this basic argument was elevated to the theory of Democratic Elitism, which also insisted that the possibility of mobility into the high-status class of managers and politicians based on merit would reduce proletarian dissatisfaction.

It is true that in Saint-Simon's industrial society, compensation would not be equal: those who contributed more to society would receive both more prestige and more material rewards. But the differential compensation would be proportional to the differential contribution, so it would not seem unjust. In one of his more remarkably naive moments, Saint-Simon portrayed the completely accepting attitude of his future proletarian in a model speech to one of the managerial class: "You are rich, and we are poor. You work with your heads, and we with our hands. From these two facts, there result fundamental differences between us, so that we are and should be your subordinates."[30] Saint-Simon was arguably one of the great founders of sociology, but he was not always a very perceptive psychologist, and he totally abandoned Adam Smith's fundamental insight that every human's satisfaction is based primarily on her or his sense of wealth and status relative to others. The managers in Saint-Simon's ideal industrial society might have been very happy, but it is extremely unlikely that the workers would have been.

Saint-Simon's Religious Turn

During the Napoleonic era, with the opening of higher education to everyone with talent, a number of young Jews, who had been excluded from traditional education, entered the *Grande Écoles* and medical schools, a number of them graduating into faculty positions. But with the restoration of the monarchy and Catholic religious authority, Jews were once again excluded from state educational institutions, and those with faculty appointments were removed. Many of these displaced Jewish intellectuals turned to their family banking concerns for employment, but they also sought to promote a wider involvement of Jews in society. To these young men, Saint-Simonian doctrines were very attractive. On the one hand, Saint-Simon granted banking and financial institutions an openly central role in society, thus dignifying their work. On the other, the Saint-Simonians sought to eradicate all considerations other than ability in determining the social roles that individuals might take up. The mathematician Olinde Rodrigues was one of the first of Saint-Simon's new Jewish disciples, and one of the most devout. Rodrigues recruited two of his cousins, Emile and Isaac Pereire (1800–1875 and 1806–80), whose *Crédit Mobiliere,* organized along Saint-Simonian lines, was to eclipse even the Rothschild fortune in the financing of European industrial expansion. Rodrigues also introduced Saint-Simon to another Jewish friend, Leon Halevy (1802–83), who had lost his job as a rhetoric instructor at the *Lycee Charlemagne.* Halevy became Saint-Simon's new secretary, giving his last works a greater literary polish than those that had come before.

In the last of his works published before his death, *Nouveau Christianisme; Dialogues entre un conservateur et un novateur* (Paris, 1825), Saint-Simon returned once again to review virtually all of the major themes of his career. Once again he rehearsed the historical pattern of progress through alternating periods of critical and organic societies, and once more he berated those critical scientists who were satisfied with partial views and refused to seek a comprehensive general theory. Saint-Simon acknowledged that there was a divine core of truth embodied in the single primitive Christian principle that "men should treat each other like brothers." Yet, in religion as in all other things there was historical progress toward greater inclusiveness, and "theological theory needs to be renovated at various times, in the same way as physics, chemistry, and physiology." So Saint-Simon's New Christianity was grounded in the updated notion that its single guiding principle must rule the actions of the temporal as well as the spiritual domain and direct both individual and social actions. As a consequence, he articulated a new, more inclusive commandment: "*The whole of society should work to improve the moral and physical existence of the poorest class; it should organize itself in the way most suited to allow it to achieve this great end.*"[31] This was shortened by his followers for dramatic effect to "*All men shall work.*"

Whether the emotional tone of *The New Christianity* simply reflected Saint-Simon's shrewd assessment that young Parisians needed an alternative to Catholicism to fill their need for commitment to something larger than themselves, or whether it reflected his own feelings that he was literally the messiah of a new creed is not clear. One of his disciples simply said that he came to realize "that it is in the name of their *sympathies* that one must speak to men, and above all in the name of their *religious sympathies.*"[32] Moreover, just a few days before he died, Saint-Simon told Olinde Rodrigues of his reasons for presenting his new religious system: "The Catholic System was in contradiction with the system of sciences and of modern industry; and therefore, its fall was inevitable. It took place, and this fall is the signal for a new belief which is going to fill with its enthusiasm the void which criticism has left in the souls of men."[33]

The new belief was, of course, his own New Christianity. Regardless of his reasons for the change in tone, there is no question that *The New Christianity* appealed more deeply and to a much broader audience than any of his earlier works. It stimulated so much interest that when Saint-Simon died shortly after its publication, his disciples established a Saint-Simonian Church and later a commune that briefly attracted more than one hundred members and sometimes up to ten thousand curious onlookers.

The Saint-Simonians and Socialism

During the last few years of the 1820s, Parisians had their choice of three major series of public lectures that focused on the relationships between science and civilization. Victor Cousin's (1792–1867) exposition of "Eclecticism" was officially sanctioned by the government and the church. It sought to demonstrate the consistency of traditional Christianity with the scientific trends of the late Enlightenment and drew heavily from the works of the Scottish Common Sense Philosophers, Thomas Reid (1710–96) and Dugald Stewart (1753–1828), in so doing.[34] Auguste Comte's *Cours de philosophic positive* (to be discussed in the next chapter) appealed to secularists and those who had a taste for intellectual rigor as well as a more than passing knowledge of the sciences. The Saint-Simonian's *Doctrine de Saint-Simon,* delivered by Saint-Armand Bazard (1791–1832), but written by Hippolyte Carnot, on the other hand, drew a more diverse and socially concerned crowd of young engineers, financiers, industrialists, and artists.

A follow-on set of lectures delivered in 1829 by ex-*polytechnicienne* Abel Transon (1805–76), *De la religion Saint-Simonienne: Aux éleves de l'École polytechnique,* drew an almost exclusively engineering student audience, but these lectures were important for a number of reasons. First, they were instrumental in establishing the Saint-Simonian Church, led by Prosper Enfantin and Saint-Armand Bazard, to which they drew nearly one hundred *polytechnicienne* converts. Second, they contained the clearest and most powerfully expressed discussion of the new Saint-Simonian emphasis on the contemporary class conflict between the bourgeoisie and the working class. Third, through an 1832 German translation, the published lectures drew a large number of German intellectuals, including Karl Marx's father and his favorite Gymnasium teacher, to Saint-Simonianism.[35]

Those who attended the Saint-Simonian lectures heard a doctrine that grew out of the thought of the master but that went beyond it in significant ways. First, his self-styled disciples refashioned the doctrines that Saint-Simon had always characterized by the term "industrialism" into a set of doctrines that focused more heavily on class conflict and on the need to abolish those forms of private property that served as means of production. These new emphases they began to identify with the term "socialism" in 1832.[36] By 1840, working-class movements throughout Europe had appropriated both the Saint-Simonian term "socialism" and key elements of the Saint-Simonian understanding of not only the historical process by which the current situation had arisen, but also the future process by which class conflict might be eliminated and the need for government abolished.

Transon's version of the present class conflict was particularly clear and widely disseminated: "As the owner of land and capital the bourgeois disposes of these at his will and does not place them in the hands of the workers except on condition that he receive a premium from the price of their work, a premium that will support him and his family. Whether a direct heir of the man of conquest or else an emancipated son of the peasant class, this difference of origin merges into the common character which I have just described; only in the first case is the title of his possession based on a fact which is now condemned, on the action of the sword; in the second case, the origin is more honorable, it is the work of industry. But, *in the eyes of the future* this title is in either case illegitimate and without value because it hands over to the mercy of a privileged class all those whose fathers have not left them any instruments of production."[37]

The discussion of the abolition of property, on the other hand, was more clearly presented in *The Doctrine of Saint-Simon: An Exposition*, which was transcribed from the first two Saint-Simonian lecture series from 1828 to 1829: "[If] it is true that mankind is moving toward a state of things in which all men, without distinction of birth . . . shall be classed by society according to their merits and be remunerated according to their work; then it is evident that the constitution of property must be changed, since, by virtue of this constitution, some men are born with the privilege of living without doing anything, namely at the expense of others, which practice is nothing but the continuation of the exploitation of man by man. From one of these facts the other can be deduced logically: the exploitation of man by man must disappear. The constitution of property which is hereby perpetuated must, therefore, also disappear."[38]

If it is the case that virtually all subsequent secular socialist movements derived their views of prior class conflict and property from the Saint-Simonians, it is also true that they divided radically over a number of other issues.

Most working-class socialist movements saw the end of class conflict occurring when only the single class of working men—the proletariat—was left. Thus they envisioned a classless society in the future. The Saint-Simonians, however, following the thought of Saint-Simon on this issue, saw a class division based on the division of labor as continuing even when the complete state of association had been reached. There would still be managers, workers, artists, and scientists, and each of these "aggregations," or classes, would see that its prosperity must be bound up with the prosperity of all others.[39]

Many subsequent socialist movements were also militant and potentially violent, since they saw a continuation of class struggle as the only effective means of progress. The Saint-Simonians, on the other hand, perpetuated their master's opposition to struggle and confrontation in any form, whether it appeared as military or political conflict, confrontation between secular and religious authority, or even economic competition. "We do not believe in any of these engines of war," they insisted. Rather, they rested confident that the law of association that continued to expand the circle of sympathy and love among all humankind would ultimately produce universal cooperation without the regressive resort to revolution.[40] It is perhaps this belief more than any other that led Friedrich Engels to derogate Saint-Simonian socialism as "utopian," and therefore, unrealistic, rather than scientific.

Toward a Managed Economy with a New Role for Banks

Perhaps the most profound consequence of their rejection of competition in all of its forms lay in the Saint-Simonian intensification of Saint-Simon's distrust of economic liberty and economic competition. The Saint-Simonians were convinced that the laissez-faire approach to economics promoted by all liberal economists was misguided in at least four fundamental ways. First, because withholding knowledge that might increase productivity from one's competitors conveys economic advantage, the protection of proprietary knowledge is important in a competitive market structure, but it functions to lower overall productivity.[41] Second, because every economic actor is focused solely on his own private gain, without concern for systemic balance, talent and capital tend to flow rapidly toward new opportunities, overcrowding some fields and leading to wasteful overproduction and to "the complete ruin of innumerable victims."[42]

Similarly, technological innovation often leads to what we now call technological unemployment. While the Saint-Simonians were inclined to be optimistic about the long-term employment increases that eventuate from productivity increases, they were aware that under laissez-faire conditions, transient unemployment could have catastrophic consequences for the displaced workers. Under these circumstances, they argued, "society ought to see to it that the conquests of industry will not be like those of war."[43] Finally, the existence of inherited wealth often—even usually—places the instruments of production in the hands of those who are either ignorant or stupid and who therefore misallocate them, rather than in the hands of either skilled workers who might use them to some good effect, or more properly, in the hands of experts with full systemic knowledge of the economy, who would be in a position to use those resources to optimize production.[44]

Given all of the weaknesses of a free market, competitive, economic system and the industrial crises to which it leads, the Saint-Simonians argued for a cooperative economic system in which the allocation of resources was made by a central institution situated to understand all needs of the economy.[45] The institution that the Saint-Simonians proposed for this crucial function was an improved commercial banking system, which would be staffed by scientifically trained experts capable of recognizing systemic economic needs. Within this system, a central coordinating bank would be the repository of all funds for production and instruments of work. This central bank would preside over specialized banks, each responsible for a particular branch of industry or a particular locality so that it might have full knowledge of local conditions or the state of a particular industry.[46] Under such a system, "capital is at the disposal of the men most able to use it profitably, and the injustices, the acts of violence, and the egotistical inclinations with which the former privileged bodies . . . were reproached, no longer need to be feared. Indeed, every industrial body is only a part, a member, so to speak, of the great social body which includes all men without exception."[47]

Of course, the Saint-Simonians admitted, even the leaders of such a system could occasionally make mistakes, because all humans are imperfect. "But," they claimed, "it must always be admitted that men of superior ability, in a position to have a general view, free of the shackles of specialization, should be exposed to the least possible chance of error in the decisions with which they are entrusted."[48]

For better or for worse, the Saint-Simonian vision of and rationale for a centrally managed economy provided the guiding vision for the Soviet Union, much of Eastern Europe, and the People's Republic of China during most of the twentieth century. F. A. Hayek, among others, has also argued that it gave a particular cast to Continental European capitalist developments during the nineteenth century as well.[49] Many European banks modeled themselves on the Pereire brothers' *Crédit Mobilier,* whose policies differed radically from those of most English and American banks. For one thing, the European banks sought to monopolize whole industries. As the Annual Report of the *Crédit Mobilier* for 1854 stated, "When we enter a branch of industry, we desire above all to encourage its development, not through competition, but through *association and fusion,* by employing forces most efficiently, and not by their opposition and reciprocal destruction."[50] For another, they showed themselves willing to forego maximum interest rates from firms in high-risk ventures to concentrate on investing in lower yield and socially important sectors of the economy, especially in utilities and transportation. The Pereire's banks, for example, built the major railway systems in Belgium and France and substantial portions of those in Austria, Switzerland, Spain, and Russia, employing Saint-Simonian engineers, including Prosper Enfantin, to manage the work.[51] Later Saint-Simonians initiated both the Suez and Panama Canal projects, although it was left for others to complete them.

Saint-Simonianism, Science, and Religion

Saint-Simonianism prided itself on being scientific and justified its stress on historical evidence in terms of its scientific value. Thus, Carnot wrote in *The Doctrine of Saint-Simon:* "If we have placed much value on the observations about humanity, it was only to place us upon ground on which the enlightened men of our epoch feel themselves so securely planted, namely that of science. We wanted to show them that if we adopted new views about the social future, that is to say about the predictions of human phenomena which are unknown to them, we are following, in justifying these predictions, the same method that is employed in all the sciences. We wanted to prove to them that our prevision had the same source and the same bases as appear in scientific discoveries in general, or, in other words, that the genius of Saint-Simon was of the same nature of that of Kepler or Galileo, and differed only in breadth and in the importance of the laws which he revealed to us."[52]

If they wanted to claim the authority of science, however, precisely what they meant by that term and how they understood the relationship between science and human ends diverged markedly from the understandings of such predecessors as Condorcet and of such contemporaries as Comte and Cousin.

The way in which these changed views were formulated owed a substantial amount to German developments that were introduced to the Saint-Simonians by Abel Transon. This German background will be developed in more detail in chapter 4; but the content, if not all of the motivating details, of the Saint-Simonian position can be understood independently.

Expanding on the trend set by Saint-Simon in his *New Christianity*, the Saint-Simonians emphasized the feelings over reason as a foundation for society: "Feeling binds us to the world and to man and to all that surrounds us. When this bond is broken, when the world and man seem to reject us, when the affection attracting us toward them is weakened and annihilated, life ceases for us. Without those sympathies that unite man with his fellow-men . . . it would be impossible to see in societies anything but aggregations of individuals without bonds, having no motive for their actions but the impulses of egoism."[53]

Given the fact that the motives for actions can only emerge out of feelings, especially sympathy, science is limited in its functions. At one level, it is purely instrumental: "Science . . . can indicate the means to be used in attaining a certain goal. But [it cannot tell us] why one goal rather than another? Why not remain stationary? Why not even retrogress?"[54] Those who, like their competitor, Comte, thought that the sciences could tell one what to do as well as how to do it, were badly misled. At a second level, science was at least capable of "'verifying' the hypotheses of feeling,"[55] that is, even though science could not create the feelings of love and association that were increasingly unifying all humankind, it could collect and assess the evidence for their existence.

One critical consequence of the Saint-Simonian understanding of the functions of science relative to those of feelings and motives, was that, for the Saint-Simonians, there could be no basic conflict between science and religion. Comte might argue that religion represented a primitive stage of knowledge that would ultimately be superseded by a positive, scientific stage. Quite the opposite, argued the Saint-Simonians: "It is not the destiny of science . . . to be the eternal enemy of religion and constantly to restrict religion's realm in order some day entirely to dispossess it. On the contrary, science is called upon to extend and constantly strengthen the realm of religion, since each of science's advances is to give man a broader view of God and of His plans for mankind."[56] As we have already seen, the Saint-Simonians were far from being orthodox Christians, but they understood themselves to be deeply religious, and they certainly decried the notion that science must be intrinsically opposed to religion.

The Saint-Simonians did not follow either Kant or Comte in claiming that because religious beliefs and feelings are inaccessible to our senses, they cannot be subject to rigorous scientific investigation. No historical event is directly accessible to our senses, yet we can confidently infer the existence and character of historical events from their evidentiary remains. Similarly, the pleasure that one gets from listening to music is not directly observable, but it can, in effect, be observed indirectly "by means of the actions it calls forth."[57] Analogically, even though one may be unable to directly sense a religious principle, one can infer from the way that people behave whether certain ideas are tending to disappear or whether "they are spreading and growing firmer at each of the great revolutions the human race undergoes."[58]

Of course, it is true that any inference about human behaviors demands that we make certain assumptions or hold certain hypotheses about the similarity between our reactions to ideas and the reactions of past actors. In this way we have to admit that our acceptance of the "fact" of religious belief depends upon a set of assumptions or provisional hypotheses. But, said the Saint-Simonians, our acceptance of all scientific facts is of the same kind: "We must note that in reasoning about observed facts, we cannot dispense with a hypothesis, no matter what its character may be, and that this hypothesis always precedes and does not follow reasoning."[59]

The Copernican hypothesis that the earth moves around the sun, for example, is capable of allowing us to coordinate an amazing amount of experience, and it underlies even the "fact" of universal gravitation. Ultimately, however, that hypothesis cannot be directly subjected to experience. Its acceptance simply permits us to make inferences that can in turn be verified by our experiences.[60] In exactly the same way, the religious belief that a God of love exists is ultimately unprovable or inaccessible to direct experience,[61] but it may allow us to make sense of social experiences that otherwise remain chaotic. Moreover, treated as a hypothesis it can be "verified" by its capacity to unify those experiences.

The Saint-Simonian understanding of science was dynamic rather than static, and science was capable only of generalizing experience, not of attaining some absolute truth. Thus, for the Saint-Simonians, every scientific knowledge claim was provisional and subject to modification when some new hypothesis showed itself capable of incorporating more knowledge than the old. Indeed, this principle was extremely important; for in the alternation between critical and organic stages of societal development, each organic phase was organized around a new hypothesis, the creation of which "is always the first necessary step toward each new coordination of facts."[62]

Since the goal of all science is to unify phenomena, or express them as general facts, it follows that the ultimate goal of science should generally be "to relate all isolated laws as far as possible to one single law."[63] Thus, the Saint-Simonians continued the opposition to the positive, specialized, sciences that Saint-Simon had initiated some twenty years earlier. They also made and supported the further claim that in organic periods science must not only not oppose religion, but that it must itself become theological or dogmatic in the sense that the specific hypotheses of all branches of science must become alternative expressions of the single general hypothesis, "which serves as dogma or as the basis of the general science and of human knowledge."[64] Moreover, that general hypothesis must not only constitute the foundation of scientific knowledge, but of religious belief and social organization as well.

Saint-Simonian Scientism and Science

Summarizing their conception of the roles, characteristics, and limits of future science at the end of the final session of the first Saint-Simonian lecture series, Hippolyte Carnot wrote:

> We foresaw a time, no longer distant, when the sciences, freed from the dogmas of criticism and viewed in a much broader and general fashion than they are today, would no longer be considered antagonistic to religion, but rather the means . . . to spread, support, and strengthen the religious sentiment. . . .
>
> On the other hand, we pointed out that scientific procedure or method always presupposed axioms and beliefs before application; that its aim was merely to classify and arrange facts according to the hypothetical conception of a relation or link existing between them, and thus to confirm this conception. In other words, we said that, strictly speaking, there existed no [unique] method for discovery, imagination, conception, or creation, and that sentiment always formed the basis of science, limited its sphere, guided its research, and determined the order of the classifications by furnishing science with a criterion for [recognizing] the differences or analogies among phenomena.[65]

These understandings of the function and characteristics of science did violence to the understanding of their own activities by most French and English mathematicians, astronomers, natural philosophers, chemists, natural historians, and physiologists and philosophers during the first third of the nineteenth century, although, as we shall see, the second paragraph might well have been acceptable to a large number of German practitioners of the various disciplines at the time and to many practitioners of the social studies of science at the beginning of the twenty-first century.

Given their understanding of the historical and future goals and characteristics of science, the Saint-Simonians unquestionably understood themselves as scientists, developing a science that comprehended the events of the social as well as of the natural world. Moreover, nearly thirty years before Friedrich Engels appropriated the term "scientific socialism" for the doctrines that he and Karl Marx had developed, Karl Grün (1817–87) had applied the term to Saint-Simonianism.[66] Yet, by 1872, Engels derided Saint-Simonian socialism as "utopian" rather than scientific. Moreover, in the middle of the twentieth century, F. A. Hayek, the distinguished advocate of laissez-faire economics, again derided the Saint-Simonian claims to being scientific, identifying their doctrines as the source of modern *scientism,* which he chose to characterize as "concerned, not with the general spirit of disinterested inquiry but with slavish imitation of the method and language of Science," and as "decidedly unscientific in the true sense of the word, since it involves a mechanical and uncritical application of habits of thought to fields different from those in which they have been formed."[67]

For many purposes, whether we call Saint-Simonianism scientific or accept Hayek's definition of scientism and his identification of Saint-Simonianism as scientistic makes little difference. As long as members believed themselves to be doing science and were able to convince others to accept their claims as grounded in the authority of science, whether they were "really" doing science or only "slavishly imitating the method and language of science" is completely unimportant. Either way, their ideas had a major impact upon nineteenth-century social movements and European economic development. On the other hand, to the extent that "science" continues to carry some authority in our culture, important political and social commitments may be at stake in choosing to accept, in part or whole, the claims of the Saint-Simonians or those of Hayek. Suppose, for example, that Hayek is correct in claiming that societies are nothing but aggregations of individuals and that their characteristics and behaviors must be understood in terms of the intentions and behaviors of the individuals who constitute them. Then the Saint-Simonian assumption that the behaviors of groups, classes, and even entire societies can be described and understood even though we do not know how those behaviors follow from the characteristics of the individual members must be wrong. Suppose also that Hayek is correct in believing that the full acceptance of the implications of methodological individualism leads one to the doctrines of laissez-faire capitalism, while the full acceptance of Saint-Simonian collectivism leads one to accept the doctrines of centrally controlled socialist economics. Then we would be ill-advised to promote any but libertarian economic doctrines.

If, on the other hand, Hayek is wrong and *he* is actually the scientistic one—he certainly does claim that without a scientific justification, humans should not attempt to manage society, and his claim that methodological individualism provides the only legitimate approach to understanding social phenomena is arguably more presumptive than the Saint-Simonian alternative—then socialist doctrines would seem to deserve a more careful hearing.

By Hayek's standards, it seems to me that Saint-Simon and the Saint-Simonians were not scientistic. For far from slavishly and uncritically imitating the dominant methods of contemporary science, they were highly critical of many of them. Moreover, they were more self-conscious about the relationship of philosophical assumptions to the character of their science than the vast bulk of contemporary natural scientists. My own position is that if we accept that *a science is any set of activities and habits of mind aimed at contributing to an organized, universally valid, and testable body of knowledge about phenomena* and that we can properly speak of "scientism" whenever *scientific attitudes, methods, and modes of thought are extended and applied beyond the domain of natural phenomena to a wide range of cultural issues that involve human interactions and value structures*—then the Saint-Simonians and F. A. Hayek were equally scientific *and* scientistic.

For purposes of the present argument, we will admit the scientific character of Hayek's economic doctrines and focus on Saint-Simonianism. The Saint-Simonians were insistent that their enterprise was aimed at seeking universally valid knowledge. They were equally insistent that their historical knowledge was verifiable or testable with reference to historical evidence, and I know of no reason to doubt their sincerity. Moreover, the fact that they admitted that their attempts at understanding incorporated hypotheses that were not independently and directly testable cannot count against the scientific character of their doctrines, because it is widely acknowledged that such hypotheses are unavoidable in the natural sciences as well. One might argue, as Auguste Comte did, that an alternative law of historical development was better than that of the Saint-Simonians because it fit the evidence better, but such an argument challenges the *adequacy* rather than the *scientific character* of Saint-Simonianism.

Saint-Simonianism was also scientistic according to my definition. Although it ultimately argued that scientific knowledge should be subordinated to "sentiment" or to values that come from outside of science, it placed a high value on the instrumental importance of science. Furthermore, it openly and critically appropriated concepts and explanatory structures from the natural science of physiology to apply to the understanding of society.

The fact that two or many equally scientific approaches to human phenomena imply radically different courses of action should not be particularly disturbing, though it should suggest a degree of caution in how we evaluate the prescriptive claims of the social sciences in particular. It should not be particularly disturbing because the fact that scientists *aim* to discover universally valid knowledge does not guarantee that they will always (or ever) be completely successful. The history of science demonstrates unequivocally that scientific knowledge claims *are* fallible and scientific knowledge *is* grounded in the acceptance of assumptions or hypotheses that are not directly testable.

In one sense, I am as scientistic as either the Saint-Simonians or Hayek, for I believe that the appropriation of methods, habits of thought, concepts, and guiding metaphors from the natural sciences by those who seek knowledge of human social phenomena offers some of the most reliable knowledge to which we have access in guiding our own actions. But I also believe that social scientific knowledge should be very carefully evaluated and used. It is extremely sensitive to the often unexamined assumptions that underlie it and that often reflect the particular interests of some group. It is vastly more difficult to "test" than most natural scientific knowledge both because of the number of salient variables that may not be controllable and because it often deals with phenomena that are difficult if not impossible to duplicate. Finally, it may even be that the widespread human perception of choice is justified and that human behavior and institutions may never be more than "softly" deterministic.

CHAPTER 3

Auguste Comte and Positivisms

Saint-Simonianism remained primarily a Parisian and almost exclusively a Continental European phenomenon, though many of its doctrines and perspectives were selectively integrated into international socialist movements. Moreover, though it continued in Paris as a minor feature of intellectual life until the end of the nineteenth century, it ceased to play a major role as a coherent movement anywhere after the revolutions of 1848. The views of another post-revolutionary Parisian scientistic prophet, Auguste Comte, drew smaller numbers of Parisian advocates before 1865. But Comte's "Positivism" developed a much greater following throughout the world and over a longer period of time. Indeed, it is arguable that Comte's Positivist philosophy and its offshoots came closer to rivaling Aristotle's philosophy, not only in scope but also in intellectual and practical consequences, than the writings of any intervening thinker.

Comte had his most important initial readership in England from about 1838 to 1860. There, although he had a few true disciples, such as the Oxford historian Richard Congreve (1818–99), his ideas were selectively appropriated by some of the most visible early Victorian intellectuals, including John Stuart Mill (1806–73), Henry Lewes (1817–78), George Eliot (1819–80), Harriet Martineau (1802–76), and Herbert Spencer (1820–1903). In France, interest in Comte's system was slower to develop, growing rapidly only after about 1865 and peaking around 1890 with the nearly simultaneous establishment of the first university chairs in sociology and the history of science, both held by self-avowed Positivists.[1] The political side of Comte's philosophy had some impact on the careers of such major French Republican politicians as Leon Gambetta (1832–82) and Jules Ferry (1832–93),[2] but it had its greatest direct impact outside of Europe, in Latin America and India from the late 1860s into the twentieth century.[3] In fact, since World War II, the financial and intellectual leadership of Positivism has come largely from Brazil, whose flag incorporates the Comtean motto, *Ordem e Progresso.*[4]

In this chapter, I consider Positivism as it was developed by Auguste Comte over the forty-year period from 1817 to his death in 1857. European responses to and appropriations of Comtean ideas will form a significant part of nearly all of the remaining chapters of this work.

The "Two Careers" of Auguste Comte and the Plurality of Positivisms

In more ways than Comte was inclined to admit even to himself, his attitudes and ideas paralleled or derived from those of Saint-Simon, who became a combination mentor and colleague to the younger Comte in 1817. On the other hand, there were crucial differences between the systems of the two thinkers that had a major bearing on their respective long-term academic and political trajectories. First, Comte was more competent and systematically trained in the sciences than Saint-Simon. He had attended the *École polytechnique* and studied mathematics with Gaspard Monge before being expelled for being among the leaders of a student protest against particularly poor teachers. Afterward he had attended the medical school at Montpellier before returning to Paris to become an examiner and tutor in mathematics at the *École polytechnique.* As a consequence of this education, though he agreed with Saint-Simon that the sciences should be unified in their goal of serving humankind, he was much more sympathetic to and knowledgeable about the differences in the methods used in the different sciences. As a result, his historical analysis of scientific developments was more sophisticated and convincing to knowledgeable readers than that of his adoptive intellectual father.

Second, although he foresaw the same long-term emergence of a two-class system based on the division between a managerial elite and the working proletariat as Saint-Simon, Comte presumed—indeed he promoted—the continuing growth of individualistic capitalism. He optimistically hoped that the owners of capital could be educated to voluntarily place their wealth at the disposal of those with expertise to use it most effectively in the general interest and in the service of others. This focus on voluntarism and the continuation of private capital made Comte's views much less interesting than those of Saint-Simon to those in working-class movements but much more acceptable to relatively wealthy and well-educated members of the bourgeoisie.

Third, because Comte believed that all societies go through the same historical sequence of developments, though not at the same time or pace, and that the character of the members of any society is determined by the features of the society, he did not believe in the intrinsic superiority of Europeans, nor did he approve of the exploitative character of colonial policies. Instead, he was convinced that Positivism could show the way toward an accelerated progress for non-European societies and he argued explicitly that though he expected his doctrines to spread first in

France and the other nations of Western Europe, in the long run they would spread "to the rest of the white race and finally to the other two great races of man."[5] For this reason, although it was in many ways less progressive than Saint-Simonian socialism, Positivism had substantial appeal among middle-class non-European leaders in the colonial world in spite of its blatantly Eurocentric perspective (i.e., its goal was to bring all societies into the European pattern).[6]

Fourth, Comte believed that in order to make it possible to use scientific knowledge to benefit rather than exploit humanity, it would be necessary to establish a religion based on the expansion of love to all humankind. But he was convinced that Saint-Simon's attempt to initiate his "New Christianity" was premature and ill advised. The success of such an attempt was certain to be dependent on the insights into human behavior that only a complete science of society could provide; and that science had not yet been established. Indeed, it was the project of establishing a satisfactory foundation for a science of society that Comte undertook in his *Cours de philosophie positive,* begun in 1826 and published in six volumes between 1830 and 1842.

Fifth, Comte, insisted that even though scientific knowledge is developed primarily for purposes of application, only if one could step back and produce a value-free, dispassionate knowledge, would that knowledge provide a completely adequate foundation for subsequent action. This point was made in a particularly clear way in an 1822 essay, "The scientific operations necessary for reorganizing society." There he wrote: "Admiration and reprobation of phenomena ought to be banished with equal severity from every positive science, because all preoccupations of this sort directly and unavoidably tend to hinder or mislead examination. Astronomers, physicists, chemists, and physiologists neither admire nor blame their respective phenomena. They observe them.... It should be the same, in this respect, in political science as in the other sciences. Such exclusion of admiration or reprobation, is, however much more needed in the former, because there it becomes more difficult and affects the investigation more deeply, inasmuch as in this science the phenomena are much more closely connected with the passions than any other."[7] In the *Cours de philosophie positive* (hereafter, *Positive Philosophy*),[8] Comte presented himself as a completely objective thinker, producing a historically driven philosophy of science that explained why scientific knowledge deserved to be considered authoritative and why the sciences developed in a certain order, culminating in social physics, or sociology, which would provide the basis for social regeneration. In addition, Comte explicated the methodological characteristics that sociology must have.

The *Positive Philosophy* attracted and stimulated a number of powerful thinkers who both were interested in the growing status of the sciences in nineteenth-century society and in creating a secular science of society. For example, John Stuart Mill, who had written on political economy and who was in the process of writing his own *Logic of the Sciences* when he was introduced to Positivism, admired Comte's work so much that he provided Comte with financial support and promoted his work widely during the 1840s. Those, such as Mill, who had been drawn to Positivism primarily by Comte's discussions of the history and philosophy of science and by the secular-scientific character of Comte's proposed sociology, subsequently came to be called "intellectual" or "scientific" Positivists. They focused on what they later called the "first career" of Comte or the "first phase" of his career; and their relationship to Comte tended to be less that of disciples to a master than that of independent thinkers who had been stimulated by an initial admiration for Comte's work to take up the questions he had written on in ways that often went beyond that of Comte, sometimes in a highly critical fashion.

A number of these scientific Positivists were so appalled at the direction that Comte's work took in the late 1840s that they sought to appropriate the term to themselves and to argue that Comte had abandoned "true" Positivism. Thus, for example, in his *Auguste Comte and Positivism* of 1865, Mill refused to consider Comte's later works because, "in their general character we deem the subsequent speculations false and misleading."[9] Some scientific Positivists even went so far as to claim that Comte had become insane and abandoned his scientific work. For example, George Sarton, the self-proclaimed Positivist founder of the International History of Science Society, argued that after 1844, "There is no doubt that Comte was crazy. . . . The prophet in him had killed the man of science. The less he knew, the more he preached."[10] Similarly, Donald G. Charlton, in an excellent history of French Positivism during the Second Empire, argues that Comte became an enemy to Positivism and that the most important "true" friends of Positivism between 1852 and 1870 were Émile Littré (1801–81), founder of the journal, *La philosophie positive*, in 1867 and the experimental physiologist, Claude Bernard (1813–78), author of a widely read and emulated *Introduction to the Study of Experimental Medicine.*[11]

Between 1844 and 1847 Comte did undergo a series of experiences that caused him to return to the subjects of social reform and the new religion of universal love—now called the Religion of Humanity—that he had projected in the 1820s. But now he approached them with greater intensity and a new, "subjective" emphasis. In 1844, Comte, estranged from the wife who had supported him through nearly two decades of bizarre behavior, became obsessed with Clotilde de Vaux, a younger woman who had been abandoned by her husband. Through his experience of "sublime selfless passion" for Clotilde, Comte felt himself finally prepared for his second career. In a letter to Clotilde from March 1846, less than a year before her death to tuberculosis, Comte wrote: "To become a perfect philosopher I lacked above all else a passion at once profound and pure that would make me appreciate sufficiently the emotional side of humanity. Its explicit consideration, which could only be accessory in my first great work, must, on the contrary, now dominate my second. . . . If you only knew the progress I have made in the last year, in the midst of these apparent perturbations, toward my principle philosophical goal, the final systematization of the whole of human existence around its true universal center: love."[12]

For those who were to become orthodox "religious" Positivists, Comte's love for Clotilde was merely the occasion that brought him back to his early project, which was to focus on the application of sociology to the restructuring of society and to the construction of a Religion of Humanity, now that the science of sociology had been placed on a secure footing. From this perspective Comte's two careers were both in the service of a single goal; and there was no major change in orientation, just a difference in degree of emphasis.[13] The *Système de politique positive,* which appeared in four volumes between 1851 and 1854, The *Catéchisme positiviste* in 1852, and the first part of the *Synthèse subjective* in 1856, simply seemed to be completing the work of the *Cours de philosophie positive.*[14]

On the other hand, there is something disconcerting about many of the specific and detailed features of Comte's Religion of Humanity—including his insistence that Clotilde de Vaux be worshiped as the symbol of the Virgin Mother of the Great Being (Humanity), and that all of his pronouncements as the High Priest of Humanity be rigorously obeyed without question—that suggests a lack of balance. Relatively few of his disciples were willing to follow him and the successive heads of the Positivist Church into all of the details of worship he proposed, although there was a small group of super-orthodox Positivists, led by the English disciple, Richard Congreve, who did so.[15]

Conflicts among different groups of self-styled Positivists often became very heated. As a historian I feel the same kind of obligation to acknowledge the legitimacy of a variety of claimants to Positivism, antagonistic as they may have been to one another, as I do to acknowledge that Catholics and Protestants alike had legitimate claims to be Christians. As a consequence, I will treat both the relatively diverse and heterodox group of scientific or intellectual Positivists and the orthodox or super-orthodox religious Positivists as legitimate "Positivists" in what follows both in this chapter and in the remainder of the work, but I warn readers that much of the secondary literature is highly sectarian and contradictory on this issue, and anyone approaching the literature on Positivism should try to be clear about what perspective the author has on the subject.

The *Cours De Philosophie Positive*

At the very beginning of the *Positive Philosophy* Comte argues that the significance of his work can only be understood in the context of a historical overview of the development of the human intellect, because "no conception can be understood otherwise than through its history."[16] Whether this claim is true for all conceptions, it certainly does seem to be true for the central conceptions of Comte's philosophy—the Law of Three Stages and the Classification of the Sciences—the sources of which would be completely bewildering without an awareness of Comte's links to Saint-Simon. Drawing from the extensive historical discussions of his mentor, but modifying and simplifying his terminology, Comte boldly announces that our historical experience provides a "solid foundation of proof" for the following fundamental law: "Each of our leading conceptions—each branch of our knowledge—passes successively through three different theoretical conditions: the Theological, or fictitious; the Metaphysical, or abstract; and the Scientific, or positive. In other words, the human mind, by its nature, employs in its progress three methods of philosophizing, the character of which is essentially different, and even radically opposed: viz., the theological method, the metaphysical, and the positive."[17]

In the first of these stages or conditions, humans suppose all phenomena to be caused by the immediate action of supernatural beings. In the second, humans replace anthropomorphized supernatural beings with abstract forces that are still presumed to have some real existence independent of the effects that they produce. In the final, positive, stage, humans abandon the search for essential and absolute causes of phenomena entirely in order to focus on "the study of their laws—that is, their invariable relations of succession and resemblance."[18] Thus, for his positive stage, Comte incorporates the explicitly anti-metaphysical and acausal understanding of positive laws that had been developed by Cabanis and the *Idéologues*. But Comte also added to the *Idéologue*'s notion with his insistence that laws could express relations of resemblance as well as relations of succession. Thus, any subject matter was capable of description by two different kinds of laws, which Comte called "dynamic" and "static." The former represented development, or "progress," over time and the latter represented structural relationships or the "order" of a system at any particular time.[19]

In further explaining the characteristics of both the Positive Philosophy and the Law of Three Stages, which was the paradigmatic dynamic law, Comte makes a series of interesting observations. Though the Positive Philosophy is not metaphysical, Comte acknowledges that it is grounded in two critical assumptions: first, that all phenomena are capable of being incorporated into invariable natural laws, and second, that it is an important goal to reduce the number of natural laws "to the smallest possible number."[20] Unlike the Saint-Simonians, however, Comte insisted that it was not his goal to reduce all knowledge to a single comprehensive law—a project that he ridiculed as "chimerical."[21] What unity could be expected was one of method—and even there, the unity of methods in different sciences was not to be understood as identity, but rather as a kind of unity characterized by the image of several branches emerging from a single trunk.[22]

The methods of the various sciences are related to one another because all sciences are ultimately aimed at two closely related goals: the satisfaction of our psychological need to dispose facts in a comprehensible order and the prediction of phenomena in order to be able to act in the world.[23] But the details of how positive scientific methods appropriate to the various domains of knowledge have developed, argued Comte, could only be discovered by observing science "in action," that is, through a detailed history of the various sciences.[24] Comte thus understood his philosophy, not as a prescriptive one, but as one that reflected the actual practices of scientists. For Comte, the Positive Philosophy had been developing since the time of Bacon and Galileo; and it was well represented in natural philosophy in the work of Newton.[25] Thus Comte, himself, seemed to justify the later intellectual Positivists' claim that Positivism should not be limited to the philosophy of Auguste Comte alone, but that it should be understood more broadly as comprehending any philosophy grounded in the study of modern scientific practices and in the rejection of theological and metaphysical assumptions and modes of explanation.

Comte did single out one purported "science" for special critical treatment at the beginning of the *Positive Philosophy*—the introspective psychology that had been developed in England and Scotland by John Locke, David Hume, David Hartley, and, more recently, by Thomas Reid and Dugald Stewart. In France it had been promoted by Etienne Condillac and more recently by Comte's unnamed nemesis, Victor Cousin. Cousin's Eclectic Philosophy, also called "Spiritualism" because of its compatibility with Christian beliefs, drew very heavily from the Scottish school, offering an interpretation of the progress of human intellectual development grounded in introspection, or psychology, that posited a domain of "internal facts."

Beginning in 1830, Cousin's works became the foundation for the official philosophy course taught in all French *Lycées*, much to Comte's dismay. Railing against the psychological foundation of Eclecticism, which he characterized alternatively as "the illusory psychology, which is the last phase of theology" and as the "pretend science" of metaphysicians, Comte offered a two-pronged critique of psychology as it had been practiced. One prong was based on logical and physiological claims and the other on a sociological consideration.

To understand the first, one must be aware that Comte, like many other contemporary scientists, was a believer in phrenology, which assigned different mental functions to different regions in the brain. At the same time, he accepted from Cabanis the notion of sympathetic influences, which implied that two functional regions of the brain could never be completely separated. Given these two assumptions, Comte wrote: "It may be said that a man's intellect may observe his passions, the seat of the reason being somewhat apart from that of the emotions in the brain; but there can be nothing like the scientific observation of the passions, except from without, as the stir of the emotions disturbs the observing faculties more or less. It is yet more out of the question to make an intellectual observation of an intellectual process. The observing and observed organ are here the same, and its action cannot be pure and natural. In order to observe, your intellect must pause from activity; yet it is this very activity that you want to observe. If you cannot effect the pause, you cannot observe: if you do effect it, there is nothing to observe."[26]

If we insist that mental activities must be sequential rather than parallel, or that if they do proceed in parallel they must somehow interfere with one another, then this criticism has some plausibility. While I know of no pre-Comtean introspective psychologist who explicitly discussed the impossibility of simultaneous independent mental processes in a single location, during the early nineteenth century, scientists in other fields were having great difficulty conceptualizing the existence of simultaneous competing processes in a single region of space. Thus, for example, it took students of heat decades after they had begun to study heat radiation to recognize that surfaces could simultaneously emit and absorb heat.

Comte's second criticism of introspective psychology was that it had been incapable of producing any consensus: "After two thousand years of psychological pursuit, no one proposition is established to the satisfaction of its followers. They are divided, to this day, into a multitude of schools, still disputing about the very elements of their doctrine. This interior observation gives birth to almost as many theories as there are observers. We ask in vain for any one discovery, great or small, which has been made under this method."[27]

In this passage Comte addresses an issue that was absolutely critical to him. If the science of sociology was to be valuable in what he viewed as the great political and moral crisis of his age, it would have to be able to overcome the intellectual anarchy that existed as a consequence of the simultaneous existence of theological, metaphysical, and positive political and moral philosophies. But it could only help to the extent that it, like other positive sciences, was capable of producing universal acceptance of its claims among those who could understand them. That is, the science of society could promote order in society only because "there is no freedom of conscience in the sciences, in the sense that the mind is not free to refuse assent to what has been proved."[28] Introspective psychologists had demonstrated themselves incapable of proving their claims to competent colleagues because they had been unable to predict consequences that could be verified; thus they were not in possession of a positive science; and no reliable sociology could be established that depended in any way on their claims.

Comte had mentioned in connection with the Law of Three Stages that different kinds of knowledge move through the three stages at different rates, such that those sciences whose phenomena are most general, simple, and independent mature first. But because the classification of sciences based on their degree of complexity and historical sequence of maturation was so important in understanding the appropriate methods to be used in each, Comte devoted an entire chapter in the *Positive Philosophy* to discussing the Hierarchy of the Positive Sciences, which became for him the paradigmatic static law. In this discussion, Comte begins by claiming that "it is clear, à priori that the most simple phenomena must be the most general; for whatever is observed in the greatest number of cases is of course the most disengaged from particular cases." Furthermore, the simplest phenomena are the easiest for us to understand, and finally, "the simple phenomena are the furthest removed from man's ordinary sphere, and must thereby be studied in a calmer and more rational frame of mind than those in which he is more nearly implicated; and this constitutes a new ground for the corresponding sciences being developed more rapidly."[29] Here then, is Comte's way of reintroducing the notion that scientific knowledge is most effectively sought from a stance of disinterest, because we can be more "rational" when we have no special perceived stake in the outcome of our investigations.

Given the need to base the classification of sciences on the simplicity of their phenomena, Comte argues that the first and most fundamental distinction that should be made is between inorganic and organic phenomena, because organized bodies, the objects involved in organic phenomena, are more complex and less general than the objects of the inorganic world and because they are to some substantial degree dependent upon the laws of inorganic phenomena. He is careful to argue that this division does not presume anything about whether the "natures" of organic and inorganic entities are different, for positive science has nothing to say about natures.[30] Inorganic phenomena are then divided into celestial and terrestrial, with celestial phenomena being the simplest and least dependent. Terrestrial phenomena are subsequently divided into physics and chemistry, the latter being more complex than the former.

Finally, organic phenomena are divided into those that deal with individual living entities—which are the subject of physiology—and those that deal with entire species—the subject of sociology. The latter are more complicated again than the former, because "in all social phenomena we perceive the working of the physiological laws of the individual; and moreover, something else which modifies their effects and which belongs to the influence of individuals over each other." This "something else" is critical to note; for Comte was insistent, in opposition to some physiologists, that social physics is not merely a branch of physiology: "The phenomena of the two are not identical, and it is of high importance to hold the two sciences separate. As social conditions modify the operation of physiological laws, social physics must have a set of observations of its own."[31] Indeed, as we move from the simplest science to the most complex in the passage from astronomy to physics, to chemistry, to physiology, to sociology, each more complicated science is simultaneously dependent upon the knowledge of the preceding science and *not* reducible to it, because new circumstances come into play.

That this set of structural relationships among the sciences is natural is confirmed by the fact of their historical emergence in the sequence given by their structural relationships to one another. The validity of Comte's taxonomy is also suggested by the fact that there is a constant decrease in the precision of the knowledge that is attainable in each science as we move from astronomy to sociology. On this issue, however, Comte has a caveat to offer that still deserves consideration today: "We must beware of confounding the degree of precision which we are able to attain in regard to any science, with the certainty of the science itself. The certainty of the science, and our precision in the knowledge of it, are two very different things which have been too often confounded; and are still so. . . . A very absurd proposition may be very precise; as if we should say, for instance, that the sum of the angles of a [plane] triangle is equal to three right angles; and a very certain proposition may be wanting in precision in our statement of it; as, for instance, when we assert that every man will die."[32] In general, different sciences continue to be marked by different degrees of precision, and practitioners of those that are more precise look down on those that are less so, but it is still true that accuracy and precision are not identical.

The issue of precision does bring up one final general issue for Comte; that is, the place of mathematics in the classification of the sciences. On the one hand, Comte viewed mathematics less as a part of natural science than as "the most powerful instrument that the human mind can employ in the investigation of the laws of natural phenomena." From this point of view, Comte saw abstract mathematics as an extension of logic. But geometry and rational mechanics were, he thought, following his teacher, Monge, ultimately based in experience.[33] Thus, they could be thought of as the first natural sciences, antecedent even to astronomy. Consequently, Comte began his analysis of the specific sciences with a long treatment of mathematics that was frequently separately translated and published.[34]

The bulk of the *Positive Philosophy* constituted a detailed historical analysis of the various sciences, which many contemporaries viewed as a great accomplishment in its own right. This material is now of interest primarily to historians of science as a founding element in their own field. There is, however, one feature of the positive method as discussed by Comte—that is, the use of hypotheses—that needs to be considered because of the problems it raised for subsequent hardline Positivists, especially in connection with their reaction to Darwin's theory of evolution by natural selection and in relation to their long-standing rejection of atomic theory and statistical mechanics in physics.

In his discussion of the development of physics, Comte reiterated a general introductory claim that in all scientific investigations we begin by making a provisional supposition regarding the objects to be investigated. This provisional supposition, or hypothesis, is absolutely necessary, and "without it all discovery of natural laws would be impossible in cases of any degree of complexity."[35] If we were free to make any arbitrary provisional assumptions, on the other hand, our intellectual work would be incredibly inefficient. So Comte insists that a single critical constraint—presented in two logically equivalent formulations—be placed on all allowable hypotheses. In its first form, Comte writes, "this condition is to imagine such hypotheses only as admit, by their nature, of a positive and inevitable verification at some future time."[36] Since Comte presumed that the verification of scientific laws depended on their ability to predict some outcome of events such that the prediction could be discovered to be either true or false, this limitation of hypotheses seemed to many of Comte's followers to exclude the hypothesis of evolution by natural selection. This was so because that hypothesis seemed to be incapable of predicting specific events and therefore of being verified.

According to its alternative formulation, "Philosophical hypotheses must always have the character of simple anticipations of what we might know at once, by experiment and reasoning, if the circumstances of the problem had been more favorable than they are. . . . If we try to reach by hypothesis, what is inaccessible to observation and reasoning, the fundamental condition is violated, and the hypothesis, wandering out of the field of science, merely leads us astray."[37] In particular, any hypotheses that posit an unobservable agent to which to refer observable phenomena "are chimerical, and can do nothing but hinder the progress of science."[38] Given the instrumentation available to nineteenth-century scientists, atoms were unobservable, so under the influence of Comtean Positivism, many outstanding chemists and physicists, including the Nobel laureate chemist, Wilhelm Ostwald, and the outstanding physicist and philosopher, Ernst Mach (1838–1916), refused to accept arguments derived from atomic theory or statistical mechanics.

Projecting the Character of Social Physics

For those who were not themselves natural scientists or analysts of natural scientific method, what Comte had to say about social physics was more important than anything he had to say about the earlier sciences, for in connection with this subject, Comte understood himself to be analyzing a field that was still dominated by theological and metaphysical concepts and to be creating new positive conceptions rather than judging, arranging, and improving conceptions that already had a positive character.

For Comte, as for virtually all of his immediate predecessors in the aftermath of the French Revolution, the two most critical conceptions in social thought were those of "order" and "progress," and the great question was how to reconcile the two. As things stood in the early nineteenth century, according to Comte, the concept of order was one that came from and that was still understood in terms of the theological and politically conservative tradition. The concept of progress, on the other hand, came from the metaphysical and politically revolutionary tradition. The notion of progress could therefore not be reconciled with order because the theological and metaphysical systems that underlay the two concepts were grounded in logically contradictory assumptions. Only by giving both conceptions new, positive, interpretations, could they be made logically consistent and be brought into harmony with one another.[39] The possibility of reaching harmony by returning to a theologically grounded social theory or by resting in a metaphysical theory was eliminated by the Law of Three Stages, which prohibited any but the most transient and unstable return to an earlier stage and which viewed the metaphysical stage as inherently transitional and unstable.

In connection with his understanding of the historical processes by which social thought had arrived at its present state, Comte again showed his direct indebtedness to his Saint-Simonian background. Thus he wrote, "The development of the sciences, of industry, and even of the fine arts, was historically the principle, though latent cause, in the first instance, of the irretrievable decline of the theological and military system."[40] Metaphysical conceptions were crucial in breaking down the conceptual hold of the old order, and they were thus progressive, but they undermined all notions of order and led to a kind of intellectual anarchy that had now outlived its usefulness. Indeed, Comte agreed with the Saint-Simonians that a whole range of notions from the metaphysical stage would have to be overturned to construct a positive science of society capable of transforming society into one that was simultaneously orderly and progressive. The "dogma" of the liberty of conscience would have to go, for instance, as would the notion of the sovereignty of the people.[41] Instead, social questions would have to be decided by "a small number of choice minds."[42] Dissent would be allowed "only with regard to opinions which are considered indifferent."[43]

Comte's readers disagreed bitterly over such claims, with liberals such as Mill and Littré opposing them violently, while conservatives welcomed them openly. But most of those who took him seriously at all—and that eventually represented a substantial fraction of late-nineteenth century intellectuals throughout the globe—concurred in admiring his discussion of the methods by which sociological knowledge should be sought, even when they were not in agreement with the doctrines he read out of their application.

Comte begins his approach to the proper methods for social science by characterizing what he sees as the principle general weaknesses of the theological and metaphysical polities. "Their attributes are the same," he says, "consisting in regard to method, in the preponderance of imagination over observation; and in regard to doctrine, in the exclusive investigation of absolute ideas."[44] The general cure must be to reverse the roles of imagination and observation and to subject scientific conceptions to the facts whose connections are to be disclosed. Before becoming more precise about method, however, Comte responded to a critical possible objection to the whole project of establishing a stable social order grounded in a science of society. Those committed to theological and/or metaphysical conceptions were (and continue at the beginning of the twenty-first century to be) inclined to believe that intellectual—and therefore social—stability can only be a product of absolute knowledge. Yet it seemed as though positive knowledge could never be more than relative, so it could never lead to a stable social order.

In response to these concerns, which had been raised by the followers of Cousin, Comte admitted that it was true that positive knowledge can never be absolute; but he insisted that the fear that relative knowledge cannot provide stability was misplaced. "The study of the laws of phenomena must be relative," he insisted, "since it supposes a continuous progress of speculation subject to the gradual improvement of observation, without the precise reality being ever fully disclosed: so that the relative character of scientific conceptions is inseparable from the true idea of natural laws, just as the chimerical inclination for absolute knowledge accompanies every use of theological fictions and metaphysical entities."[45] Thus, relativism follows because knowledge production is an iterative process in which the recognition of "facts" must always depend upon some prior theory whose modification is always demanded by the new organization of facts that it produces.[46] Even when a subject of study reaches the positive stage, we must expect continuing and endless change in our knowledge as new arrangements of facts are recognized.

On the other hand, we have the strongest empirical evidence that this relativistic knowledge is capable of producing consensus:

> All antecedent experience shows that in other departments of natural philosophy, scientific ideas have not become arbitrary by becoming relative; but have, on the contrary, acquired a new consistence and stability by being implicated in a system of relations which is ever extending and strengthening, and more restraining of all serious aberration. There is therefore no fear of falling into a dangerous skepticism by destroying the absolute spirit, if it is done in the natural course of passing on toward the positive state. Here as elsewhere, it is characteristic of the positive philosophy to destroy no means of intellectual coordination without substituting one more effectual and more extended; and it is evident that this transition from the absolute to the relative offers the only existing means of attaining to political conceptions that can gradually secure a unanimous and permanent assent.[47]

Having asserted the unique ability of scientifically acquired relative knowledge to produce assent and, presumably, acquiescence, Comte turned to the question of what specific methods were appropriate to sociology, assuming the universal tenets of the positive philosophy—that is, "that social phenomena are subject to natural laws, admitting of rational prevision," and that those laws will be limited to statements about static and dynamic relationships.[48] For Comte, as for the *Idéologues* who preceded him, the single most important fact about social phenomena was that they involved a level of complexity and interconnectedness that made it pointless to consider the different elements in a social system as if they could exist separately. Consequently, social systems must be studied using methods more like those of the physiologist than like those of the physicist:

> [The] master-thought of universal social interconnection becomes the consequence and complement of a fundamental idea established, in our view of biology, as eminently proper to the study of living bodies. Not that the idea of interconnection is peculiar to that study. . . . It is, in fact, true that wherever there is any system whatever, a certain interconnection must exist. The purely mechanical phenomena of astronomy offer the first suggestion of it; for the perturbations of one planet may sensibly affect another, through a modified gravitation. But the relation becomes closer and more marked in proportion to the complexity and diminished generality of the phenomena, and thus, it is in organic systems that we must look for the fullest mutual connection . . . the animal interconnection being more complete than the vegetable, and the human more complete than the brute. . . . The idea must therefore be preponderant in social physics, even more than in biology.[49]

As a consequence, "the methodological division of studies that takes place in the simple inorganic sciences is thoroughly irrational in the complex science of sociology and can produce no results."[50]

One critical consequence of the interconnectedness of social phenomena is that there is no significant role for experiment. Totally aside from the practical problems that would be involved in mobilizing the resources to undertake large-scale social experiments, there is a theoretical reason to believe that experimenting with social systems could not lead to valuable knowledge, for "any artificial disturbance of any social element must affect all the rest, according to the laws both of coexistence and succession; and the experiment would be deprived of all scientific value, through the impossibility of isolating either the conditions or the results of the phenomena."[51]

Fortunately, according to Comte, the analogy between physiology and sociology does suggest alternatives to experimentation. In physiology, pathological cases offer an insight into the normal operation of biological systems. In the same way, and to an even greater extent, claimed Comte, social pathologies, or cases in which the ordinary laws of harmony or succession are disturbed, offer insight into society.[52] Even more important than the study of natural social pathologies such as political revolutions, was the use of comparative studies, both in terms of comparisons between human societies and those of other animals and of comparisons among human societies separated from one another in time and/or space.

If the complex interconnections among elements of society make for some difficulties relative to the simpler sciences, however, they also offer some advantages, for the interconnectedness leads to its own Law of Harmony, or Law of the Unified Character of Society, which can be expressed in the following way:

"There must always be a spontaneous harmony between the whole and the parts of the social system, the elements of which must inevitably be, sooner or later, combined in a mode entirely conformable to their nature."[53] In practice this claim means that if one can see changes over time in any single social relationship, one can use the structural relationships within societies to predict changes in other social relationships without observing them directly.[54]

The comparison of coexistent societies separated in space can help in establishing static social laws, but the historical comparison of societies that succeed one another in time offers the possibility of generating both static and dynamic laws. So the historical method became the most central method for sociological study for Comte. This is particularly true because what most distinguishes human societies from animal ones is the fact that both what is learned and institutions that are created or modified by one generation can be passed on to the next, "accumulating continuously until it constitutes the preponderating consideration in the direct study of social development."[55] For Comte, then, most social evolution can be said to be directly produced by cultural factors such as the use of language, technological innovation, and so on. On the other hand, he does admit the possibility of "the gradual and slow improvement of human nature" through the process proposed by Lamarck—that is, through the inheritance of characteristics acquired through "homogeneous and continuous exercise."

Comte was convinced that both individual human intellect and altruism have been increasing over time in highly civilized societies and he was sure that they could similarly be improved in members of those societies that were less advanced in their social evolution.[56] This notion of a progressive social evolution, through both cultural and biological means, was verified through historical investigation, and it was a notion that was particularly important in stimulating the sociological investigations of Herbert Spencer, though, as we will see, Spencer was not convinced that altruistic tendencies were increasing over time. Through both Spencer's writings and more directly through Harriet Martineau's translation of *The Positive Philosophy*, the idea that social evolution was driven by increasing altruism entered Charles Darwin's writings in *The Descent of Man.*

I want to explore just one final consideration from the *Positive Philosophy* before we turn to Comte's attempt to use sociological knowledge to change society; for this final consideration reflects one of the greatest differences between the *Positive Philosophy* and the views of his Saint-Simonian predecessors. Moreover, it suggests why some of the intellectual Positivists had such great difficulty in taking Comte's later work seriously. Ultimately, Comte argued against all materialist perspectives on history, claiming that intellectual and moral change must always precede significant social and institutional change in society. The bulk of his attention throughout his first great work was on intellectual development, but the question of the relationship between knowledge and morality absorbed his attention toward the end of the *Positive Philosophy.*

The Saint-Simonians had argued that in the final analysis, knowledge had to be subordinated to sentiment. The universalization of love that underlay moral development, while it could be described scientifically, could not be produced through knowledge in any way. After the transformation he underwent in connection with his love of Clotilde, Comte seemed to have come back to this position; but in the *Positive Philosophy,* Comte insisted that in the positive state of society, our powerful affective lives must be guided by a naturalistic morality derived from our sociological understanding. He did not believe that this new naturalistic morality could be derived from Benthamite Utilitarianism, because Utilitarian doctrines were based on a false psychological perspective and because they were too focused on materialistic considerations. Indeed, Comte had relatively little to say about the new morality except that it would be grounded in sociological knowledge, that it will be a unifying force, and that it would be promulgated through a basically secular system of education and discipline:

> In our present state of anarchy, we see nothing that can give us an idea of the energy and tenacity that moral rules must acquire when they rest on a clear understanding of the influence that the actions and the tendencies of every one of us must exercise on human life. There will be an end then of the subterfuges by which even sincere believers have been able to elude moral prescriptions, since religious doctrines have lost their social efficacy. The sentiment of fundamental order will then retain its steadiness in the midst of the fiercest disturbance. The intellectual unity of that time will not only determine practical moral convictions in individual minds, but it will also generate powerful public prepossession. . . . The instrumentality will not be merely the influence of moral doctrine, which would seldom avail to restrain vicious inclinations; there would be first the action of a universal education, and then the steady intervention of a wise discipline, public and private, carried on by the same moral power which had superintended the earlier training. The results cannot even be imagined without the guidance of the doctrines themselves.[57]

The *System of Positive Polity*

When Comte returned to print in 1851 in the first volume of the *Système de politique positive, ou Traité de sociologie instituant la religion de l'Humanité* (hereafter, the *Positive Polity*), which was directed at the application of the ideas developed in the *Positive Philosophy,* several things had changed. First, the notion that a positive science of sociology was in its infancy and that its most important findings were still far in the future was gone. Without claiming that sociology was complete, he now argued, "Our system of scientific knowledge is already so far elaborated that all thinkers whose nature is sufficiently sympathetic may proceed without delay to the problem of moral regeneration, a problem that must prepare the way for that of political reorganization."[58]

Second, the sense that positive knowledge should somehow govern the emotions was abandoned in favor of "the ascendancy of the heart over the intellect." Thus, Comte wrote, "In the first [treatise], where the process of scientific preparation is carried to its fullest limit, I have carefully kept the objective method in the ascendant, as was necessary where the course of thought was always preceding from the world in the direction of man. But the fulfillment of this preliminary task, by the fact of placing me in the true universal point of view, involves henceforth the prevalence of the subjective method as the only source of complete systematization, the procedure now being from man outward toward the world."[59] In this connection, Comte returned to another notion that had been developed by the Saint-Simonians. Whereas in the *Positive Philosophy* Comte had complained that sociology had heretofore been crippled by the failure to subordinate imagination to reason, in the *Positive Polity* he touted the importance of imagination and claimed that, "Positivism is eminently calculated to call the imaginative faculties into exercise."[60]

Finally, Comte had reconceptualized the spiritual and moral authority in the positive society as fundamentally religious. He now insisted that the leadership of the new Religion of Humanity be constituted in a hierarchical priesthood modeled very closely on that of the Catholic Church, with Comte himself functioning as "high priest" or Pope of the new Positivist Church. Given all of these basic changes in orientation, it should hardly be surprising that some of those attracted by the *Positive Philosophy* were disturbed by the tone of the *Positive Polity.*

There were, however, important continuities between the two works, especially in terms of the obvious political implications, which promised a middle ground between the most retrogressive apologists for the old feudal order and those drawn to radical and increasingly violent working-class movements. On the one hand, like conservatives such as Burke and De Maistre, Comte stressed the limited sense in which conscious human action can be expected to improve circumstances: "We are powerless to create: all that we can do in bettering our condition is to modify an order in which we can produce no radical change. Supposing us in possession of that absolute independence to which metaphysical pride aspires, it is certain that so far from improving our condition, it would be a bar to all development, whether social or individual."[61]

On the other hand, Comte argued that the most important function of the Positive Philosophy was to point out those domains in which directed human action *could* be effective in making changes; and he argued that "those who look wisely into the future of society will feel that the conception of man's becoming, without fear, or boast, the arbiter, within certain limits, of his own destiny, has in it something far more satisfying than the old belief in Providence, which implied our remaining passive."[62] While the political possibility of improving social conditions without disrupting society made explicit in Positivism continued to attract many, the emphasis on emotion and religiosity with a scientific rather than traditional Christian foundation certainly attracted new converts as well, for it purported to offer a set of doctrines that could satisfy people's cravings for emotional experience without insisting that they moderate their rationality.

At the very beginning of the *Positive Polity,* Comte argued that it would be essentially impossible to convert men of the upper classes of society to his ideas, because "they are all more or less under the influence of the baseless metaphysical theories and aristocratic self-seeking. They are absorbed in blind political agitation and in disputes for the possession of the useless remnants of the old theological and military system." As a consequence, he argued, "It is among women, and among the working-class that the heartiest supporters of the new doctrine will be found."[63] Given this belief, Comte chose to focus special attention on the meaning of Positivism for workers and for women.

In connection with the working class, Comte organized his discussion primarily in the form of a comparison between his own doctrines and those of the communists (to be discussed in more detail in chapter 5). On the one hand, Comte praised communism for articulating many features of the great social problem of working-class poverty and for the "generous sympathies by which it is inspired."[64] He even acknowledged the truth of its insistence that property is a social rather than an individual matter.[65] But, he argued, communism, being "ignorant of the true laws of society," offered a series of unworkable solutions to the problems it so ably posed. Positivism alone could solve those problems. Positivism offered "the only doctrine which can preserve Western Europe from some serious attempt to bring communism into practical operation." And Positivism alone would be able "to satisfy the poor and, at the same time, to restore the confidence of the rich."[66]

From this point on, though he constantly assured the reader that he had the needs of the poor in mind, Comte often seemed much more concerned with restoring the confidence of the rich—not the old landowning aristocracy linked to monarchical government, but the rising rich of the capitalist bourgeoisie. Thus, for example, he insisted that the distribution of wealth and power is beyond our direct conscious control and that all attempts to redistribute them are certain to be a waste of time and energy. Instead, we should try to get that wealth used for the broad benefit of humanity. Moreover, whatever regulations are established to encourage the proper use of wealth should be moral rather than political. Finally, "Those who accept them [the moral guidelines] will do so of their own free will, under the influence of their education."[67]

The following passage, which addresses the issue of the concentration of capital, will serve to illustrate his approach to a large range of issues:

> It is evident that under the present system of industry there is a tendency to a constant enlargement of undertakings; each fresh step leads at once to still further extension. Now this tendency, so far from being opposed to the interests of the working classes, is a condition that will most seriously facilitate the real

> organization of our material existence, as soon as we have a moral authority competent to control it. For it is only the larger employers that the spiritual power can hope to penetrate with a strong and habitual sense of duty to their subordinates. Without a sufficient concentration of material power, the means of satisfying the claims of morality would be found wanting, except at such exorbitant sacrifices as would soon be found incompatible with all industrial progress. . . . We are not to encourage the foolish and immoral pride of modern capitalists. . . . But at the same time we must be careful not to underrate the immense value of their function, or in any way obstruct its performance.[68]

There can be little doubt that Comte truly saw himself as sharing the progressive goals of the communists and socialists of the mid nineteenth century. Nor can there be much doubt that he felt that the moral regeneration of society was a prerequisite to any lasting social regeneration and that moral regeneration depended on the separation of moral and spiritual authority from temporal authority. As a consequence, for him, the great problem was to establish a moral order in which capitalists would *want* to use their wealth to benefit everyone, rather than to abolish private capital. At the same time, it should be easy to understand why more cynical—or realistic—members of working-class movements were inclined to see Positivism as an apology for capitalism rather than as a progressive force.

Much the same story can be told about the relationship between feminism and Positivism. Comte had a deep and abiding admiration for the roles that women played in society and a desire to see women treated well. He even made the worship of feminine characteristics a central feature of his Religion of Humanity. But he had nothing but disdain for the notions of sex equity promoted by Enfantin and the Saint-Simonians or for the communist attacks on marriage as a primarily economic institution in which women are exploited and dominated by men. Such views, he characterized as "subversive schemes that are growing every day more dangerous to all the relations of domestic and social life."[69]

In connection with what Comte identified as "the most essential attribute of the human race, the tendency to place social above personal feeling," he insisted that women are superior to men.[70] Thus, he argued that in marriage, a man "enters into a voluntary engagement of subordination to woman . . . he finds his highest happiness in honorable submission to the ennobling influence of one in whom the dominant principle is affection."[71] Furthermore, he insisted that the most important part of education, "the spontaneous training of the feelings," should be done in the home by a child's mother, so Comte argued against instituting any form of public education for children.[72] At the same time, he argued that in all other things, "whether physical, intellectual, or practical, men are naturally superior to women," so women should have no significant political standing or economic roles outside the home.[73]

There were important precedents, from Rousseau through the *Idéologues*, to support Comte's claims that women had a more loving and sympathetic nature than men. There were equally important precedents, especially within the Scottish tradition of philosophical history, for his claims that marriage and the family provided critical stages in the socialization of all human beings.[74] But Comte took such arguments far beyond any prior mainstream social theorist when he claimed that while sexual urges may produce the occasion that initiates marriage, its primary goal is not procreation, but rather, "raising and purifying the heart."[75] Given this critically high goal of marriage, Comte insisted that in positive society it be "exclusive and dissoluble." Divorce would be impossible in the regenerated society. But more than that, the constancy of love symbolized in the positive state of marriage would be such that it should be held to endure beyond death, so the institution of "perpetual widowhood" was to be established as was the institution of "union in the tomb."[76] The overall consequence of such doctrines was that, although women were to be venerated, they would also find themselves even more excluded from intellectual life, political life, and nondomestic economic life than was typically the case in nineteenth-century society. And their one-and-only husband's control over their lives would extend even from beyond and into the grave.

The Religion of Humanity

The institutions of perpetual widowhood and union in the tomb bring us back to the issue of the character of the Religion of Humanity that Comte sought to institute. Its goals, like that of Saint-Simon's New Christianity, were to promote the strengthening of social passions over self-love and to encourage the growth of universal love toward all of humanity. But Comte's expectations with respect to the growth of benevolence and altruism were more limited than Saint-Simon's, for Comte was convinced that it would be neither possible nor desirable to totally eradicate the individualistic and self-oriented passions that serve our animal natures. Social passions are the highest human attributes, but the most that they will ever be able to do is to moderate and oppose egotism.[77] As a consequence, there will never come a time when the universal love of humanity has triumphed over self-love and personal ambition.

If the tension between egoism and altruism can never be eliminated, however, there are conditions under which the balance between the two can be shifted toward allowing altruism to direct a greater fraction of our actions, and Comte's discussions of those conditions are interesting both in their own right and because they allow us to see in a particularly clear way the extent to which his entire theory is grounded in biological notions, especially those associated with Bichat, Lamarck, and Cabanis. Comte began his discussion of religion with Bichat's notion that humans have three basic kinds of needs, associated respectively with feeling, thought, and action. As a result, any acceptable religion would have to relate to all three dimensions of humanity. Next, he argued that religion must emphasize the subordination of humans to "some external Power possessed of superiority so irresistible as to leave no sort of uncertainty about it."[78] This claim, he insisted, "is at bottom, merely the full development of that primary notion of sound biology, the necessary subordination of every organism to the environment in which it is placed." He pointed out that this general principle of the subordination of every organism to its environment was true because that environment provides the food an organism needs and the stimuli that initiate both physical activity and mental activity. As a consequence, "a sound theory of biology thus furnishes the positive theory of religion with a foundation wholly unassailable; for it proves the general necessity for the constant supremacy of an external Power as a condition of unity for man, even in his individual life."[79]

This superior external power or "Supreme Power" of Comte's Religion of Humanity was nothing but the entire physical and social universe, the character of which is known through the positive sciences. By replacing traditional theology with the sciences, the Religion of Humanity offered satisfaction of the intellectual needs of humanity and brought the objects of religious veneration out of the supernatural and metaphysical domains and into the domain of nature. It was largely this promise of offering religion without "superstition" that drew the attention of many nineteenth-century figures for whom traditional Christianity had become impossible to believe, but who still felt a need for some kind of religious life.

Moving more directly to the relations between egoistic and altruistic orientations, Comte argued that the reason that individualistic tendencies tended to predominate in our mental activities is that we are constantly stimulated by "physical wants." If those physical wants were not so pressing, our minds would be more influenced by forces arising from our social contacts and we would cultivate our social instincts more. One consequence of the increased exercise of our social instincts would be their gradual Lamarckian strengthening from one generation to the next.[80] A second consequence of the satisfaction of our material needs would actually be an inversion of the emphasis placed on intellectual vis-à-vis emotional life, for to a great extent our intellectual activity is called forth to satisfy our material needs. Science could become less important in education and emotional development could receive increasing attention.[81]

There would be a third critical outcome of the lessening of our material needs as a consequence of the idea, emphasized by Cabanis and the *Idéologues,* that we have a spontaneous need for activity and self expression, "quite apart from any external object." When there are material needs to be met, much of our activity is directed toward their satisfaction. But what happens to the drive for activity and self exertion if the material needs are met? Comte responded: "The only consequence is that its exercise becomes artistic, instead of being technical. . . . In a word, actions become games, which are not mere preparations for active life but simple modes of exercise and expansion. This transformation would become particularly obvious in the case of the activity of the social body—which, being no longer absorbed by material undertakings, would give itself up to festivals, whereby to express and develop the common affections of the society. The artistic character would predominate in practical as well as speculative life."[82]

Given that both our emotional and aesthetic lives would be enriched if we could be liberated from our material needs, Comte offered a special Positivist version of political economy culminating in what he called "the positive theory of accumulation, without which all . . . progress from selfish to unselfish toil would be impossible."[83] The central theme of this political economy was that capital accumulation makes possible such dramatic increases of productivity—through both the division of labor and technological innovation—that it promises virtually to eliminate material want. If there were no altruistic instincts that could be strengthened as material wants were increasingly met, Comte admitted that capital accumulation would only lead to a world of ever more self-centered individuals. Fortunately, he argued, such is not the case. Once again Comte could find comfort in promoting capitalism as a precondition for the material and moral betterment of the human condition.[84]

At this point, having established the essential goals of the Religion of Humanity and the conditions favoring its growth, we come with Comte to the character of its worship, which is aimed at "regulating the direct cultivation of our sympathetic instincts,"[85] so that we may live for others. This worship is most clearly discussed in the *Catéchisme positive* of 1852, which Richard Congreve translated as *The Catechism of Positive Religion* in 1858. One key to understanding Positivist worship is to note that Comte said that it was intended to focus on the subjective rather than the objective life. Since even Comte recognized that such a claim needed explanation, he offered as an illustrative example the way in which we use the commemoration of the lives of dead heroes and heroines to improve our own lives. This example is particularly important because most rituals of the Religion of Humanity were organized around such commemorative acts.

"During the objective life," Comte writes, "the dominion of the outer world over the world of man is as direct as it is unbroken. But in the subjective life, the outward order becomes simply passive, and no longer prevails except indirectly, as the primary source of the images we wish to cherish. Our beloved dead are no longer governed by the rigorous laws of the inorganic order, nor even of the vital . . . the existence which each one of them retains in our brain . . . is composed essentially of images, which revive at once the feelings with which the being snatched from us inspires us and the thoughts which he occasioned. Our subjective worship is reduced, then, to a species of internal evocation."[86] The point of this passage seems to be that our dead ancestors literally continue to live subjectively in us as remembered images that cause us to reflect on those moral attributes that they embodied in a particularly powerful way. Our commemoration of their lives thus reinforces and exercises our own commitments to those virtues and allows us to improve our own lives.

In carrying out a commemorative act we also engage in a crucial process of idealization through which we remove from our memories of the dead all of those imperfections that might otherwise have interfered with our evocation of their positive characteristics.[87] Just such a process of idealization allowed Comte to transform his own mother, who had not been an objectively emotionally nurturing person, into the symbol of universal motherhood, to transform the abandoned Clotilde de Vaux into the symbol of universal love, and to transform the daughter of a friend into the symbol of daughterly love.

After these general reflections, Comte went on to stipulate in excruciating detail such details of private worship as the number of daily prayers and the relative lengths of each, as well as the characteristics and timing of nine personal sacraments—presentation, initiation, admission, destination, marriage, maturity, retirement, transformation, and incorporation.[88] With respect to public religion, he offered even greater detail. He stipulated, for example, that all Positivist churches should face toward the single source of their doctrines, Paris.[89] He laid out a series of eighty-one major annual festivals as well as a positivist calendar of thirteen twenty-eight-day months (plus one extra day to be the festival of the dead), incorporating the names of individuals to be commemorated on each day, and with special sequences for leap years. Furthermore, he stipulated what particular icons or symbols should be used in Positivist churches and worship. Finally, he set up a prescribed training for the Positive Priesthood as well as a fourfold hierarchy of priests.[90] With respect to all of these details, only the most rigid of his disciples, such as Congreve, insisted upon absolute submission to Comte's will.

PART II

Science and Culture in Germany and Britain

CHAPTER 4

Naturphilosophie, Romanticism, and Nationalism

In most of its central features the conception of science that was articulated in Auguste Comte's *Positive Philosophy* is one that has dominated worldwide science and scientisms from the mid-nineteenth century into the early twenty-first century. According to this view, positive, or scientific, knowledge forms the most reliable basis we have for action in the world. This is so because scientific investigators attempt to achieve a disinterested or objective attitude in which they submit to nature in order to discover its laws, rather than imposing on nature some scheme calculated to serve particular human interests. The scientific approach avoids making "metaphysical" claims about essences and necessary causes because they cannot be tested and it limits itself to stating laws describing "general" facts that come directly from our sensory experience of the physical world. Furthermore, it avoids making claims about mental states of human beings based on introspection because they are unreliable and unconfirmable. In spite of all attempts to exclude knowledge without an adequate empirical foundation, however, positive knowledge is contingent and imperfect, rather than absolute and perfect, because as our experience is extended we recognize new and more inclusive general laws that supersede those we had held provisionally before.

There was a radically different understanding of the nature of science that existed as part of a significant movement in the natural sciences and as a dominant feature of the social sciences in German-speaking lands during the early nineteenth century, however. That vision not only had a major impact on German science and culture at the time, but it has also had long-term consequences, especially for Continental political and intellectual life. To anticipate just a few major features of this understanding, which came to be associated with the term *Naturphilosophie,*[1] natural scientific knowledge, to be worthy of the name, had to be capable of being certain. Instead of trying to do away with metaphysical notions in science, this philosophical position assumed that only a science fully grounded in adequate metaphysical concepts could be true. Moreover, far from avoiding the notion of causation, it presumed that scientific laws must be causal.

This way of thinking assumed that, in a very important sense, humans impose these causal laws on nature rather than derive them from nature. It was openly spiritual (usually Christian) rather than secular. Finally, this way of understanding set the human sciences apart in a radical way, arguing that the very possibility of historical knowledge, for example, depends on introspective knowledge and on our ability to empathize with past actors, whose mental functioning must have been identical with our own. In this insistence that human sciences, such as history, have an introspective foundation, as well as in other ways to be discussed below, this vision of science placed subjective rather than objective considerations at the center of the sciences.

Because the understanding of science involved in *Naturphilosophie* is so different from that which most English language readers today can accept or sympathize with, we will spend some time discussing how it came about and how it produced a number of startling results—especially in physics and physiology—during the nineteenth century. Then we will look at its broader relationships to German culture, especially in connection with the Romantic movement and the promotion of an intense cultural and political nationalist ideology among German intellectuals. In the next chapter, we will focus on the processes by which the *naturphilosophisch* tendencies in German science and culture came to be transformed or abandoned in Germany in favor of various versions of positivism and "materialism" during the mid nineteenth century, starting in the 1830s.

Newtonian Science, the Experience of Freedom, and the Kantian Philosophy

In order to have any hope at all of understanding how the *Naturphilosophen* and other early nineteenth-century German intellectuals could have operated with a conception of science so foreign to our own, we need to begin with a look at the understandings of science that emerged out of the writings of two of the giants of late-eighteenth-century German culture, Immanuel Kant (1724–1803) and Johann Wolfgang Goethe (1749–1832). With respect to Kant, we will focus on a few key features of a general philosophy that was capable of incorporating three crucial elements: Newtonian mathematical natural philosophy; the emphasis on "reason" and rationalism that had been central to the system of the great German philosopher, Gottfried Wilhelm Leibniz (1646–1716), and our experience of human freedom, which guaranteed that the human sciences could not be perfectly deterministic, as the natural sciences were. Although Kant's ideas on these key issues changed over time,[2] for our present purposes it is enough to focus on the way they appeared in his mature works of the so-called critical period, from roughly 1781 to 1793. In fact, since Kant's ideas were not always crystal clear to his readers, what we really need to consider is how most of his readers interpreted Kant's ideas regarding these key issues in his critical works, even though subsequent scholarship may have shown that some of those interpretations were almost certainly not ones intended by Kant.[3]

Immanuel Kant

Kant was born in 1724 at Königsberg, Prussia, the fourth of nine children of a strongly but not fanatically Pietist harness-maker. He rarely left Königsberg, and, so far as we know, he never left the boundaries of eastern Prussia, whose economic conditions were as yet virtually unaffected by even a hint of industrialization. As a consequence, he showed little of the interest in the relationships between scientific knowledge and material change that dominated the concerns of his English and French contemporaries. When he spoke of "practical" considerations, these were practical in the older, Aristotelian, sense of being related to human life in communities, that is, to morality and ethics. Always small—about five feet tall as an adult—slightly deformed, and a man of incredible intellectual appetites and ability, he was nonetheless popular and sociable. When he became a professor at the University of Königsberg, he was an extremely popular teacher, whose classes were often so large that they overflowed the largest rooms available. Johann Herder (1744–1803), one of his many outstanding pupils, left a marvelous description of Kant as he was in the 1760s: "Though in the prime of life, he still had the joyful high spirits of a young man, which he kept, I believe, into extreme old age. His open brow, built for thought, was the seat of indestructible serenity and gladness. A wealth of ideas issued from his lips, jest and wit, and good humor were at his bidding, and his instructional lecture was also the most fascinating entertainment. . . . The history of mankind, of nations and of nature, natural science, mathematics, and his own experience were the well-springs which animated his lectures and his everyday life. He was never indifferent to anything worth knowing. . . . He encouraged and gently compelled people to think for themselves. . . . I recall his image with pleasure."[4]

After attending the local Protestant college, Kant entered the University at Königsberg in 1740. There he studied with Martin Knutzen, one of the first German academics to teach Newtonian natural philosophy. Forced for financial reasons to leave the university in 1746, Kant served as private tutor in two families for seven years while he finished his Ph.D. equivalent. Then, in 1755, he returned to Königsberg as a *Magister,* or lecturer in philosophy. In 1770, after receiving offers of triple his Königsberg salary from several other universities, he was finally promoted to professor of theoretical philosophy. In 1786, he became rector of the university.

Though Kant lectured and wrote on a huge range of topics, including anthropology, natural history, and physical geography, as well as religion and aesthetics, the central concerns of his early academic life were associated with mathematics and natural philosophy. His first publication, "Thoughts on the True Estimation of Living Forces" (1747), was a contribution to debates between Leibnizians and Newtonians in the Prussian academy of sciences regarding what was then called *vis viva,* and is now called kinetic energy. His dissertation was on the nature of fire, and in 1755 he published the first major theory of cosmological evolution under the title *A Universal Natural History and Theory of the Heavens, or an Attempt to Explain the Composition and Mechanical Origin of the Universe on Newtonian Principles.* From the very beginning, however, Kant was especially interested in broad philosophical issues raised by mathematics and science, and his reading of David Hume's arguments undermining both reason and experience as adequate sources of certain knowledge turned the focus of his attention to what are now called metascientific questions, or to the philosophy of science.

The German philosophical tradition of Leibniz and Christian Wolff (1679–1754) held that scientific knowledge had to be about necessary rather than contingent relationships. Since sensory experience was unavoidably contingent, this meant that reason and logic alone could produce scientific knowledge and that scientific knowledge had to be rationally derivable without the need for sensory experience, although sensory experience *did* have to be consistent with truly scientific knowledge. This point of view is held by almost no one today, and it was already almost certainly a minority view among French and English intellectuals by the early eighteenth century. But it was not only prevalent in Germany, it was still widely held within the community of mathematical natural philosophers in France during the late eighteenth century and was the view of Jean D'Alembert (1717–83) and Joseph Louis Lagrange (1736–1813), among others.

Kant shared the belief that no knowledge that was incapable of certainty deserved the name of science. He insisted, nonetheless, that both geometry and Newtonian natural philosophy provided examples of true sciences. This left Kant with a huge problem, since neither traditional rationalist accounts nor empiricist accounts could explain how such sciences could be possible. His self-imposed task then came to be determining the conditions under which certain, scientific, knowledge could be achieved. *That* it was achieved was not for him a question, for both geometry and the law of universal gravitation provided existence proofs in his view. The question was, How?

Kant's Critical Philosophy

This question was to have vastly greater ramifications for Kant and for subsequent German intellectual life than one might expect, for Kant argued that once one discovered the conditions that make mathematics and natural philosophy possible, then this discovery could be used to deal with a whole range of questions, including the nature of metaphysics, morality, religion and aesthetics. In a series of works published between 1781 and 1793, Kant attempted to analyze the foundations of virtually all domains of human thought. These works form what is often called the Critical Philosophy, and they established what many have viewed as the equivalent of a Copernican revolution in philosophy. The *Critique of Pure Reason* (1781) dealt with general problems of cognition and experience at a very abstract level. It proved so difficult to readers (even Kant said that it was "dry, obscure, opposed to all ordinary notions, and moreover long-winded")[5] that Kant provided a vastly abridged and simplified version in 1783. This work, *A Prolegomena to Any Future Metaphysics,* began to draw widespread attention to Kant's work. In 1785 he published both *The Metaphysical Foundations of Natural Philosophy,* which had a major impact on German physics and chemistry for the next fifty years, and *Fundamentals of the Metaphysics of Morals,* which tried to explore how reason is related to moral activity. The year 1788 saw publication of the *Critique of Practical Reason,* which continued the discussion of the roles of reason in ethics and morality. In 1790 Kant published *The Critique of Judgement,* which is primarily about aesthetics, but also contains some crucial considerations relating to the life sciences. Finally, in 1793 he published *Religion within the Limits of Reason Alone,* which completed his system.

In order to follow the basic thrust of Kant's arguments, one needs to recognize two basic ways of organizing knowledge that Kant brought to the Critical Philosophy. One was a logical classification of statements imported from the Leibnizian tradition, and the other was a distinction between knowledge derived from thought and knowledge derived from "intuition," which he imported from his *Inaugural Dissertation* given in 1770 when he became Professor of Philosophy.

According to the first of these classifications, every statement is either analytic or synthetic—that is, the predicate of the sentence is contained in the subject as a part of the concept of the subject (analytic), or the predicate is something added to the concept that serves as the subject (synthetic). Thus, "A rainy day is a wet day" is analytic, because the concept of wet is contained in the concept of rain and a denial of the statement would be self-contradictory. On the other hand, the statement, "This rainy day is a hot day" is synthetic, because the concept of heat is not contained in that of rain and the denial of the statement is logically possible, even though it might be wrong. All statements are also either a priori, that is, logically independent of particular experiences, or a posteriori, dependant on particular experiences. All a posteriori statements are ineluctably contingent because of their dependence on experience.

Given these two dichotomies, all statements must be one of four kinds, analytic a priori, analytic a posteriori, synthetic a priori, or synthetic a posteriori. Since Kant denied that any contingent statement could be the basis for certain knowledge, all a posteriori statements had to be excluded from science.[6] Leibniz had discussed analytic a priori statements and had concluded that they were the foundation of all scientific knowledge, because, he argued, there are no synthetic a priori statements. If such statements did exist, however, they would be extremely important, because they would allow us to make claims that are necessary and that simultaneously extend knowledge beyond what we put explicitly into the concepts that we use.

Kant claimed that we do, in fact, have examples of synthetic a priori statements, one of which is that the sum of the interior angles of a triangle is 180 degrees. This statement is certainly synthetic, since the concept "triangle" contains only the concepts three, line segment, intersection, enclose, and plane. On the other hand, it is a priori, because given the definitions of Euclid's *Elements of Geometry*, it is true even if there are no existent triangles to which the statement refers, so it must be independent of particular experience. In order to understand how such synthetic a priori statements can exist, we also need to accept Kant's distinction, imported from the *Inaugural Dissertation*, between the faculty of understanding, which deals with concepts and their relationships to one another, and the faculty of sensibility, which produces intuitions. Cognition, or the production of knowledge, depends upon uniting these two faculties, and thus upon the laws that regulate each. The faculty of understanding stipulates all of the logical forms of relationships between concepts and all of the categories into which concepts might fit (Kant identifies twelve such categories). In this way it sets certain limits on our possible experiences by insisting, for example, that they incorporate the law of noncontradiction—that is, a claim cannot be simultaneously true and false—and that they fall into certain categories, such as substance, cause, existence, and so on.

The forms and categories provided by the faculty of understanding would be empty without some objects to apply them to, and the faculty of sensibility provides those objects as intuitions. But even here, the faculty places constraints on our intuitions; or rather it provides the categories into which intuitions must fall. In particular, the faculty of intuition organizes all external appearances spatially and all internal experiences temporally, so space and time, or extension and succession are preconditions of our experience.

At this point we are in a position to understand how geometrical propositions such as "the sum of the internal angles of a plane triangle equals a straight angle, or 180 degrees" may be synthetic a priori, and in doing so, we begin to recognize the extent to which subjective elements—that is, features of our own mental faculties—structure our experiences and our knowledge. The figures that geometry deals with are ideas constructed in our faculty of intuition, which is capable of containing them because they have spatial form, which is given as an a priori condition for experience. Into these figures we put everything that is necessary for the deduction: not just the definitions of the concepts, but also such constructive processes as the extension of line segments, which are necessary to Euclid's proof of the proposition. Geometry is not totally divorced from sensations for Kant. It is from sensation that we get the raw materials for cognitions.

But it is what the mind adds that allows us to know things about triangles that we could not possibly sense and that go beyond the definitions.

Extending this notion in the *Prolegomena,* Kant also argues that universal natural laws—especially the law of gravitation—can be understood as following from the laws of our understanding and intuition. In one of the most famous passages of the *Prolegomena,* he concludes: "It thus at first sounds strange, but is nonetheless certainly true, if with regard to the latter [universal natural laws] I say: *The understanding does not extract its laws from, but prescribes them to, nature.*"[7]

This last statement deserves some careful consideration. First, Kant has argued that only universal natural laws are capable of being given a priori determinations, thus only rational mechanics, among natural sciences, could be constructed as a science without reference to experience.[8] Most kinds of natural knowledge did have empirical sources and were thus contingent rather than strictly scientific, but even these forms of knowledge were governed by the general constraints of the laws of reason and intuition. Second, for Kant, though the laws and categories of the understanding and faculty of intuition bound our experiences of external objects (what Kant calls *Phänomena* [hereafter, phenomena]), they say nothing about the "things as they are in themselves" (what Kant calls the *Noumena*) being represented through experience.[9] So even though we must experience the phenomena of nature as organized spatially, temporally, and causally, that does not legitimate any claim that the noumenal world is so organized. Indeed, the fact that our internal experience includes an equally necessary a priori notion of freedom suggests that the noumenal world cannot be fully causal.

Finally, and very importantly for Kant, though subjective, the categories of thought and intuition were by no means either arbitrary or imaginative. In a sense they can be understood as "hard wired" into the brain. Most of Kant's readers, as we will see, were inclined to extend the notion of subjectivity in science far beyond Kant's intent. We will soon follow that extension into German *Naturphilosophie,* but it is worth noting that a number of important non-German scientists also read out of Kant an authorization for a much more extensive role of imagination in scientific activity than other understandings of science allowed. The Irish mathematical physicist, William Rowan Hamilton, for example, explained what the physicist does in a Kantian inspired and anti-positivistic way, writing: "In all . . . physical science, we aim not only to record but also to explain appearances; that is, we aim to assign links between reason and experience; not merely by comparing some phenomena with others, but by showing an analogy to those phenomena in our own laws and forms of thought, 'darting our being through earth, sea, and air. . . .' And this appears to me to be an essentially imaginative process; although I do not deny that it must be combined with a diligent attention to the appearances themselves in their most minute details, and with rigorous reasoning on the hypotheses the scientific imagination has suggested."[10]

Much as geometry is possible only because our faculty of intuition is set up to organize experiences spatially, natural science is possible only because our faculty of intellect is set up to structure experience in conformity with such categories as causation and substance. In connection with the first, we are impelled to hold that "every natural event is determined by a cause according to constant laws" and in connection with the second, we are impelled to hold that "substance is permanent."[11] For now, we assume that the causal principle is unproblematic, but we do need to explore some of the ramifications of the concept of substance, especially of that particular kind of substance called matter.

Kant discusses matter primarily in the *Metaphysical Foundations of Natural Science* in connection with the concept of force. He defines matter as "[that which] is moveable in so far as it fills space," where "to fill a space means to resist everything moveable that strives by its motion to press into [that] space."[12] Now, to resist something that moves means to repel that thing, and that means to change its motion. But by definition, that which changes the motion of any object is a force. As a consequence, "matter fills space, not by its mere existence, but by a special moving force."[13] Thus we recognize that Newton was wrong in saying that forces do not belong to matter by necessity. In fact, repulsive force constitutes the very condition for the possibility of matter.

If there were nothing but repulsive force associated with matter, however, all parts of matter, each of which is itself material, would simply disburse to infinity. That does not happen, so we can also know that "the possibility of matter requires a force of attraction, as the second fundamental force of matter."[14] All matter is constituted by a balancing of repulsive and attractive forces, such that what we call the boundary of a body consists of that surface at which the two forces are in equilibrium. Inside that boundary, repulsive forces dominate, while outside that boundary, attractive (gravitational) forces predominate. This so called "dynamical" understanding of matter had already been suggested by the Jesuit natural philosopher, Roger Joseph Boscovich (1711–87), in his *Natural Philosophy* of 1747, but Kant's formulation was more general and it seems to have stimulated the rise of what subsequently came to be called "field theories," which focus on the spatial distribution of forces.[15]

For our purposes, one additional argument from the discussion of matter in *The Metaphysical Foundations of Natural Science* is crucial. Kant demonstrates not simply that both attractive and repulsive forces are necessary to our notion of matter, but also that they are the only kinds of forces that can be conceived. Consequently, "all the forces of motion in material nature must be reduced," to these two.[16] We cannot help considering nature under the category of unity, and thus we unavoidably consider the phenomenal world of nature to be a single unified world. For this reason, we seek to unify any partial understandings that we might achieve. This notion was to underlie the *Naturphilosophen*'s emphasis on the interconnectedness of apparently different kinds of phenomena. Coupled with Kant's insistence that all forces in nature could be reduced to attraction and repulsion, this led to the expectation that the different forces connected with electricity, magnetism, chemical reactions, and so forth, might all be converted from one form into another.

The notion of unity is also critical in understanding a set of issues discussed by Kant in the *Critique of Judgement* that had major bearings on the development of *Naturphilosophie* and of German Idealist philosophy in general. The problems discussed in the *Critique of Judgement* are all associated with how to unify different kinds of knowledge, for his earlier works had left a seemingly unbridgeable chasm between the noumenal world of human freedom and morality and the phenomenal world of nature, as well as a troublesome disjunction between the nature understandable a priori through rational mechanics, and the subject matter of the empirical sciences, also presumed to be "nature." Kant argues that there is no a priori way to know that the infinite variety of empirical laws relate to a single, unified nature. To respond to all of these problems of disjunction, Kant posits a faculty of judgment, mediating between intuitions and thought. This faculty imposes a subjective unity on our experiences that may not be necessary to make experiences possible, but that is a precondition for our understanding of our experience as integrated into a single whole. In Kant's terminology, the faculty of judgment has a regulative rather than a constitutive function relative to experience. The way it regulates experience is by presupposing the notion of purposiveness, which allows us to unify our experiences by seeing them all as directed by a single plan to some specific end. For Kant, there is no guarantee that there is such a purposiveness outside of ourselves; we simply must formulate our knowledge as if there were if that knowledge is to contribute to a unified conception of the world.

As applied to the general problem of unifying the deterministic world of phenomena and the noumenal world that includes moral choices or freedom, this attempt was widely held by Kant's near contemporaries to have been a failure. Moreover, its unconvincing character stimulated a number of young students of Kantian philosophy, including Johann Gottlieb Fichte (1762–1814), Friedrich Wilhelm Joseph Schelling (1775–1854), and G. W. F. Hegel (1770–1831) to take up the problem in ways that led to what has been called German Idealism and to its offshoot, *Naturphilosophie.* But as applied to the understanding of organized nature, Kant's analysis of the way in which we must understand organisms had a wide impact on the life sciences, even among those who did not accept the idealisms of Fichte, Schelling, or Hegel. Kant argued that something is said to be naturally purposive if it is both an end and a means, or, "both cause and effect of itself."[17] To illustrate this notion, he uses the example of a watch, which is not living, and compares and contrasts its features with those of a living being:

> In a watch, one part is the instrument that makes the others move, but one gear is not the efficient cause that produces another gear; and hence, even though one part is there for the sake of another, the former part is not there as a result

> of the latter. That is also the reason why the cause that produced the watch and its form does not lie in nature, but lies outside nature and in a being who can act according to the ideas of a whole that he can produce through his causality. It is also the reason why one gear in a watch does not produce another; still less does one watch produce other watches, by using (and organizing) other matter for this production. It is also the reason why, if parts are removed from the watch, it does not replace them on its own; nor, if parts were missing from it when it was first built, it does not compensate for this lack by having other parts help out, let alone repair itself on its own when out of order. Yet all of this we can expect organized nature to do. Hence organized nature is not a mere machine. For a machine has only *motive* force. But an organized being has within it *formative* force [*Bildungstrieb*], and a formative force that this being imparts to the kinds of matter which lack it (thereby organizing them).[18]

The notion that every part of an organism should be understood in terms of what it contributes to the whole had long been a central claim of physiology, but the Kantian insistence that organisms incorporate some kind of nonmechanical formative force (a concept probably borrowed from Johann Blumenbach) helps to account for the special strength and persistence of so-called vitalism within early-nineteenth-century German biology and organic chemistry at a time when it had virtually disappeared from France and England.[19]

Goethe's "Romantic" Science

Johann Wolfgang von Goethe (1749–1832) is certainly much better known in the twentieth century as a poet, playwright, and novelist than as a scientist, and literary scholars rarely view him as a "Romantic," yet much of his intellectual effort went into the study of natural phenomena in a way that he understood to be scientific, though certainly not Newtonian. At least during part of his life, he viewed his work on the science of colors, *Zur Farbenlehre,* rather than *Faust* as his greatest single work. He left massive collections of scientific instruments as well as mineralogical and biological specimens at his death. And recent historians of science have generally characterized Goethe's science as Romantic.[20] Beginning around 1771, he became a follower of Kant's student, Johann G. Herder (1744–1803), and after 1790, during the period of his mature scientific works, Goethe interpreted his work through a filter of Kantian philosophy, but many of his most characteristic notions seem to have emerged before their Kantian expressions.

After attending the University of Strassbourg in the early 1770s and taking courses in anatomy, surgery, and chemistry, as well as more typical liberal arts subjects, Goethe obtained a license to practice law; but his plans to get a doctorate in law foundered when his dissertation was rejected for being too radical and anti-Christian. In 1774 he published his first novel, the immensely successful *Sor-*

rows of Young Werther, soon to be followed by a play, *Goetz von Berlichingen,* and a series of essays and poems. His literary successes brought him to the attention of Duke Karl August of Saxony-Weimar (1757–1828); and in 1775 the duke appointed him as an administrator, initially with responsibility for cultural affairs, but with rapidly expanding responsibilities for a variety of economic, fiscal, and educational issues. The next year Goethe brought his old friend, Herder, to Saxony-Weimar to become the minister of education, and the two men remained in close touch with one another through the remainder of Herder's life.

Even before 1775, Goethe had expressed both a passionate love of nature and a fascination with descriptions of natural settings. Then, beginning in 1776, he became increasingly interested in geology in connection with a series of trips he took with the duke and several mining engineers into the mountains of Thuringia, with the intent of reopening medieval copper and silver mines. His interests were further stimulated when Abraham G. Werner (1744–1817), the famous "Neptunist" geologist was hired to teach in the Mining Academy of Saxony in Freiberg. A 1784 essay, "On Granite," was Goethe's first major effort in this field, illustrating a very personal and passionate style, and introducing a notion that would be central to much of his scientific work, the notion of "type." For Goethe, granite formed the "type" of all rock, which somehow underlies all others and forms, "the basis of all geological formation."[21] Soon he was attempting to develop a unified geological theory to unite those theories, such as that of Werner, which focused on the effects of a cataclysmic flood, and those, such as that of Geuttard, which focused on volcanic activity. This theorizing, which emphasized gradual geological change through a variety of processes including both externally driven processes, such as weathering, and internally driven "metamorphoses,"[22] continued throughout the remainder of his life.

Three general features of his geological work typify Goethe's approach to all scientific topics and contribute to its characterization as Romantic. First, though his geological theorizing may have initially been stimulated in connection with commercial interests in mining, all technological interest is both implicitly and explicitly expelled in his mature science, which seeks "to interpret nature, not serve technology."[23] This nonutilitarian perspective was particularly important in shaping Goethe's attitude toward Carl Linnaeus (1707–78), whom Goethe acknowledged as one of the most important influences on his thought. Goethe found the Linnaean system anathema because it was aimed more at use in identifying plants than in understanding nature. "As I tried to take up his sharp and suggestive distinctions, his expressive, *useful,* but frequently arbitrary laws," Goethe wrote of Linnaeus, "there arose in me an inner conflict: what he tried to hold asunder, tended, according to the innermost demands of my nature, to be united."[24]

This statement raises the other two general characteristics of Goethe's Romantic science. His science made no effort whatsoever to be objective. It was, instead, subjective and imaginative, both in acknowledging that much of what we demand of scientific knowledge comes from inside us (e.g., it was the innermost demands of *Goethe's* nature, that demanded unity), and in projecting onto nature not simply human reason, but human passion as well. Thus, for example, Goethe saw the dispassionate and highly quantitative approach to the world associated with Newton's *Opticks* as deadening and alienating, and in *Zur Farbenlehre,* which offered an alternative approach to the phenomena of light, he explained that he wanted to interpret colors as "the deeds and sufferings of light." Finally, his comment on Linnaeus points up the special emphasis on unifying phenomena that was central to all of Goethe's science and all of Romantic science. We have seen that this search for unifying themes and laws is a central feature of all science, but for Goethe and the Romantics it became an obsession.

Goethe's second significant subject of scientific investigation illustrates all of these features of Romantic science in a context in which his work actually entered into state-of-the-art scientific discussions. As a student, Goethe had become fascinated with anthropological questions in connection with Herder's teachings. He agreed with Johann Blumenbach (1752–1840), among others, that humans were a single species because they satisfied the criteria established by Buffon—that is, all races of humans were capable of interbreeding with one another—but he was interested in the relationship between humans and other species. Linnaeus had suggested that facial bones were relatively easy to use in distinguishing among species, and it was generally believed that one of the most clear-cut distinctions between apes and humans was that the former had an intermaxillary bone in their upper jaw, whereas the latter do not. Goethe, convinced that all animals were unified somehow (later, as we will see, he developed the notion of type or archetype to account for such unification), felt that there should be evidence of the transition from ape to man in the form of some residual evidence of an intermaxillary bone in the human upper jaw.

In 1784, Goethe published his discovery of the intermaxillary bone in humans, which, he claimed, is generally invisible because it has become fused with other bones. It can, however, be identified in embryological studies before it has become fused, and it is also evident in the pathological conditions associated with cleft palate. This discovery was immediately communicated to a number of leading comparative anatomists and the existence of the intermaxillary bone in humans was widely debated, though acknowledged by very few.[25] Perhaps more importantly, this study provides a theme for much of Goethe's subsequent science, which focused on the modification of various natural structures, or the science of morphology.

This science began from the notion of "archetype" or "type." It seems fairly clear that Goethe's notion of type was confused in his early writings. Sometimes it was intended to designate some empirically existing primitive entity that might be related to others through the discovery of intermediate forms. At other times it was understood to be a kind of ideal creation of his own mind that could be used to structure empirical knowledge and to provide a foundation for the comparability of different entities. But after meeting the Kantian, Schiller, in 1794, and discussing with him the status of his ideas about the metamorphosis of specific plants out of some archetypal or Primal Plant, Goethe became increasingly explicit in regarding his archetypes as pure concepts containing only the most fundamental and universal attributes of those classes to which they gave structure, rather than as physically existing objects out of which others evolved.[26] The archetype thus established the basic outline for the actual forms that alone have empirical existence as modifications of this basic structure. Such modifications, or transformations, might occur either in response to external environmental forces or in response to internal forces.

In "The Metamorphosis of Plants," first published in 1790, Goethe focused on transformations driven from within. Beginning with the notion of a leaf "Type," he accounts for six stages of the growth of a plant in terms of transformations of the basic leaf structure driven from within by the nourishment provided by sap. Each successive stage of this process becomes more complex. The cotyledons or seed leaves are most simple, the next stage is slightly more complex, and the more mature, later leaves, become "enlarged, indented, divided in points and parts."[27] "The Metamorphosis of Animals," which generated much less interest, focused on adaptation to environmental circumstances as the driving force for transformations.

Goethe's biological writings focused attention on the concept of type or archetype, which came to play a major role in early-nineteenth-century German biology, and which played a major role in stimulating Charles Darwin's early thinking about the transformation of species.[28] His work on magnetism and optics, which began in the early 1790s, focused attention on the notion of polarity, which became another central organizing principle for the *Naturphilosophen.* Goethe set up a magnetic laboratory and studied magnetic phenomena quite extensively. At one level, these experiments, while done very carefully, led to few interesting results because Goethe was unwilling to separate what we would call the "noise" in his experiments produced by imperfections in his equipment, by changes in the local magnetic fields due to the presence of objects that were accidentally introduced, and so on, from the fundamental phenomena he was studying. As a radical empiricist, he viewed the "noise" as an integral part of the phenomena. On the other hand, his conceptualization of magnetism became important because he perceived magnetism as an archetypal phenomenon for all of nature in the way that his *Urpflanze* was for botanical entities and his *Urtier* was for zoological phenomena: "The magnet represents an archetypal phenomenon that needs but to be stated in order to explain itself. Hence, it becomes a symbol for all other things."[29] The sense in which it modeled all other natural phenomena was in its embodiment of the concept of polarity. It stood for all of the polar oppositions that pervade the natural world, the opposition between light and dark, between the subject and the object, between the body and the soul, between spirit and matter, between God and the world of sensory experience, between thought and extension, and between the ideal and the real.[30]

Of all the polarities, that which absorbed the greatest amount of Goethe's attention was that between light and darkness, which he discussed in *Zur Farbenlehre* of 1810, after working with optical phenomena for nearly twenty years. This work embodies virtually all that Goethe and the *Naturphilosophen* saw as most important in his work. Far from remaining dispassionate, it was passionate in its opposition to the objective methods of the Newtonian opticians:

> Friends, avoid the darkened prisons
> Where they pinch and tweak the light
> And in pitiful decisions
> Bow to rays distorted quite,
> Worshipers most superstitious
> Thronged in plenty down the year.
> Leave in hands of teachers vicious
> Spectres, madness, cheats, and leers![31]

Moreover, *Zur Farbenlehre* was insistent that colors be understood in terms of the subjective experiences of observers, so it emphasized physiological and psychological phenomena associated with color as well as the merely physical. Finally, in connection with the physical aspects of color, Goethe insisted that white was not a mixture of colors as Newton had suggested, but that it was pure light, just as black was pure darkness. Colors were then produced by the archetypal polar conflict between light and dark when it took place in a turbid (opaque or translucent) medium.[32]

F. W. J. Schelling and *Naturphilosophie*

The most fundamental problems that the generation of Fichte, Schelling, and Hegel saw emerging out of Kantian philosophy were almost universally motivated by moral and ethical considerations rather than by concern for natural philosophy. But in 1797, F. W. J. Schelling, after obtaining a theology degree and immersing himself for just one year in the study of medicine, physics, and mathematics, turned once again to the themes raised in Kant's works on the foundations of the natural sciences. Now, however, he brought to these topics a background that included a knowledge of Goethe's early scientific works and of Fichte's rejection of Kant's distinction between the noumenal and phenomenal worlds. The result was *Ideas for a Philosophy of Nature,* a text that had a major impact on scientific practices in Germany for the next forty years.

We have already developed the notions that Schelling either appropriated or modified from Kant and Goethe, but we need very briefly to consider the material that he adapted from Fichte. Fichte had become a self-avowed Kantian in the early 1790s because Kantian philosophy seemed to offer a serious way to justify the existence of free will. Over time, however, he became disturbed by the notion of a thing-in-itself distinct from a thing as experienced. He thus argued that there ought to be a way to avoid this Kantian version of dualism. Two potentially consistent options seemed available. Either one could absolutize the object—creating a form of materialism, making intelligence and consciousness merely epiphenomena, and making human feelings of freedom completely illusory—or one could absolutize the subject, in which case, the external world becomes epiphenomenal and the experience of freedom is validated. Given his primarily ethical interests, Fichte understandably chose the second option.

At least in his early writings, up to the first edition of *Ideas for a Philosophy of Nature,* Schelling agreed with Fichte on the primacy of the subjective. But that left him with the problem of accounting for phenomena without any things-in-themselves to serve as a ground for experience. After considering a whole series of possible ways to account for nature, given the primacy of the thinking subject—including those of Leibniz, Benedictus Spinoza (1632–77), Christian natural theologians, and Fichte—Schelling argued that there must be something more fundamental than either the subjective or the objective world. In much the same way that the Cartesian "I," which recognizes itself both as a thinking being and as a physical being, is logically prior to either of its manifestations as mind or body, there must be something prior to nature and the thinking mind.[33] Identified in the 1803 edition of *Ideas for a Philosophy of Nature* as the "Absolute," this higher principle manifests itself simultaneously as mind in us and as nature outside of us,[34] displaying the omnipresent polar structure that pervades the world. At the same time that the polar character of the Absolute somehow explains the difference between our inner experiences and those we experience as outside of us, the fact that our minds and external nature are manifestations of the same Absolute explains why the structures in nature and of our cognitive capabilities are matched to one another.

Before turning to the implications Schelling draws from this notion for the development of the natural sciences, I want to point out one of its implications with respect to the difference between our knowledge of nature and our knowledge of human activities. For Schelling, as for Kant and all other German Idealist philosophers following from Kant, there is a radical difference between our knowledge of nonhuman nature, which is deterministic, and our knowledge of self, which must hold our consciousness of freedom to be central. But how are we to understand other human beings: as objects in nature, or as free moral agents? Practical considerations, Schelling asserts, force us to accept the latter approach:

> I must in practice be *compelled* to acknowledge beings outside me, who are like me. If I were not compelled to enter into the company of people outside me and

> into all the practical relationships associated with that; if I did not know that beings, who resemble me in external shape and appearance, have no *more* reason to acknowledge freedom and mentality in me than I have to acknowledge the same in them; in fine, if I were not aware that my moral existence only acquires purpose and direction through the existence of other moral beings outside me, then left to mere speculation, I could, of course doubt whether humanity dwelt behind each face and freedom within each breast. All this is confirmed by our commonest judgments. Only of beings external to me, who put themselves on an even footing with me in life, between whom and myself giving and receiving, doing and suffering, are fully reciprocal, do I acknowledge that they are spiritual in character.[35]

Because we must treat other humans as uniquely endowed with agency, and because that notion of agency must be presumed to be the same as our own, all of our attempts to understand human actions, including the critically important attempts to understand human history, will depend upon trying to place ourselves as much as possible in the same position as the human whose actions we are trying to understand and then asking ourselves what we might have done in those circumstances. Given the noncausal character of human activity, we have no alternative to such an empathic strategy in trying to understand humans. On the other hand, given the causal character of nonhuman nature, no empathic strategy for seeking natural knowledge could possibly be of any value. This fundamental claim provided one foundation for the special character of the German *Geisteswissenschaften* during much of the nineteenth century.

Turning now to Schelling's discussion of approaches to nature, he appropriates Kant's dynamical analysis of matter and its universal properties virtually unchanged, concluding that, "attractive and repulsive forces constitute the *essence* of matter itself."[36] But Schelling goes beyond Kant in using this conclusion not simply to account for the existence of gravitational attraction, but also to account for elasticity, chemical interactions, light, heat, electricity, and magnetism, all of which are understood as qualities of matter that rest "wholly and solely on the intensity of its basic forces."[37] Both light and heat are understood, for example, to be particular intensities of repulsive force. Light is, "the highest degree known to us of expansive force,"[38] which can only be sustained by that part of the atmosphere that Schelling calls "vital air" or oxygen. Heat, on the other hand, is a degree of expansive force that can be manifested by all forms of matter and that can be communicated from one body to another until equilibrium is achieved.[39] Moreover, heating a body may produce what we call a change of state when the degree of expansive force achieves some critical level, as when water is changed to steam.[40]

For German physics, Schelling's discussions of the relationships between magnetism, electricity, and chemical activity turned out to be particularly significant. Following Goethe, Schelling identified magnetism as one of the two foundational phenomena in nature, and one that was identified with the polar principle of differentiation that transformed universal matter into the particular kinds of existing matter. Thus, he writes, "By virtue of gravity the body is in unity with all others; through magnetism it picks itself out and gathers itself together as a particular unity. Magnetism is therefore the universal form of individual being in itself."[41] If this is true, magnetism must be a universal characteristic of bodies. Why then is it not observed in all bodies? Here is Schelling's response: "Magnetism is as universal in Nature generally as sensibility is in organic Nature, where it belongs even to plants. Only for the *appearance* is it [magnetism] abolished in particular substances, in the so called nonmagnetic substances. What in magnetic substances is still distinguished as magnetism is immediately lost upon contact, as electricity; just as in plants, what in the animal is still distinguished as sensation is lost immediately in contractions."[42] It seems as though certain kinds of forces can be manifested in different ways—we now say they are transformed into one another—and, according to Schelling, under the appropriate conditions (e.g., in nonmagnetic substances) magnetism can be manifested as electricity in much the same way that the forces that are manifested as sensations in animals are manifested directly as motions in plants.

This strategy of arguing through analogy and generalizing from unproven specific assertions is characteristic of Schelling and other *Naturphilosophen.* It seemed justified by the Kantian emphasis on the unity of nature from the *Critique of Judgement,* which suggested that in a unified cosmos every phenomenon of nature should reflect the same pattern and that any pattern discovered in connection with one set of phenomena should be replicated in others. Schelling thus felt no need to offer specific empirical evidence for the transformation of magnetism into electricity in nonmagnetic materials. Rather, he inferred its necessity from his fundamental proposition that magnetism is a universal force of differentiation, coupled with the observation that it cannot be detected in some materials and with his observation in plants that sensation—also a fundamental force *in living entities*—seems to be transformed into tropic motions in plants.

If magnetism can sometimes appear as electricity, we might reasonably ask what electricity is. Schelling initially answers this question in an extremely abstract way: "It [electricity] is the dynamic or identity-nisus of two bodies entering into cohesion with one another."[43] In the process of trying to clarify and illustrate this definition, Schelling suggests that when two bodies cohere, it is generally the case that the surface of one expands into the region that had been occupied by the other, while that of the other contracts. In this case we define the body that is expanding (i.e., which evidences a condition of reduced internal cohesion) as in a state of positive electrification, while we define that which is contracting (i.e., which is experiencing a state of increased internal cohesion) as being in a state of negative electrification. As a consequence, the general law expressing the electrical relation between two bodies can be stated as follows: "*That one of the two which enhances its cohesion in opposition to the other will have to appear negatively electric, and that one which diminishes its cohesion, positively electric.* It is evident from this how the electricity of every body is determined, not only by its own quality, but equally by that of the other. . . . The bearing which the electric relationship of bodies has upon their oxidizability is [thus] intelligible, since this too is determined by cohesion relationships."[44]

The final sentence of this passage is particularly important for its suggestion that chemical and electrical phenomena might be intimately connected with one another (oxygen is identified by Schelling as a cohesion-intensifying principle),[45] while the first part suggests how it can be the case that even though all metals are electropositive relative to oxygen, some can be electronegative relative to others. In suggesting a fundamental relationship between electricity and chemistry, Schelling's first edition of *Ideas for a Philosophy of Nature* preceded the experimental work of Alessandro Volta (1745–1827), who first demonstrated in 1800 that electricity could be produced in a "pile" constructed out of alternating disks of silver and zinc brought into effective contact through brine-soaked disks of cloth or paper. But in the various supplements incorporated into the second edition, Schelling drew on Volta's work to illustrate his fundamental ideas. He was among the earliest to emphasize the significance of Volta's studies, and his views unquestionably encouraged interest in electrochemistry in Germany during the early years of the nineteenth century.

Just as Schelling adapted Kant's doctrines to establish his dynamical view of matter and Goethe's views to envision magnetism as the universal principle of differentiation in nature, he also appropriated their common view regarding the need to understand living organisms as ends as well as means and to understand the parts of organisms in terms of how they contribute to the functioning of the whole. With regard to the existence of some special life-force, however, Schelling certainly qualified the views of Kant, arguing that if there is a life-force, it cannot exempt living bodies from ordinary physical and chemical laws. Instead, it must be understood as somehow standing on the same level but acting in opposition to other chemical and physical forces. If this is the case, then there must be some higher, nonphysical, principle, lying outside the limits of empirical research, "which keeps this conflict going and maintains the work of Nature in this conflict of alternately prevailing and submissive forces."[46] This higher principle had to be mental rather than physical and it was manifested in the overall plan or design that directed the growth and activity of the organism—what one might call its "soul." In *Von der Weltseele, einer Hypotheses der hoheren Physik zur Erklarung des allgemeinen Organismus* (*On the World Soul*) of 1798, Schelling went on to suggest that the universe as a whole was one single living organism whose "soul" consisted of an archetypal polarity, opposition, or "conflict" of opposed forces, which manifested itself in all of the particular polarities found in nature.[47]

The special way in which Schelling formulated his understanding of organisms demands just a bit more exploration because of the way in which it—and closely related formulations by Fichte and Hegel—influenced German biological, social, and political thought in the early nineteenth century. We saw in chapter 1 that French theorists turned increasingly to organicist models for understanding society at the end of the eighteenth century because they understood the complexity of organisms to preclude the usefulness of more analytic models derived from the exact sciences. In doing so, they did not pretend to be able to have access to any inner essence associated with society; rather, they rested content with articulating general facts or laws. When German theorists turned to organic models and metaphors, however, their focus was on different issues connected with Kant's notions of purposiveness, Goethean notions of archetypes and metamorphoses, and/or Schelling's notion of the nonmaterial "soul," or spirit, which guided the organism's growth through a series of polar conflicts that might combine internal and environmental elements. Moreover, for the German organic theorist, the whole point of his investigation was to discover the idea or spirit or archetype that was manifested in the concrete phenomena rather than to rest comfortably in a mere description of events and objects. These features of German organicist thought should be kept in mind for future reference.

Naturphilosophie and German Physics in the Early Nineteenth Century

In most ways *Naturphilosophie,* grounded in the rationalist assumptions of Kant, the Romantic passion of Goethe, and the Idealist philosophy of Schelling, must seem bizarre to twenty-first century readers—even German readers—whose notion of science is much more directly related to Positivism.[48] There is something especially peculiar about Schelling's vision of dynamical nature, which insists that all phenomena must be understood as manifestations of attractive and repulsive forces of differing intensities, but that wants nothing to do with quantification, which alone can deal directly with differences in intensity. Charles Coulston Gillispie, a distinguished American historian of science, expressed a widespread view that was certainly shared even by some of Schelling's immediate colleagues, in writing that in spite of the deep interests and deep feelings associated with the Romantic view of nature, "it is the wrong view for science."[49]

If this statement simply means that it is a viewpoint not shared by the vast majority of modern scientists and not consistent with current scientific practices, I am in complete agreement. But it *was* a point of view shared by a substantial number of persons who viewed themselves as scientists and who held academic positions as natural philosophers and physiologists in early-nineteenth-century Germany. Moreover, the practicing scientists who identified to some substantial degree with *Naturphilosophie* did publish in scientific journals, and their views became the subject of important discussions in the fields of physiology, optics, electrochemistry, crystallography, and electromagnetism. For a brief period of time, *Naturphilosophen* even dominated the philosophy and medical faculties at the University of Berlin.[50]

Among academic physicists holding university positions between 1790 and 1840, almost all wrote textbooks grounded in Kant's dynamical understanding of matter, including extensive sections on Kantian or Idealist epistemology.[51] Only a relatively small number, however, including Wilhelm Ritter (1776–1810) at Jena, Hans Christian Oersted (1777–1851) at the University of Copenhagen, Christian Samuel Weiss (1780–1856) at Leipzig and then Berlin, and Thomas Seebeck (1770–1831) and Georg Friedrich Pohl (1788–1849) at Berlin, followed Schelling into his speculations concerning chemical, electrical, and magnetic forces. Even fewer followed Goethe in his anti-Newtonian understanding of colors. But the appetite for Romantic physics among students was disproportionate to the number of professional physicists who were sympathetic to it. Thus, for example, when Leopold Henning lectured on Goethe's theory of colors at Berlin starting in 1821, he drew enrollments of from forty to sixty, two to three times the average enrollments of introductory natural philosophy courses at the time, and he also drew from the town artists, secondary school teachers, and military officers as auditors.[52]

Within the sciences, the most important selective appropriations from Schelling's *Naturphilosophie* came from experimentalists, who found in his emphases on polarity and claims for the close connections among light, heat, electricity, magnetism, and chemistry, a rich storehouse of suggestions to stimulate their experimental work. Thus, in his introductory textbook, published in 1811, Hans Christian Oersted wrote that, "Schelling created a new *Naturphilosophie,* the study of which cannot but be important to the experimental investigator of nature; it not only may excite in him many new ideas, but also cause him to repeat tests of much that was previously regarded as settled."[53] This view was prophetic for the work that made Oersted famous, his demonstration that electrical currents can produce magnetic effects. As early as 1805, Oersted explicitly followed Schelling's suggestion that magnetism, electricity, and chemical processes should be viewed as the construction of matter out of attractive and repulsive forces corresponding to the three forms of line, surface, and volume. As part of this discussion, he insisted that "the same forces that express themselves in electricity also express themselves in magnetism; although in another form."[54] By 1812 this notion of the connection between many phenomena through their unification as different manifestations of fundamental attractive and repulsive forces had led Oersted to the conclusion that the different forms could "pass over into one another," and in the following year he concretized this notion in his *Recherches sur l'identité ses forces chemiques et électriques,* arguing, "It must be tested whether electricity in its most latent state has any action on the magnet as such."[55] No magnetic effects could be found that were produced by static electricity, but after years of seeking a magnetic effect of electricity, Oersted finally demonstrated that an electrical *current* does cause a magnetic needle to orient itself perpendicular to the direction of motion of the current and in opposite directions above and below the wire through which current is flowing.[56] This constituted the first empirical evidence for the interconnectedness of electricity and magnetism, and there is no serious doubt that it had been produced in connection with expectations that arose out of *Naturphilosophie.*

Oersted's interest in *Naturphilosophie* had almost certainly come about as a result of his visits to and correspondence with Schelling's Jena colleague Johann Wilhelm Ritter, who also drew from Schelling's *Ideas* to guide several important experimental research programs. Ritter, for example, sought to explore the significance of the electrical polarity that was one manifestation of the polar principle of the World Soul throughout all of nature, and especially in living nature. Thus, he carried out a huge range of experiments in the electrophysiology of both plants and animals. He demonstrated, for example, that the optical system of animals is affected differently when connected to the positive pole of a voltaic battery than when it is connected to the negative pole and that changing the intensity of the charge on the eye will change the color that one perceives.[57]

Ritter's fascination with polarities and with the interconnections among various phenomena led him to what many have identified as the discovery of ultraviolet light. In 1800, William Herschel (1738–1822) had demonstrated that at wavelengths longer than those associated with the visible spectrum there was a form of radiation that heated bodies. Indeed, for sources at temperatures that are easily reached by bodies heated in ordinary fire, these invisible rays conveyed even more heat than the visible spectrum. Ritter had become interested in the relationship between light and chemical reactivity. He identified Herschel's infrared rays as a source of "oxygeniety" because they seemed to stimulate oxidation reactions (we now know that almost all chemical reactivity increases with temperature), and then, assuming the general principle of polarity, he started to search for light beyond the violet end of the visible spectrum that might be a source of "hydrogeneity," that is, which would stimulate reduction reactions. In 1801 he announced the discovery of the expected rays in an article on the "Chemical Polarity of Light," in the Erlangen *Litteratur-Zeitung.*[58] Subsequent investigators separated the discovery of light beyond the visible violet from Ritter's photochemical speculations to identify this as the discovery of ultraviolet radiation.

Similar stories can be told of Christian Samuel Weiss, who came to know Schelling while he was a medical student at Leipzig, and whose important work in crystallography showed strong *naturphilosophisch* elements; of Thomas Seebeck, who lived in Jena and was personally acquainted with Schelling, Ritter, and Weiss, and whose experimental work on thermoelectric phenomena was also stimulated by the search for interconnectedness among different kinds of phenomena; and of the theoretician and professor at Berlin and later at Breslau, Georg Friedrich

Pohl, who came to *Naturphilosophie* after 1805 without personal connections to Schelling or Ritter, but who alone among German physicists, identified himself as an unqualified *Naturphilosoph.*[59] Many later German physicists appropriated some elements of *Naturphilosophie* without buying into all of its doctrines.

Naturphilosophie and German Physiology before 1840

In physiology, as in physics, *naturphilosophisch* tendencies persisted among German academics during the first four decades of the nineteenth century. Moreover, as in physics, theoreticians who tended to accept virtually all aspects of *Naturphilosophie,* including its idealist metaphysics, were much rarer than experimentalists, whose early exposure to *Naturphilosophie* often played a particularly important role in suggesting fruitful lines of experimental investigations even after they had rejected its more speculative elements.

Of the theoreticians, far and away the most important was Lorenz Oken (1779–1851), a south German peasant who managed to gain an education and to get an appointment as assistant to the professor of anatomy at Jena in 1807. An open disciple of Schelling's philosophy, Oken sought in his first major publication, *Die Zeugung* (*On Generation*) of 1805, to account for the relationships between simple and complex organisms by arguing that all must be formed out of the same primitive living units, or *infusoria.* Oken argued that within a primordial, undifferentiated slime, primitive spherical vesicles arose. These simplest living things joined together into communities or aggregations to form ever more complex organisms. Thus, he writes, "All flesh may be resolved into infusorians. We can invert this statement and say that all higher animals must be formed from constitutive animalicules. These we call Primitive Animals, and note that they constitute not only the fundamental materials of animals but also of plants. . . . In a larger sense they may be called the primitive matter of all which is organized."[60]

Though historians of biology are uncomfortable in counting such speculative comments, unconfirmed by direct observational evidence, as of major significance, there is little doubt that Oken's comments on the necessity that some small, primitive, but nonetheless living unit should be found as the fundamental building block of animal and plant tissues stimulated the microscopic observations of Matthias Jacob Schleiden (1804–81), who is generally acknowledged to be the founder of cell theory. Such a downplaying of Oken's role began with Schleiden, himself, and is easy to understand in part because Schleiden was a philosophical follower of the anti-Schellingian, Jacob Fries.[61]

Oken continued to write on physiology and natural history well into the 1840s, and his *Lehrbuch der Naturphilosophie* seems to have been widely read, but he was also important in the institutional development of the sciences in Germany. As we will see shortly, *naturphilosophisch* notions became widely associated with anti-French sentiment and with German cultural and political nationalism during the period following 1807. Oken was particularly caught up in nationalist fervor. Thus, in 1822 he was the primary moving force behind the establishment of the *Versammlung deutscher Naturforscher und Artze* (Association of German Natural Scientists and Physicians), which was intended to bring together scientists and physicians from all over the Germanies for annual meetings that would explore common interests. In addition, he published the scientific journal, *Isis,* which invited German language articles from advocates of many different scientific perspectives and in a variety of fields, much as *Philosophical Magazine* did with English language articles in Britain. He also participated in the German-unification festival sponsored by student fraternities near Jena in 1817. This last act led in 1819 to the loss of his academic position at Jena. For a time he became a professor at Munich, but again he got into political trouble for his nationalist activities, so in 1832 he accepted a post in the more liberal environment at Zurich, where he remained until his death.

One of the distinguished experimental physiologists to draw on *naturphilosophisch* ideas—though only selectively and partially—was Karl Ernst von Baer (1792–1876), who studied physiology from Schellingian sympathizers, Karl Burdach (1776–1847) at Dorpat and Ignaz Döllinger (1770–1841) at Wurzburg, and then went on to the professorship in physiology at Königsburg, where he and Burdach became colleagues. Typically for men of their generation, both Burdach and Döllinger had begun their careers with a grand enthusiasm for Schelling's speculative philosophy, but Döllinger had gradually been repelled by what he saw as the speculative excesses of such *Naturphilosophen* as Oken and Nees von Esenbeck (1776–1858). Consequently, he focused on the need to construct physiology on the foundation of precise observations that might then become the material for philosophical reflection.[62] At the same time, he continued to utilize many of the organizing notions from his earlier days as heuristic devices for suggesting experimental approaches, and it was this set of heuristics, linked to an empiricist approach, that von Baer appropriated.

At Döllinger's suggestion, von Baer took up the central problem of how fertilized eggs grow into adults. Following a Kantian tradition that Timothy Lenoir has identified with the term "telio-mechanism," von Baer rejected preformationist theories, according to which the structure of the adult organism is already present in the fertilized egg, in favor of the epigenetic doctrine that a special developmental force, which he also at times identified with Schelling's Idea,[63] controls the development of the fertilized egg into the adult. In trying to understand the particular mechanisms by which the regulation of embryo growth was regulated, von Baer often appealed to the presumption of forces linked to electromagnetism and to the primacy of polarities. Thus, for example, he speculated that the asymmetrical alignment of materials in a fetus might be associated with electromagnetic effects and suggested that experiments might be devised to determine whether this was the case; moreover, in trying to understand why the organs of ingestion and excretion develop at opposite sides of the embryo he uses the standard *naturphilosophisch* language of polarity and metamorphosis.[64]

In his multivolume *Über die Entwicklungsgeschichte der Thiere* (*On the Developmental History of Animals*), published from 1828 to 1837, von Baer offered minutely detailed descriptions of the developmental stages of chick embryos and many mammals, focusing on the extent to which the embryological development of different organisms runs in parallel until specific divergences occur. Von Baer explicitly disagreed with Johann Meckel's (1781–1833) 1821 expression of the ontogenetic claim—suggested by the Kantian notion that there is a close parallelism between the way that a tree produces itself and the way it reproduces its species—that "the development of the individual organism obeys the same laws as the whole animal series; that is to say, the higher animal, in its gradual evolution, essentially passes through the permanent organic stages which lie below it."[65] Instead, following Cuvier and Carl Rudolphi (1771–1832), he identified four basic types, or ideas, each of which directed the development of a major group of organisms. Within each of these groups, however, von Baer did admit that some form of evolution, manifesting itself in ever greater degrees of specialization, does take place and that the embryological development of the more specialized forms does recapitulate the developmental history of the less specialized forms in its initial stages. One of the major consequences of the descriptive comparative embryological work of von Baer and his contemporaries was that German biology was particularly open to Darwinian evolutionary perspectives when they entered the scientific scene in 1859. A second, and in the long run, less positive consequence was that because they tended to take some form of ontogenetic analogy very seriously, German biologists had a very difficult time separating the study of inheritance, or genetics, from the study of embryological development. Only in places such as the United States, where no single theory was expected to account both for individual growth and for evolutionary transformation, was genetics able to grow rapidly at the beginning of the twentieth century.

In terms of contemporary reputation, the most distinguished physiologist to express *naturphilosophisch* tendencies was probably Johannes Peter Müller (1801–58). The son of a wealthy Coblenz shoemaker, Müller studied medicine at Bonn under Schellingian tutors, including the botanist Nees von Esenbach, and wrote a *naturphilosophisch* doctoral dissertation on the motion of animals in which he identified bending and stretching as the two poles of life and compared them with the unopened buds and withered flowers of plants respectively. "In both," he wrote, "night prevails; but between them moves life."[66]

From Bonn, Müller went to study with the experimentalist Carl Rudolphi, who was unsympathetic to the speculative works of Schelling and Oken. As a consequence, Müller, like Döllinger, seems to have become a careful experimentalist, renowned for his development of new instruments and procedures. His early work was, nonetheless, addressed to questions raised by Goethe, Schelling, Ritter, and other Romantic scientists. Beginning from the physiological side of Goethe's *Zur Farbenlehre,* Müller devoted much of his early career to a comparative study of sensory physiology in humans and other animals, which appeared in 1826 as *Zur vergleichenden Physiologie des Gesichtsinnes des Menchen und der Thiere* (Toward a Comparative Physiology of the Senses in Men and Animals). Later, when his emphasis turned to comparative embryology, he continued to view his work in Goethean terms, as, "uncovering the metamorphosis of the organs and organism in its finite development as Caspar Friedrich Wolff and Goethe recognized and demonstrated."[67] Especially in his studies of the development of the urogenital systems in men and animals, Müller's work reached a level of precision that was unprecedented, and his discovery of the convoluted manner in which elements from different embryological sources come together to form the genital system seemed to confirm the widespread German belief that no purely mechanical and nonpurposive process could possibly guide the growth of organisms.[68] Throughout his career, Müller continued to view his scientific task as fundamentally "spiritual." And, what will become equally important for our story in the next chapter, his conservative political views, which emerged during his idealist youth, never changed.

The Relation of *Naturphilosophie* to the German Romantic Movement

Romantic or "Romantisch" is primarily a literary term whose use in Germany began during the 1760s in the writings of C. M. Wieland (1733–1813). Its initial use was intended to point up certain characteristics of popular imaginative literature that was intended for entertainment rather than for didactic purposes and that emerged as important in the European Middle Ages, though it was seen as having earlier Oriental sources.[69] These characteristics included an acceptance of elements that were imaginary, or unreal, often magical or mystical; a fascination with exotic places and persons; an intensification of emotions beyond the normal; and, as time went on, a special interest in the conflict and resolution of presumably opposed elements; that is, of sentimental and erotic love, of good and evil, of beauty and ugliness, of subject and object, of the natural and the artificial, of the comic and tragic, and so on.

Prior to the mid 1790s, Romanticism had little if any contact with philosophy in general or natural science in particular. Furthermore, it was self-consciously and intentionally apolitical. In complete opposition to the view of art and literature that was to be promoted by the Saint-Simonians in early-nineteenth-century France, Romantic artistic productions were not intended primarily to motivate and direct collective social action. To the extent that they were intended to "do" anything other than entertain, it was to promote the development of the individual reader's openness to a wide range of experiences that extended beyond the pedestrian and realistic.

Largely through the dual efforts of Georg Hamann (1730–88) and Johann Herder, key features of Romanticism came to be associated with the "poetic imagination," which Hamann, like Giambattista Vico (1668–1744), identified with the particularly creative youth of a culture. He saw this prephilosophical youthful creativity as being undermined and destroyed by the critical rationality associated with the Enlightenment.[70] For Herder, the creative element in each people or *Volk*—its special genius—was expressed particularly clearly in the folk music and folktales of that people, and there was an important sense in which the shared stories of each particular language community formed a bond that distinguished members of that *Volk* from all others. Though not explicitly nationalistic in the writings of Hamann and Herder, such notions were available for appropriation by nineteenth-century German Romantics in the construction of an ideology of "cultural nationalism" that was to gain in political importance through the mid nineteenth century and that had a resurgence in the late twentieth century with the breakup of Soviet-controlled governments in Eastern Europe.

Romanticism became a self-conscious and programmatic movement in Germany beginning at Jena during the 1790s, at a time when Fichte, Schelling, Ritter, Novalis (1772–1801), and Friedrich and Auguste Schlegel (1772–1829, 1767–1845) formed the core of a group of interacting intellectuals. When it did so, it took on a philosophical and even a "scientific" dimension. Thus one of the chief theorists of the Romantic movement, Friedrich Schlegel, wrote: "The whole history of modern poetry is a continuous commentary on a short text of philosophy: all art should become science, and all science, art: poetry and philosophy should be united."[71] Trying to explain what this statement meant to the Romantics, the later-Romantic historian, H. A. Korf, wrote: "Reason cannot get along without imagination, but neither can imagination get along without reason. The wedlock of the two is, however, of such a peculiar kind that they wage a life and death struggle, while they are capable of attaining their greatest achievements only in close cooperation. The result of this combat is called science when reason plays in it the leading part; it is poetry when the leading part has been taken over by imagination; it is, finally, philosophical intuition when reason and imagination cooperate harmoniously."[72] On a very important level this statement is simply another expression of the Romantic fascination with polarities and with their resolution in some new form of unity that had predated any attempt to make it programmatic. Beginning around 1797, however, the synthesis of opposites became an obsession, and it was often formulated in the language of *Naturphilosophie.* In his notebooks, for example, Friedrich Schlegel consistently discussed "the Romantic" as a synthesis of disparate elements, using the term *Indifferenzpunkt,* in the manner of Schelling and Ritter, to indicate the point of balance among conflicting elements.[73] This interest in the point of indifference or balance, in which opposites somehow mixed without destroying one another and in such a way as to produce a higher synthesis, was often expressed in optical images. Novalis, for example, has his protagonist Heinrich von Ofterding explain his preference for twilight, saying: "Who would not like to walk in twilight, when the night is broken up by light and light broken up by night into higher shadows and colors."[74] And in his novel, *Godwi*, published in 1801, Clemens Brentano (1778–1842) focuses on the medium between the subject and object, symbolizing the effect of the mediation by considering the reflecting and refracting effects of a bowl of water held between a visual object and an observer.[75] In a more playful mood, Novalis explores the interaction of plant and animal, male and female, spirit and matter, in a way that reminds one of Ritter's vision of animals as the inverse of plants: "Are the plants perhaps the products of female nature and male spirit, and the animals the products of male nature and female spirit? The plants perhaps the girls, the animals the boys of nature?"[76]

If the concept of polarity, or the synthesis of opposites into some kind of higher unity, is one central concept that links Romantic aesthetic notions with the *naturphilosophisch* formulations of natural science, there is a second, almost equally important, common ground in an emphasis on the organic as opposed to the mechanical. In an essay on Giordano Bruno, Schelling had explained how these two are connected in a discussion of the tension between the infinite and the finite or the unlimited and the limited, which many Romantic writers saw as a fundamental polarity to be resolved in works of art: "The way in which the finite is linked up with the infinite, is most clearly paralleled in the world of known and visible things by the way the parts are linked up with the whole in an organic body."[77] This notion was transferred to the world of art by Friedrich Schlegel, who argued that the success of an artistic production in resolving all of the conflicts with which it deals can be judged by the extent to which it constitutes an organic entity in which every part both reflects and serves the whole, and in which every portion can be understood equally as end and means.[78] Thus, Schlegel predicated his criteria for criticizing art upon the conditions established by Kant, Goethe, and Schelling for the study of living organisms.

Organicism, Romanticism, and Nationalism

There is one use of the Romantic emphasis on organicist thought that played a particularly significant role in German political and cultural developments. That is the incorporation of the state or nation into the organicist understanding of history that had been initiated in the 1770s and 1780s by such German late-Enlightenment historians as Herder and August Ludwig Schlözer (1735–1809).[79] I have already suggested that for Herder's generation, the *Volk* manifested its creativity primarily through its cultural productions in music, literature, and art. For Herder in particular, the political state or nation was usually seen as a destructive rather than as a constructive institution.[80] But this negative view of the state, which pervaded Romanticism in its early years, was radically changed in the aftermath of Napoleon's defeat of the Prussian armies at Jena and Auerstadt in early October of 1806, his occupation of Berlin in late October of 1806, and the final defeat of Prussia in July of 1807.

Although the Romantic notion of freedom was quite different for most purposes than that of French liberals, it was equally opposed to the restriction of thought and action by alien forces. Thus, the Romantics were all but universally opposed to French hegemony in German lands. This was especially true because Napoleonic France was understood by German Romantics as the carrier of everything that they hated about the French Revolutionary ideology. That is, it stood for universalism in opposition to the belief in the unique character of each *Volk;* it stood for secularism in opposition to religiosity; it stood for a peculiar form of rationality that turned its back on history; and it stood for egoism and pure self-seeking in opposition to cooperation and altruism. As a consequence, Romantic theorists, especially Johann Fichte in Berlin and Auguste Schlegel in Vienna, took the lead in promoting an anti-French sense of German unity. This sense of unity, building upon the cultural nationalism of Herder and earlier German historical thinkers, focused attention on a national political unification that would allow a successful rebellion against French influence.

When they focused on German unification, they cast the German nation as an organism in the sense discussed by Schelling above. That organism was a single entity guided by a unique idea or spirit that differentiated it from other nations. It was to be understood as developing or metamorphosing over time through a process that was at once internally driven and manifested in conflicts that sometimes involved external influences. Above all, it was understood to be an entity that had its own peculiar ends above and beyond the protection of the individual citizen's property. Indeed, Germans, as parts of the organic German nation, should understand themselves as both created by and for the nation—as contributors to a greater synthesis, rather than as self-seeking individuals for whom the nation was nothing but a necessary evil justified solely by the need for protection of property.

Schlegel's entrée into these issues was primarily historical. Returning to the period of the Holy Roman Empire, he saw a deep spirituality as the distinguishing spirit of the German people—one that set them apart from the French in particular, and one that persisted through all of the political transformations the Germans experienced. Furthermore, he saw in the Germans a special preference for a feudal-aristocratic social structure, rather than a democratic one. Fichte also viewed the German nation in terms of a pattern of historical development. He too saw German uniqueness as linked to religion and a political system dominated by an aristocratic elite. Thus, in defining the guiding spirit of the Teutonic race, of which the Germans are the most important branch, he wrote: "Its mission was to combine the social order established in ancient Europe with the true religion preserved in ancient Asia, and in this way to develop in and by itself a new and different age after the ancient world had perished."[81]

In a series of fourteen *Addresses to the German Nation* given in the occupied city of Berlin during 1808, Fichte drew upon the organic character of the German *Volk* to explain why subordination to Napoleonic visions of a *universal* monarchy must be resisted:

> Only when each people, left to itself, develops and forms itself in accordance with its own peculiar quality, and only when in every people each individual develops himself in accordance with that common quality, as well as in accordance with his own peculiar quality-then and then only, does the manifestation of divinity appear in its true mirror as it ought to be. . . . Only in the invisible qualities of nations, which are hidden from their own eyes—qualities as the means whereby these nations remain in touch with the source of original life—only therein is to be found the guarantee of their present and future worth, virtue, and merit. If these qualities are dulled by admixture and worn away by friction, the flatness that results will bring about a separation from spiritual nature, and this in turn will cause all men to be fused together to their uniform and conjoint destruction.[82]

He insisted that, "it is only by the common characteristic of being German that we can avert the downfall of our nation which is threatened by its fusion with foreign peoples, and win back an individuality which is self-supporting and quite incapable of any dependence upon others." He pled with his audience both to think of Germany as an "organic unity in which no member regards the fate of another as the fate of a stranger," and to "see the whole in each part."[83]

The German nation's passion for freedom from alien influences had played a critical role in world history when it had successfully resisted Roman domination.[84] Now Fichte called for the German nation to repel the "Rome of to-day." "Think that in my voice there are mingled the voices of your ancestors of the hoary past, who with their own bodies stemmed the onrush of Roman world dominion," he cried. "They call to you: 'Act for us; let the memory of us which you hand on to posterity be just as honorable and without reproach as it was when it came to you, when you took pride in it and in your descent from us.'"[85] But Fichte was aware that in the present state of affairs, military opposition to the French was pointless. What form then should resistance take? Eventually, Fichte did look forward to political unification of the German peoples and to the creation of a uniquely Germanic state, which would be vastly different from the liberal state that was focused on the protection of individual liberties and property. But before that could occur, he saw the need for a recovery of German cultural identity through a transformation of education: "In a word, it is a total change of the existing system of education that I propose as the sole means of preserving the existence of the German nation."[86]

Though Fichte did not lay out the full content of the curriculum of his new education, he made it clear that it would focus on philosophy, science, and German literature and he made it clear that its details would depend upon a reformed professorate and literary community. On the one hand, it would place new emphasis on scholarship, for in a very real sense it was scholarship that would drive the continual growth of culture: "The person who is not a scholar is destined to maintain the human race at the stage of culture it has reached, the scholar to advance it further according to a clear conception and with deliberate art. The scholar with his conception, must always be in advance of the present age, must understand the future, and be able to implant it in the present for its future development. For this purpose, he needs a clear survey of the previous condition of the world, unlimited skill in pure thought, independent of phenomena, and, in order that he may be able to communicate his thoughts, control of language down to its living and creative root."[87]

The last phrase in this passage indicates that scholars would not only have to focus on being knowledge producers, but they would also have a role in the broader community of "men of letters," and as men of letters, especially of *German* letters, they would have an additional obligation: "The noblest privilege and the most sacred function of the man of letters is this: to assemble his nation and to take counsel with it about its most important affairs. But especially in Germany this has always been the exclusive function of the man of letters, because Germany was split up into several separate states, and was held together as a common whole almost solely by the instrumentality of the man of letters, by speech and writing."[88] Many of the features of Fichte's proposed educational reforms were implemented beginning in 1810, when Wilhelm von Humboldt (1767–1835), Prussian minister of education, established the University of Berlin and appointed Fichte as professor of Philosophy. At Berlin, and then spreading throughout Germany, university reform—the first stage of overall educational reform—was so successful that the network of roughly twenty German universities became the envy of the rest of the world by 1870. One of the most important features of this reform involved the focus on advanced scholarship and the specialization of academic life that demanded. But a second, and for our purposes, equally important, aspect involved the expectation that even the most specialized scholar had an obligation to contribute to the broader culture by reflecting on and communicating widely his understanding of the relationships between his specialized knowledge and the interests of the nation as a whole.

Throughout the remainder of the nineteenth and the early twentieth centuries, whereas a very small number of academic natural and social scientists sought to interject themselves into lay cultural affairs in England, France, and the United States, for example, there was virtually no major scientist in Germany who did not. Thus, the popular works of scientists, including Alexander von Humboldt (1769–1859), Emil Du Bois-Reymond (1818–96), Hermann von Helmholtz (1821–94), Wilhelm Ostwald (1853–1932), Jacob Moleschott (1822–93), Ludwig Büchner

(1824–99), and Ernst Haeckel (1834–1919) were among the most widely read of all literary productions, as we will see in subsequent chapters.

Mathematics, natural science, philosophy, and German literature were also added to the curriculum of the *Gymnasiums* that functioned as university preparatory schools, and under von Humboldt, the old emphasis on vocational *Realschule* for ordinary citizens was downplayed, opening up the *Gymnasiums* to a larger segment of the population. As a consequence, a philosophically based and specifically German-oriented education did become much more widespread, providing an increasingly extensive sense of national identity grounded in intellectual life.

Fichte had finally insisted that the state should take up responsibility for providing the kind of education that he envisioned, and under the Humboldtian reforms this also took place. The *Gymnasia* had traditionally been sponsored by religious organizations and had been staffed by theologically trained teachers or, more often, with persons who had never finished any university-level degree. Now the state took over support of elementary and secondary education, requiring that all *Gymnasium* instructors have degrees from the philosophy faculty of a university. With respect to the state support of education, Fichte offered an interesting and prophetic insight when he argued that for the purpose of educational reform, the disunity of German political entities was a potential advantage: "What has so often been to our disadvantage may perhaps in this important national business serve to our advantage. The rivalry of several states and their desire to anticipate one another may perhaps bring about what the calm self-sufficiency of the single state would not produce."[89]

At least in connection with the reform of university education, and within that reform, in connection with the growing emphasis on and support for scholarship in the sciences in the middle decades of the nineteenth century, Fichte was certainly correct. Though it was initially most important in humanistic disciplines such as philology and history, competition for educational and cultural leadership drove up the status and remuneration of all university faculty. Even more importantly, it drove up funding for laboratory facilities and research support in the natural sciences beginning around 1825, but increasingly after 1848.[90] For those who were truly committed to research and scholarship, there had probably never been a more nurturing environment in world history than the German university around 1860.

That Fichte and other German nationalist educational reformers viewed the state as the entity that was most appropriate to take on support of education brings us back to the pressures for the political unification of Germany that their organicist views produced. For Fichte and virtually all of his idealist followers, the truly natural boundaries of states emerged out of the linguistic unities of peoples: "Those who speak the same language are joined to each other by a multitude of invisible bonds by nature herself, long before any human art begins; they understand each other and have the power of continuing to make themselves understood more and more clearly; they belong together and are by nature one and an inseparable whole."[91] During the Middle Ages, the German people did have a unified political life as the Holy Roman Empire, but with the split in Christendom, that unity was destroyed, and Germany became the tragic battleground in which the dynastic conflict between Bourbon and Habsburg forces played itself out. "If only the German nation had remained united, with a common will, and common strength," Fichte wrote, "then, though the other Europeans might have wanted to murder each other on every sea and shore, and on every island too, in the middle of Europe the firm wall of the Germans would have prevented them from reaching each other."[92] Once again, in 1807, French and Austrian forces were battling one another on German soil, and Fichte hoped that the revival of a unified Germany might provide for another turning point in European history.

Though Fichte recognized that the institutionalized state was probably a necessity to provide the circumstances for the continuation of the nation in its broader cultural sense, the elaboration of both the character and importance of the state was largely the work of a series of Fichte's colleagues at the newly established University of Berlin between 1810 and 1830. Of these, G. W. F. Hegel, a younger student friend of Schelling's who was deeply influenced by Schelling's *Naturphilosophie,* was undoubtedly the most important.

Starting from Schelling's doctrine of the World Soul, which accounts for the structure of nature through its polar character, Hegel incorporated a more explicitly temporal and developmental notion under the concept of history. For Hegel, "World history in general is the development of *Spirit in Time,* just as nature is the development of the *Idea in Space.*"[93] The polar character of this Spirit manifests itself in an internally driven process in which any given moment of historical existence (designated by Hegel as the "thesis") holds within itself the source of its own destruction (the "antithesis") in such a way that the working out of the conflict between thesis and antithesis leads to a new condition (the synthesis), which then becomes the thesis for a new developmental round. Just as Schelling's nature is to be understood as an organism, and thus as an entity that has a purpose, so too does Hegel's history have a purpose, which is that Spirit might gradually come to disclose itself in self-consciousness and self-creativity (freedom). This process goes through a number of critical stages associated first with inorganic processes subject only to universal laws, then with "chemism" or processes that create order without life, then with organic developments, and finally through the stage of humanity in which self-consciousness finally emerges.

It is here, at the stage of the development of humanity and in the transformation of mere self-consciousness into true freedom, that the state becomes absolutely critical in two different and conflicting ways. First, it provides the structure of laws and social arrangements within which the individual subjective will becomes linked to and subordinated to the objective universal will, or in which man's animalistic, self-absorbed, nature gives way to his social subordination to a common good: "The State is the definite object of world history proper. In it freedom achieves its objectivity and lives in the enjoyment of this objectivity. For law is the objectivity of the Spirit; it is will in its true form. Only the will that obeys the law is free, for it obeys itself and, being in itself, it is free. In so far as the state, our country, constitutes a community of existence, and as the subjective will of man subjects itself to the laws, the antithesis of freedom and necessity disappears. . . . The objective and subjective will are then reconciled to form one and the same harmonious whole."[94]

From this point of view, those who chafe against the laws and customs of their own state and who set their personal notions of what are right and good against "things as they are" are likely to be misguided, though, "in asserting good intentions for the welfare of the whole and exhibiting the semblance of good heartedness, [they] can swagger about with great airs." Age and philosophical reflection lead us rather to the insight that "the actual world is as it should be."[95] On this level, Hegel's notion of the state served the conservative political ends that had led the Prussian minister of education to bring him from Heidelberg to Berlin in the first place; and it was emphasized by the so-called Hegeleans of the right.

On the other hand, for Hegel, no particular state at a particular time embodied the final realization of the Spirit. The laws and morality of each state remained imperfect, and there were certain world-historical individuals through whom the Spirit acted to transform the present into the future. Such men, Hegel insisted, are rarely, if ever, conscious of being the agents of the Idea. They are never academics. Rather, they are practical and political men, who bring about the new stage of the self revelation of the Spirit by transforming the state (Hegel explicitly cites Alexander the Great, Caesar, and Napoleon as examples). The state is, thus, not simply the locus within which individual citizens become linked with the universal through obedience to the laws, it is also the locus of conflict in which new syntheses are created in the Spirit's progress toward self revelation. From this point of view, which appealed to the young Hegelians, or Hegelians of the left, the state was more important as the primary site of historical change than as the structure within which morality emerged.

Though Hegel's understanding of the state was certainly more heavily politicized than that of earlier Romantic thinkers, it retained strong cultural overtones, for it was characterized as much by customs as by formal legal structures. During the third decade of the nineteenth century, however, as it became specialized like other academic disciplines, German history at Berlin moved away from the cultural emphasis that remained in the views of Fichte and Hegel, and turned to an emphasis on the history of governmental actions, especially on diplomatic history. Ironically, the chief agent of this change and the founder of the modern professional discipline of history, Leopold von Ranke (1795–1886), was a close student of Schelling's *Naturphilosophie;* and his reasons for shifting focus to states as formal legal entities drew heavily from both Schelling and Hegel.

Perhaps more than any other early-nineteenth-century German scholar, Ranke managed to fuse the artistic and scientific in the way that Schlegel had proposed. In him a fanatically careful and critical approach to source materials and a commitment to the rationalist understanding of historical processes that informed the writings of Hegel were fused with a consummate artistry as a storyteller. Reflecting on the writing of history, Ranke wrote: "History is at once art and science. It has to fulfill all the demands of criticism and scholarship to the same degree as a philological work; but at the same time it is supposed to give the same pleasure to the educated mind as the most perfect literary creation. One might feel inclined to assume that beauty of form is only achieved at the expense of truth. If this were the case, the idea of combining science with art would have to be abandoned and shown to be false. But I am convinced of the contrary, and think that interest in form may even stimulate the passion for research."[96]

Like his immediate predecessors, Ranke understood history as a process through which the Spirit expresses itself over time, but unlike them, he did not consciously and openly seek to stand outside the historical process, making judgments and using the past to inform present actions. Instead, Ranke tried to see historical developments from the inside. "The historian is merely the organ of the general spirit which speaks through him and takes on real form," he wrote. Then, speaking very personally, he continued, "I desire that I could obliterate myself and only let the things talk, to allow the powerful forces to appear, which in the course of the centuries have arisen with and through one another and now stand opposed to one another, placed in conflict."[97]

When Ranke considered the sources of the conflict that drove historical processes, instead of identifying them with internally driven metamorphoses within culturally defined nations, he identified them with the external struggles between states. Taking the notion of the state as an organism more literally than most of his predecessors, he argued that any internal struggle was more likely to be pathological than constructive. Whatever internal drives exist within a nation are set at the origin and are automatically expressed when the conditions are right. The expression of such drives can help to understand the growth of an individual state; but it cannot help in understanding the role that particular state plays in the overall pattern of history. To understand this greater development, one must focus on the external conflicts between states, that is, to *Aussenpolitik.* Long before Social Darwinist imagery came along to legitimate international conflict as a source of progress, then, Ranke and his colleagues had provided a view of history that made German academics and statesmen particularly open to its militaristic tendencies and that appealed to the Prussian Junker class. As Roger Smith has argued, with Ranke, "academic history became the scholarly core of nationalist opinion."[98]

Summary Reflections

There is an important sense in which all of the most important elements of *naturphilosophisch* thought were already present in a nascent form in the Critical Philosophy of Kant and in Goethe's discussions of types and magnetism. Thus, persons such as the Schlegels, Fichte, Hegel, and even Ranke might have come to them without ever hearing of Schelling or *Naturphilosophie*. But as a matter of contingent historical fact, it was primarily through Schelling's writings that these notions were elaborated and formed into a single intellectual system. All of the figures mentioned were personally linked with Schelling. And all except for Fichte, whose intellectual rivalry with Schelling precluded an admission of indebtedness, admitted their admiration for his works and adapted or appropriated many of his ideas as their own.

Perhaps the most important way in which key conceptions associated with the Romantic appropriation of *naturphilosophisch* ideas came to have broad cultural significance was connected with their role in reshaping the linguistically grounded cultural nationalism of Herder into a political and state-focused nationalist ideology. The first step in this transformation was initiated through educational reforms carried out in Prussia by Wilhelm von Humboldt that incorporated the educational philosophy of Fichte. These reforms directed increased attention to German philosophy, literature, and science and deemphasized the study of classical languages, making cultural identity a chief focus of education. Equally importantly, they placed a high value on scholarship and raised the status of the scholar in German life.

One consequence of the new emphasis on philosophy and scholarship was the creation and rise to importance of the historicist doctrines of Hegel, Ranke, and their various disciples and the placement of those doctrines at the center of higher education in Germany. For both of these men, history became the process of the self-disclosure of Spirit in Time, as Schelling had understood nature as the disclosure of the Idea, or "World Soul" in space, and for both of them, though in different ways, the organic state became the central vehicle through which the later stages of this historical process were carried out. We will see in the next chapter that at least the Hegelian version of historicism was capable of being turned to liberal and progressive uses. But the overwhelming impact (largely unintended by founders such as Herder or Hegel) of organicist nationalism was to promote a sense of German national and racial identity that set Germans apart as mentally and spiritually superior to other persons and that strengthened the power and authority of religious institutions and of traditional, aristocratic, ruling elites. That is, at least in the early years of the nineteenth century, the Romantic view of the nation state as a teleologically unified organism served primarily conservative interests.

CHAPTER 5

The Rise of Materialisms and the Reshaping of Religion and Politics

The Romantic nationalism that fired the audiences for Fichte's 1807 *Addresses to the German Nation* and Auguste Schlegel's 1809 *Appeal to the German Nation* led immediately to the establishment of many organizations intended to prepare Germans for the day when they would be liberated from French hegemony and become free again to develop as Germans. The year 1808 saw the establishment of the *Tugenbund,* or Society for the Practice of Civic Virtue, in Königsberg, with branches throughout Germany. In 1810 the *Deutsche Bund* was established in Berlin to spread throughout German-speaking lands. In 1811 the first *Turngesselschaft,* or Gymnastics Society, intended to develop both the bodies and minds of youth for the coming struggle was initiated; and after 1810 new Patriotic Singing Societies sprang up in virtually every significant town. When Napoleon began to retreat from Russia in the spring of 1813, Frederick William III of Prussia issued a call to rise up against the French, and young middle-class intellectuals associated with these organizations flocked as volunteers into the various militias that served alongside the regular army. Then, once the fighting was over, the students and young academics established *Burschenschaften,* or "fraternities," which promoted Romantic idealist views and looked forward to ever greater glories for the German nation.

Alas, the political, economic, and academic realities of post–Napoleonic Germany diverged rapidly from student expectations. At the Congress of Vienna in 1815, the political life of the Germanies became dominated by the socially and religiously conservative Austria, led by Prince Klemens von Metternich, who was the architect of the German Confederation. Though each of its thirty-nine member states retained great autonomy, the Confederation, under Metternich's guidance, had a conservative and anti-democratic impact. For the most part, the traditional nobility was confirmed in its feudal rights over peasants, public lands, and local courts; and with rare exceptions that nobility opposed any form of nationalism that would constrict its local authority. The clergy were granted state pensions; and freedom of the press, speech, and assembly were limited.

Even in Prussia, where there was a strong liberal tradition within the government bureaucracy, the 1820s saw the reestablishment of noble control over local government.

After the 1817 student festival near Jena, state concern over the activities of the *Burschenschaften* grew; and in 1819 Metternich brought together a group at Karlsbadruhe in Bohemia to draft a set of decrees that shut down the fraternities, tightened press censorship, made gymnastic societies illegal, and established police surveillance in the universities. This move led to the dismissal of numerous outspoken academics, even when, as in Oken's case, they tended to be conservative themselves. Again in 1832, after twenty-five thousand people attended a German unification rally at the Hambach Festival in May, Metternich reacted repressively, this time getting the German Confederation to forbid all public meetings, to restrict the right of local sovereigns to extend the rights of citizens, to regulate universities more closely, and to place suspicious political figures under surveillance.

At this point, a substantial number of liberal-leaning faculty members, especially in law and the cameral sciences (named for the room in which the prince's advisors met), were dismissed from their positions.[1] As a consequence, there was an increasing politicization and polarization of academic life, with some faculty members, such as Johannes Müller, becoming increasingly reactionary; while most students and other faculty members were increasingly radicalized. One group, the so-called Gottingen Seven, who were dismissed from academic positions for their political views, banded together in 1837 to establish the newspaper, *Deutsche Zeitung*, which pushed the limits of allowable liberal expression.

A few years later, in 1842, a more radical group in Cologne established *Die Rheinische Zeitung für Politik, Handel, und Gewerbe* (Rhenish Newspaper for Politics, Trade, and Industry). This group hired a young philosophy graduate named Karl Marx as editor. Within a few months the Prussian cabinet shut the paper down and Marx went into exile, first in Paris, then in London. With rare exceptions, because idealist philosophy, traditional religion, and *naturphilosophisch* science were all associated with political conservatism, younger liberal and radical students tended to move away from all three and to develop intellectual perspectives that combined materialism, extremely heterodox religious views, and liberal to radical political perspectives.[2]

At the same time, economic conditions in the Germanies were changing rapidly under external pressures and as a consequence of liberal reforms promoted by camerally trained bureaucrats. One consequence was that although productivity, especially agricultural productivity, increased rapidly, so that aggregate per capita income was rising in most regions of Germany during the first half of the nineteenth century, a concentration of wealth in small segments of the population meant that there were wide pockets of severe economic distress, especially among agricultural workers and the traditional artisanal classes.

Throughout Germany between 1805 and 1835, population increases averaged about 40 percent in a region that was overwhelmingly rural and agrarian. (In 1850, less than 5 percent of the Prussian and less than 4 percent of the Bavarian population lived in towns of twenty thousand or more, whereas more than 34 percent of the British population did.)[3] At the same time, economic reforms led to the effective concentration of landholdings in smaller numbers of hands and to increased productivity. One result was that the Germanies became major exporters of foodstuffs, but another result was the creation of a huge army of landless underemployed agricultural wage laborers whose real income declined precipitously between 1820 and 1847.

Some of those workers moved into the towns, where liberal reforms were gradually undermining the traditional guild practices that had, for example, limited the number of persons entering most artisanal professions. The creation of a relatively unrestricted labor market undoubtedly drove production costs down, and in those few factories that were established in the first half of the century, some of the increased profits were passed on to labor; so factory wages seem to have risen faster than the cost of living. Factory production was slow to come to Germany, however; and in almost all traditional crafts, the chief result of the influx of new workers was to increase unemployment and to drive incomes down. The rapid economic decline of a once-prosperous and well-educated artisan class led to growing organized unrest among German workers beginning in the 1830s.

The economic decline of the previously comfortable artisanal class had a critical immediate impact on German culture because this class was not only highly literate, but it had also produced a substantial fraction of the student bodies and faculties of *Gymnasia* and the universities—recall that Kant's father had been a harness-maker, Johannes Müller's father was a shoemaker, Fichte's father had been a silk ribbonmaker, Oersted's was a pharmacist, and the poet, Johann Schiller's was a baker. As long as the financial stability of their world seemed relatively assured, such sons of tradesmen and craftsmen could focus their attention on Romantic literature or idealist philosophy without much worry about how their activities related to the world of material production and consumption. But as the livelihood of their families became increasingly problematic, the attention of the sons of the burgher class became increasingly turned toward the material conditions of life.

The social and economic considerations mentioned above served to selectively amplify the importance of specific developments that were occurring within the academic disciplines represented in the universities. As we shall see, the most important initial developments probably occurred within the humanistic and theological disciplines, but these soon spread outward to change the way in which the natural sciences were understood and the status that they enjoyed both within the academy and within the broader community.

Natural Science and Religion from Kant to David Strauss

We look first at some of the developments in theology that emerged in connection with Kantian philosophy. On one level these developments are particularly important because the conservative German state of the early nineteenth century drew heavily from religious authority to support its own legitimacy. Moreover, the whole meaning of the uniqueness of the German nation was bound up for the idealists and Romantics with its special religious role. Thus anything that challenged traditional understandings of religious authority also challenged the foundations of the state. At the same time, however, Kantian philosophy contained within itself features that were bound to challenge critical aspects of traditional religious authority and to shift the major locus of intellectual support for religion away from the natural theological traditions that had dominated eighteenth-century religious discourse throughout Europe.[4]

Given his distinction between the spatiotemporal and causal world of natural phenomena and the moral-choice-laden noumenal world, Kant was forced to argue for a radical disjunction between scientific knowledge, which was the legitimate and certain creation of speculative reason, and theology, which was not. In the *Critique of Pure Reason,* he wrote, "I maintain that all attempts to employ reason in theology in any merely speculative manner are altogether fruitless and by their nature null and void, and that the principles of its employment in the study of nature do not lead to any theology whatsoever. Consequently, the only theology of reason which is possible is that which is based upon *moral laws,* or seeks guidance from them. . . . [With respect to religion,] I have therefore found it necessary to deny knowledge in order to make room for faith."[5] In making this claim Kant was clearly not intending to denigrate either natural science or theology. He was simply saying that one could not legitimately infer claims regarding either domain from arguments that were appropriate to the other.

For the nineteenth-century German student of theology, then, two central questions emerged out of Kantian philosophy. First, in the absence of a rational, natural theological warrant, what kind of warrant exists for religious beliefs? Second, and perhaps even more crucial for many of those seeking to cope with the social and economic chaos of early-nineteenth-century Germany, in the absence of a natural theological account, how was one to understand the relationship between God and the material world within which human actions take place? For Kant's philosophy not only deprived religion of the support of natural science, it also deprived the material world of moral significance.

In 1837, David Friedrich Strauss (1808–74) identified three fundamental groups of German theologians who were struggling with these issues during the first half of the nineteenth century. First were what he called the "believers": traditionalists who were basically uninfluenced by Kant's philosophical arguments because they contended that religious faith had never depended on the ability to provide rational knowledge about God or his Word or World in the first place. For the most part, such figures tended to understand the miracles of the Bible and the doctrine of salvation through Christ as mysterious and supernaturally produced. Indeed, according to many in this group, our chief warrant for believing the divine character of Christ was his ability to perform miracles. Second, there were the mediators, whose views were associated largely with those of Schelling and Friedrich Schleiermacher (1768–1834). Most members of this group tried to combine biblical and philosophical elements in a moderate stance aimed at holding German Christians together in the face of a series of intellectual and political pressures toward schism. Finally, there was the speculative or philosophical school, with whom Strauss identified. The views of this group were derived largely from the philosophical perspective of Hegel.

In the remainder of this section, we will be concerned primarily with the views of the speculative theologian, David Strauss. But since Strauss's doctrines were largely a response to those of Schleiermacher, we characterize his views briefly first. Schleiermacher accepted the Kantian notion that religion cannot be grounded in phenomenal knowledge. But like many others, he could not accept Kant's claim that there is an autonomous human moral sense, or faculty of practical judgment, that grounds religion. Instead, he developed Schelling's notion that we have a feeling of absolute dependence on the Absolute as spiritual beings in much the same way that we recognize our absolute dependence on the physical universe as physical beings. It is this feeling of absolute dependence that is the foundation for all religion. Religion is thus born out of precognitive, prevolitional feelings (*Gefühl*) for Schleiermacher.

Virtually all of modern liberal Protestant theology derives from Schleiermacher's emphasis on affective experience as the source of religious impulses, but Hegel found this notion appalling because it seemed to him to preclude the possibility of any cognitive component to religion. If Schleiermacher were right, Hegel fumed, "a dog would be the best Christian, because it has this feeling [of dependence] most intensely."[6]

One consequence of Schleiermacher's views was that he took a vastly different view of biblical miracles than the believers at the same time that he accepted the divine source and historicity of the Bible. For Schleiermacher, whether or not one believed in miracles was irrelevant to whether one was a Christian. God's presence could be felt in ordinary natural events. It did not need to be proven supernaturally. Indeed, it could not be "proven" at all by the argument that because some event was not understandable, it therefore had to have a supernatural, divine cause.

Though Schleiermacher made a significant break with traditional Christian theology, his religion remained orthodox in many ways. Throughout his *Speeches on Religion* of 1799, his *The Christian Faith* of 1803, and in his frequently repeated

Berlin lectures on "The Life of Jesus," delivered during the 1820s and 1830s, he continued to identify the historical Jesus of Nazareth with the saving Christ, even though he refused to commit himself to either naturalistic or supernaturalistic explanations of the supposed miracles reported of Christ.

The man who more radically transformed theological discourse in his own *Life of Jesus,* published in 1835, just three years after he had attended Schleiermacher's lectures at Berlin, was David Strauss. Strauss, born in 1808, the son of a downwardly mobile merchant family in Wurtemberg, entered the school of theology at Tubingen in 1825 to complete his theological training. There, because he and several friends found the philosophy faculty dull and inept, he established a study group to read contemporary German philosophy, beginning with Kant's critical works. But Kant seemed to him perverse in his emphasis on the extremely limited range of human knowledge. Fichte's abstract Idealism had little more appeal. On the other hand, Schelling's *Naturphilosophie* and its extension in Hegel's *Phenomenology of the Spirit* excited the young Strauss so much that after receiving his doctorate from Tubingen and serving (successfully) for about a year as pastor to a small congregation, he left for Berlin to study with Hegel.

Strauss arrived in Berlin in late 1831, just weeks before Hegel's death from cholera, so instead of studying with Hegel, he began to attend Schleiermacher's lectures. But Strauss brought to them a profoundly different approach, grounded in his study of Hegel. For Hegel and for Strauss, religion and philosophy have the same content, which is focused on the participation of humanity in divinity.[7] But religion must clothe the content in representation [*Vorstellung*] while philosophy deals with the content directly as concept [*Begriff*]. Like Schleiermacher, Strauss was inclined to believe that there had been a historical Jesus; and like Schleiermacher, he was inclined to discount the importance of miracles as warrants for belief. But Strauss went far beyond Schleiermacher, both in denying the credibility of all of the miracle stories and in denying the historicity of almost all of the particular events reported of Jesus' life. In fact, in his *Doctrine of Faith,* published in 1840, Strauss extended his attack on biblical literalism in interesting ways by drawing on Karl Burdach and Gustav Carus's *naturphilosophisch* understanding of biological change to undermine the Genesis story of Adam and Eve as the original parents of humankind: "According to science, all organic beings are produced originally from the inorganic. In particular, there is no doubt that our planet has acquired its present state gradually, that it was uninhabitable for organic beings in primitive times, and that these have originated gradually without having had ancestors, that is, through dissimilar reproduction."[8] Humans, like every other species, had evolved gradually under the pressure of the internal formative force (*Bildungstrieb*), and could not have emerged from a single pair of "original" parents.

From Strauss's Hegelian perspective, even if the biblical stories of the creation and of Jesus and his miracles were factually false, they could legitimately "represent" true aspects of the union of God with the universe and with man. They were fictions, or myths, created in a particular historical context to represent or symbolize a particular stage of that union for participants in a particular culture. In his later years, Strauss believed that because the historical process in which the Spirit continued to express itself in human self-consciousness was ongoing, the old, biblical, myths could no longer represent the highest religious consciousness. Instead, he turned to Darwinism and natural science to discover his religious truths.[9] But in his early and most influential works, Strauss still understood himself to be a Christian and he continued to hold that religious truth was represented in the biblical stories.

Indeed, the application of critical techniques to the biblical narratives by Strauss were intended, in Fred Gregory's words, "to force theological attention where it belonged, on the spiritual truth behind the life of Jesus,"[10] by demonstrating that traditional literal interpretations of the Gospels were impossible and that the biblical stories must, then, be understood as mythic symbols or imaginative representations of philosophical truths rather than as the truths themselves. Because of such views, Strauss found himself effectively ostracized from the evangelical Lutheran theological community. He lost his appointment at Tubingen and he became increasingly embittered and increasingly critical of traditional Christianity; so that in his later works, he no longer considered himself Christian at all.

The "Sensualist" Materialism of Ludwig Feuerbach

One of the most important stages in Strauss's growing estrangement from Christianity grew out of his interest in another young Hegelian theologian-turned-philosopher, Ludwig Feuerbach (1804–72), whose works were to transform the thrust of German philosophy during the 1840s and to point the way to a number of materialist attacks on all forms of Idealism. Strauss wrote of Feuerbach immediately after his publication of *Toward the Critique of Hegelian Philosophy,* in 1839, *The Essence of Christianity* in 1841, and *Preliminary Theses for the Reform of Philosophy* in 1842, "To-day, and perhaps for some time to come, the field belongs to him. His theory is the truth for this age."[11]

Feuerbach was born in 1804, son of the distinguished liberal Bavarian jurist Anselm Feuerbach and his wife Wilhelmine Troster. By age sixteen he showed an intense interest in religion, taking private Hebrew lessons during his time in the *Gymnasium* at Ansbach. From there he went to study theology at Heidelberg. Like Strauss, Feuerbach was introduced to Hegelian philosophy as a theology student, and like Strauss he made the pilgrimage to Berlin to study with Hegel. But Feuerbach arrived in 1824, early enough to study with the master as well as to attend the theology lectures of Schleiermacher.

At this point, Ludwig, along with his brothers, was (probably wrongly) suspected of being a member of a secret organization. As a consequence, he came under police observation and he was denied formal admission to the university for several months. His older brother, Karl, was actually put in prison, where he attempted suicide. This experience, along with his concurrent fascination with the eighteenth-century French egalitarian materialist philosopher, Claude Adrien Helvetius, seems to have turned Feuerbach's politics increasingly toward radical democracy.

For financial reasons, Feuerbach had to leave Berlin for Erlangen in 1826. There, while he completed a basically Hegelian doctoral dissertation, Feuerbach began to study anatomy, and this study seems to have raised questions in his mind regarding whether Hegelian philosophy treated the natural world, or world of phenomena, fairly. For Hegel, as for Fichte, the identity between the laws of nature and the laws of thought followed from the fact that nature was ultimately created out of thought. Schelling had been bothered by this asymmetry, so he had posited an Absolute that was neither thought nor thing and was prior to both. The two manifestations of the Absolute then interacted in such a way that both Nature and Spirit evolved out of their interaction. Feuerbach was dissatisfied with this answer as well, but for nearly ten more years he continued to write as a Hegelian Idealist in spite of his growing concerns.

In 1829, Feuerbach became a *Privatdozent* in philosophy at Erlangen, with the expectation that he would move up the academic ladder to professor, but his publication of *Thoughts on Death and Immortality* in 1830 put an end to those hopes. Though it was published anonymously, Feuerbach was widely known to have been the author of this work, which challenged the traditional Christian notion of an otherworldly Heaven. Heaven and the notion of immortality, Feuerbach suggested, were representations or symbols of the search for the perfection of life in the here and now.[12] By persisting in the illusion of literal immortality, humans actually held themselves back from a complete submission to God. "True and complete surrender to and submersion in God," he wrote, "is possible only when the human recognizes death as true, real, and eternal."[13] Most importantly, in holding off their expectations of perfection for an afterlife, humans were relieved of the responsibility to make the best of their present lives. Thus, Feuerbach wrote that his overarching purpose in *Thoughts on Death and Immortality* was "to cancel above all the old cleavage between this side and the beyond in order that humanity might concentrate on itself, its world, and its present with all of its heart and soul."[14]

It should hardly be surprising that Feuerbach's attack on the doctrine of immortality produced a serious response in the political and theological communities, for it was widely held that the possibility of endless punishment in an afterlife was crucial as an instrument of social control. Without that inhibitor, who knows what kinds of immoral, illegal, and socially disruptive acts people might engage in? After his authorship of *Thoughts on Death and Immortality* became officially known, the authorities at Erlangen, and throughout the Germanies, made it clear that Feuerbach would never get a regular academic appointment.

Throughout the mid 1830s the increasingly embittered Feuerbach published a very successful series of works on the history of philosophy from a Hegelian perspective, but then in 1837 he took up anatomical and physiological studies again, and his philosophical writings began to show an increasing interest in empirical natural science. In his *Philosophical Fragments,* under the title, "Bruckberg 1836–41," for example, he wrote: "All abstract sciences stunt man; only natural science reinstates him as an integral being, makes use of all the capacities and senses of the whole man."[15] In an 1837 study of Leibniz, he wrote: "Empiricism has facilitated the freedom and independence of thought—it has delivered us from the bonds of belief in authority by referring us to the holy, inalienable natural right of autopsy and self-examination."[16] From this time on he refocused attention from the realm of the rational ideal to the realm of concrete experience. In Kantian terms, Feuerbach increasingly focused on the notion that sensory intuitions are prior to intellectual cognitions in the formulation of knowledge, so that the idealist tendency to posit the existence of extra-sensory "objects" had no adequate foundation any more than Kant's positing of the noumenal thing-in-itself.

Feuerbach's new empirical emphasis was at the very heart of *The Essence of Christianity,* a work that rocked the foundations of German philosophy. In the preface to that work he insists: "I unconditionally repudiate *absolute,* immaterial, self-sufficing speculation,—that speculation which draws its material from within. I differ *toto coelo* from those philosophers who pluck out their eyes that they may see better; for *my* thought I require the senses, especially sight; I found my ideas on materials which can be appropriated only through the activity of the senses. I do not generate the object from the thought, but the thought from the object; and I hold that alone to be an object which has an existence beyond one's own brain. . . . I attach myself, in direct opposition to the Hegelian Philosophy, only to *realism,* to materialism in the sense above indicated. . . . I am nothing but a *natural philosopher in the domain of mind;* and the natural philosopher can do nothing without instruments, without material means."[17]

Though Feuerbach proclaimed himself now a materialist, his views incorporated a strange and ultimately unstable amalgam of classical materialist, Hegelian, and Humian emperico-positivist views that took on the form of a kind of inverted Hegelian idealism. Feuerbach continued to use the Hegelian distinction between religious representations and philosophical conceptions that had informed his own *Thoughts on Death and Immortality* and Strauss's *Life of Jesus,* focusing on the claim that religious dogmas ought to be understood as symbols of truths rather than the truths themselves. In addition, he continued to accept the Hegelian notion that the laws of nature are the laws of our thinking, but whereas for Hegel this assertion was true because thought imposed its structure on nature, for Feuerbach it was now to be understood as true because nature imposed its structure on thought. For Hegel, cosmic history and its later stage, human history, was constituted through the gradual self-revelation of the Idea, or God, whereas for Feuerbach, God was constituted by the gradual self-revelation of Man through his interaction with the cosmos.

Even the structure of *The Essence of Christianity* was Hegelian, in the sense that it began in part 1 by exploring the truths represented through Christianity, then it turned in part 2 to exploring the contradictions contained in Christian theology, and it concluded with a new religious synthesis in which man comes to realize "that there is no other essence which man can think, dream of, imagine, feel, believe in, wish for, love, and adore as the *absolute,* than the essence of human nature itself."[18] Ultimately, for Feuerbach, there is a union of the human and the divine. But whereas for Hegel this union was achieved because in some sense God gradually created self-conscious humanity in his image, for Feuerbach, it was achieved because humanity, made gradually self-conscious, has finally realized that God is nothing but a representation of its own attributes and aspirations—thus his insistence that the most fundamental philosophical truth underlying religion is that "Theology is Anthropology."[19]

The occasion for Feuerbach's change of perspective was undoubtedly his reconnecting with the empirical science of physiology. The reasons, however, are almost certainly much more closely related to his feeling, already present in 1830, that idealist philosophy and the religion that it supported had lost touch with the experiences of ordinary humans and with the moral focus that Reformation Lutheranism and Kantian theology had both insisted upon. This general disillusionment with contemporary Christianity was no doubt amplified by the personal problems he had faced as a consequence of the complicit relationships between state and church in nineteenth-century Germany. These issues became increasingly important in the face of the economic crises facing many Germans during the 1830s and 1840s and the perceived insensitivity to those crises on the part of church and state authorities.

Those materialist philosophers whom Feuerbach admired, especially Helvetius, had been highly critical of the failure of Christian authorities to act positively to improve the living conditions of the mass of humankind and had bemoaned the way in which political and clerical authorities acted together to protect their interests against the interests of the great majority. In his 1846 introduction to the first volume of his collected works, Feuerbach highlighted both the moral and the political issues and the significance of his own sensual-materialist position: "[The question is] not whether God is a creature whose nature is the same as ours but whether we human beings are to be equal among ourselves; not whether and how we can partake of the body of the Lord by eating bread but whether we have enough bread for our own bodies; . . . not whether we are Christians or heathens, theists or atheists, but whether we are or can become men, healthy in soul and body, free, active, and full of vitality. . . . I deny God. But that means for me that I deny the negation of man. In place of the illusory, fantastic, heavenly position of man which in actual life necessarily leads to the degradation of man, I substitute the tangible, actual, and consequently also the political and social position of mankind."[20] That is, Feuerbach insisted that any account of religious beliefs had to begin from an assessment of the concrete material and social conditions of human lives. For Feuerbach, as for Schleiermacher, religion was based on man's recognition of his dependency. But according to Feuerbach, Schleiermacher had failed to recognize that theology went deeply wrong when it posited something over and above the natural world and one's fellow humans as the object of that dependency. In their imagination, theologians created a transcendent God, placing humankind in an unnecessary state of humility and subordination.

Feuerbach's was a position that had tremendous appeal to liberal and radical young people during the 1840s, when social and economic distress and inequity was an increasingly obvious fact of life and when religious institutions were often functioning as arms of the state to enforce intellectual conformity and economic injustice. Thus the young Karl Marx wrote admiringly in 1843 that Feuerbach's critique of religion "ends in the teaching that, *man is the highest being for man,* it ends, that is, with the categorical imperative to overthrow all conditions in which man is a debased, forsaken, contemptible being forced into servitude."[21] And reflecting back in 1888, Friedrich Engels wrote that after the *Essence of Christianity* appeared, "we all immediately became Feuerbachians."[22]

The failure of the 1848–49 uprisings in Germany and the reestablishment of Prussian military dominance by April of 1849 left most liberal and radically oriented intellectuals deeply discouraged. But those who built upon Feuerbach's materialist philosophy generally refused to acknowledge defeat. They constituted a cadre of optimistic opponents of traditional authority and the social status quo who almost welcomed persecution by the authorities as a sign of their own significance.

Scientific Materialism and Political Liberalism

During the early decades of the nineteenth century, the materialist philosophical perspectives associated with Laplace among physicists, Helvetius and d'Holbach among social theorists, and Cabanis, among physicians, had relatively little appeal among German academics, in part because of their identification with French culture. But immediately after the publication of Feuerbach's *Essence of Christianity,* a variety of materialistic philosophies began to develop in Germany, especially among medical and biological scientists and among social thinkers. And each form of materialism seemed to have its own political agenda and implications.

The first of these to mature, and the one most closely associated with Feuerbach's own political views, was what Frederick Gregory has called "scientific" materialism. It is virtually certain that the principle motives for the materialist turn taken by the major scientific materialists of the 1850s, Carl Vogt (1817–95), Jacob Moleschott (1822–93), and Ludwig Büchner (1824–99), derived from the humanistic religion and from the antagonisms to an arbitrary political and religious power structure that they shared with Feuerbach, rather than from their scientific activity. All three had fathers who were liberal physicians. Vogt's father lost his academic post at Giessen in 1834 because of his association with radicals, ending up teaching at a Hochschule in Zurich. Moreover, three of Vogt's maternal uncles were either imprisoned or forced to flee Germany because of their politics. After studying with Justus Liebig (1803–73) and the physiologist, W. G. Valentin, Vogt worked for five years as an assistant to Louis Agassiz, whose increasing religiosity Vogt deplored. When Agassiz left for America, Vogt, who was persona non grata in Germany because he had helped a fellow student escape from the police, moved to Paris, where he survived by writing science articles for *Allgemeine Zeitung* and where he began writing popular science texts. In Paris he formed close associations with the Russian anarchist Michael Bakunin (1814–76) and gradually adopted a materialist scientific perspective that he identified with that of Pierre Cabanis.[23]

Moleschott's father was an agnostic Dutch physician who sent his son to a German *Gymnasium* and then to the University of Heidelberg, where he became associated first with Feuerbach's writings and then with Feuerbach himself, before he began his materialist career, criticizing Liebig's acceptance of vital forces.

Büchner's older brother, George, was a politically active dramatist, who eventually became a communist, while his younger brother, Alexander, was charged with treason when he met with exiled German dissidents while on a trip to England. In 1845, Ludwig helped establish a student political society at Giessen, promoting Vogt's election to the Frankfurt Parliament in 1848, before he wrote a materialistically oriented medical dissertation on reflex actions.[24] It seems, then, that the materialism of the major scientific materialists involved a set of commitments that they brought to their science as a consequence of broader social and political considerations. Yet it is equally certain that each presented his antagonisms to idealist philosophy, traditional Christianity, and monarchist politics as emerging out of scientific activities and knowledge claims.[25] Additionally, they collectively managed to convince themselves and many of their readers—both those who sympathized with their perspectives and those who bitterly opposed them—that materialism was a natural consequence of scientific activity.

The overwhelming majority of German scientists by the 1840s had almost certainly become fed up with religious and philosophical controversies and simply wanted to get on with their work without committing to any particular metaphysical or religious position. They had no more use for the secular dogmatism of materialism than for the religious dogmatism associated with Christian conservatives. But from the perspective of those outside the sciences, the polemical zeal of the materialists often made it seem as if they spoke for the scientific community as a whole.[26]

If the status of science had not been rapidly on the rise in Germany during the 1840s, the materialist's appeal to scientific authority in the name of humanistic religion and liberal politics would have had little general impact, but such was not the case. At least two different trends in German culture tended to raise the status of science across the spectrum of political and intellectual positions during the period between about 1810 and 1870. First, the notion of *Bildung* (self-cultivation), which had been so deeply linked to scientific scholarship (*wissenschaft*) by Humboldt, Fichte, and other early-nineteenth-century academic reformers, persisted within the educated elite of Germany, leaving even those whose professions were associated with law and theology with a sense of the central importance of scientific methods and content. For this group, which seems to have been most powerful during the early part of the century, science was understood as something to be pursued purely for its own sake, and theoretical and observational sciences tended to be more important than laboratory and applied clinical sciences.

Alongside this largely anti-utilitarian notion of science, a second perspective, usually associated with the prominence of economic liberal bureaucrats within a given state, grew rapidly after about 1830. The key to this perspective was at least in part the growing dominance that German translations of J. B. Say's *Traite d'economie politique* had in the teaching of the cameral sciences within the German university structure. First translated as *Abhandlung Über die Nationaloekonomie, oder einfache Darstellung der Art und Weise, wie die Riechtthumer enstehen, verheilt und versehrt warden*, by L. H. Jacob in 1807, Say's work, along with that of Adam Smith, guided most economic theorizing in Germany for nearly fifty years.[27] For our present purposes, Say's insistence on the importance of industrialization and on the roles of both scientific knowledge and a technically educated cadre of entrepreneurs and workers for increasing productivity are central. Well before there was a substantial market-driven pressure toward applied science in Germany, camerally trained bureaucrats under the influence of Say were promoting practically oriented scientific education.

This emphasis occurred in part at the level of vocationally oriented *Realwissenschaften,* so that German workers would be prepared for the transition from agriculture to manufacturing, which was "the path now being taken by Civilization as a whole."[28] With one eye on the industrial advances occurring in England, for example, the Baden minister for education in 1833, C. F. Nebenius, lobbied for the construction of a technical high school in Karlsruhe.[29] But the turn to applied sciences also occurred at the university level. As a consequence, argues Arleen Tuchman, "The construction of large research institutes, financed by state governments, occurred before science and industry formed the link that came to define much of Western culture, providing an essential, if not *the* essential, driving force for the economy."[30] During the last third of the nineteenth century, science-based industries did play a central role in Germany's rapid economic growth, but the expectation that this must occur preceded the event, and science began to *appear* as a crucial feature of German economic life nearly half a century before there is much evidence that it had become so.

As a consequence of the growing emphasis on science within both vocational and university education, the appetite for scientific knowledge grew rapidly among middle-class Germans, and during the late 1840s and 1850s, a number of new popular scientific journals came into existence. Though scientific materialists published in virtually all of these journals, two of the most popular, *Die Natur: Zeitung zur Verbreitung naturwissenschaftlicher Kenntniss und Naturanschauung für Leser alle Stande* (*Nature: a journal for disseminating natural scientific knowledge and attitudes among readers from all classes*), founded in 1852, and *Das Jahrhundert, Zeitschrift für Politik und Literatur* (*The Century: A Journal of Politics and Literature*), founded in 1856, had strongly materialist editorial policies.

Perhaps even more important for the spread of scientific materialism, the public developed a huge appetite for popular books in natural science, which the materialists exploited very self-consciously. Writing of the circumstances under which his brother Ludwig came to write *Kraft and Stoff. Empirisch-naturphilosophische Studien* (*Force and Matter*) in 1855, Alexander Büchner describes his conversation with Ludwig when Ludwig brought the manuscript to him. Alexander suspected his brother of having written a Romantic novel. Ludwig tells him that he had, instead, written a work on natural science, and then he explained why: "This kind of thing has strong appeal these days. The public is demoralized by the recent defeat of national and liberal aspirations and is turning its preference to the powerfully unfolding researches of natural science, in which it sees a new kind of opposition against the triumphant reaction. Look at Vogt, Rossmässler, and Moleschott, all of them are finding good publishers."[31]

At Alexander's recommendation, Ludwig sent *Force and Matter* to the publishing house of Meidinger in Frankfurt. Within three years it had gone through five editions. By the end of the century it had gone through nineteen German editions, and in addition, it had been translated into seventeen different languages, becoming one of the great literary successes of the century.

Force and Matter

Among themselves, the big three German-educated scientific materialists, Vogt, Moleschott, and Büchner, published no fewer than sixty-three popular scientific monographs in addition to hundreds of articles in popular scientific journals. Because Büchner's *Force and Matter* was both the most popular and most comprehensive of these works, I will let it stand as an illustration of the whole genre, though one should be aware that there were slight differences in perspective within the group and that over time, even Büchner backed away from the extreme self confidence regarding knowledge that he displayed in *Force and Matter.*

Scientific materialism contained a strong populist and anti-intellectual element that undoubtedly enhanced its appeal to a broad audience. Thus, in the preface to the first edition of *Force and Matter,* Büchner insists, "We shall seek to present our views in a generally intelligible form and to base them on known or easily comprehensible facts, and in doing so shall avoid all those philosophical technicalities the use or rather abuse of which has justly brought all theoretical, and especially German, philosophy into discredit in the eyes of both the learned and unlearned. *It is part of the very nature of philosophy to be intellectually the joint property of all.*"[32]

Laying out the fundamental claim of the entire scientific materialist movement, Büchner promises the destruction of traditional religion and idealist philosophy alike: "Starting from the recognition of the indissoluble relation that exists between force and matter as an indestructible basis, the view of nature resting upon empirical philosophy must result in banning every form of supernaturalism or idealism from what may be called the hermeneutics of natural facts, and in looking upon these facts as wholly independent of the influence of any external power dissociated from matter. There seems to us to be no doubt about the ultimate victory of this realistic philosophy over its antagonists. The strength of its proofs lies in facts, and not in unintelligible and meaningless phrases."[33]

Linked to an insistence that the materialist perspective must ultimately triumph, in the minds of the scientific materialists, was a tremendous sense of optimism about the extent of knowledge that had already been attained, about that which remained accessible to their methods, and about the positive societal benefits of that knowledge: "We shall be able all the better to understand [the world] and to control it, the more we endeavor to know it in its infinite fineness and its incredible energy and capacity, by means of observation, of investigation, of experiment. Experience has spoken here with sufficient clearness. The scientists, unfairly decried as materialists, have not only made it possible for our mind to penetrate by thought into the All and to obtain scientific certitude on questions and things which appeared forever sealed to it; but we also owe it to them that the human race is more and more borne upwards in the mighty arms of matter, known and controlled through its laws, and that we can perform by it works and acts, which in former times seemed possible only to giants and magicians."[34]

Büchner begins his major conceptual argument by denying the Kantian identification of matter with attractive and repulsive forces. It is true, he admits, that force and matter are indissolubly connected with one another, so that no matter can be found without force and no force can be found without matter. But he goes on to draw from Hermann von Helmholtz's mechanical formulation of the principle of the conservation of energy and from the widespread mid-nineteenth-century interpretation of heat as constituted by the microscopic motions of the particles of matter, to argue that force and matter are two distinct but intimately connected entities. "Force," he writes, "may be defined as a condition of activity, or a motion of matter or of the minutest portions of matter or a capacity thereof."[35] This slightly awkward definition follows from the mathematical expression for energy used by Helmholtz for any system of particles, $\Sigma m_i V_i^2 + \Sigma\Sigma\, Y(r_i, r_j)$, where m is the invariant mass of each particle, v is its velocity, and Y(r) is some function, now know as a potential function, which depends upon the distance, r, between each pair of particles. The principle of the conservation of energy, which had been independently articulated in slightly different forms by Julius Robert Mayer (1814–78), James Prescot Joule (1818–89), and Helmholtz between 1842 and 1847, could thus be expressed by saying that the energy of a system consisted of an invariant expression that included the summation of the motions of the particles of a system plus the potential or capacity for those particles to move by virtue of their spatial relationship to one another. Since the German word, *Kraft,* is translated by both the English term, energy, and the English term, force, the Helmholtz equation can be restated in English as the claim that the sum of the forces, or motions and capacities to move, of all bodies in a closed system is conserved throughout any change in the system. According to this principle, which was a key element in all of classical physical theory after the 1840s, then, matter and energy, or matter and force are separately and independently conserved in all phenomena.

From this basic principle Büchner immediately concluded that the universe could not have been created ex nihilo as Christian theologians claimed. Neither force nor matter can be created out of one another, so if one conceives of a preexistent creative force as the source of the universe, there is no way that it could produce matter. Nor can either force or matter come into existence out of nothing according to the conservation of energy principle. As a consequence, Büchner insists, "The universe, or matter, with its properties, conditions, or movements, which we name forces, must have existed from eternity, or,—in other words—the universe cannot have been created."[36] Büchner could not even imagine the possibility that the law of conservation of energy was created with the physical universe and thus could not serve as a preexistent constraint on the act of creation.

In an interesting short chapter entitled, "The Value of Matter," Büchner argued that it was principally Pauline Christianity that turned people in the ancient Mediterranean world toward "that foolish conception which looks upon matter as a crude, dismal, inert something, hostile or opposed to spirit."[37] Materialism, he argued, was much more evenhanded. It did not denigrate the idea of spirit, indeed it also venerated the concept. It simply argued that "spirit can only exist on a substratum of organized matter, and that not a shadow of a proof can be brought forward to show that spirit can attain to an independent existence outside of matter."[38]

Drawing from Vogt, who in turn drew from Cabanis, Büchner offered an interpretation of spirit, or psyche, that made it an emergent property of organized matter. Thus, he continued, "The simple solution of the problem [of the relationship between matter and spirit] lies in the fact that not only *physical* but *psychical* energies inhere in matter, and that the latter always become manifest wherever the necessary conditions are found, or that, whenever matter is arranged in a certain manner and moved in a certain way in the brain or nervous system, the phenomena of sensation and thought are produced in similar fashion as those of attraction and repulsion are under other conditions."[39]

Decrying all dualistic theories that set matter and spirit in any kind of opposition to one another, Büchner promoted a philosophical monism that he sometimes described as being neither idealistic nor materialistic, but rather, "simply natural."[40] In doing so, he prepared the way for the Monist religious movement of Ernst Haeckel and Wilhelm Ostwald, which was formally organized at the beginning of the twentieth century.

Turning to the issues of determinism and moral responsibility raised by Kant, Büchner begins by reiterating a position superficially similar to that taken by the Kantians. That is, he insisted on the necessity and immutability of the laws of nature. Büchner, however, argued that such an insight was not a precondition of scientific knowledge, but rather a consequence of the extended study of nature.[41] Given the immutability of natural law, there can be no room in nature for morality or compassion. Büchner quotes Feuerbach on this issue, insisting that "nature answers neither the questions nor the plaints of man; she inexorably flings him back upon himself."[42] Natural science, then, offers no warrant for the belief in a personal God who might respond to human desires or prayers by intervening in events. Indeed, it effectively precludes such a belief.[43]

Next, he proceeds to undercut the Kantian argument for a noumenal world, writing, "We are only busying ourselves with that world that is accessible to our means of intelligence, and can find no scientific reasons compelling us to believe that behind this world there is another, a higher one, independent of the influence of the laws of Nature, and perhaps arranged in an entirely different fashion. . . . *Let every one believe whatever and as much as he likes, and let his fancy have free scope where science forsakes him!*"[44]

Having turned the world of traditional religion and morality into a creation of the human fancy, Büchner proceeds to consider another key Kantian question: How does it happen that our minds are suited to an understanding of nature? Here he takes the line developed by Feuerbach, claiming that since the mind was produced by natural law it is consequently able to recognize it. "Could it be otherwise," he asks, "considering that Nature's laws have created the mind and that the same forces are at work in it which rule the world and Nature? Hence the laws of thought and of our minds must be in unison with the most recondite principles of the laws governing in Nature, and hence the laws of thought are also the laws of the universe."[45]

In what had become one of the longest chapters by the fifteenth German edition of *Force and Matter* (published in 1884), Büchner confronted the issue of teleology or design, which was so important to both natural theologians and to Idealist and Kantian, as contrasted with materialist, biologists. Well before Darwin and Haeckel, Büchner had offered an evolutionary and natural selectionist critique of teleology grounded in Lamarck's notion that organisms can transmit to their progeny characteristics that they acquire in the process of adapting to their environment; and he never abandoned this Lamarckian view, though he increasingly focused on natural selection.[46] Second, he drew on the work of Helmholtz on the human eye to point out that careful analysis of our organs of vision suggest that they are really very clumsy and imperfect optical instruments.[47]

As a counter to the purposelessness of the universe, Büchner turns once again to Feuerbach, "the philosopher, *par excellence* of emancipated and self-contained humanity," who finds purpose in the human attempt to realize its own ideal self.[48] Anthropology, or the study of man, thus becomes the highest materialist science, and the final half of *Force and Matter* is completely devoted to exploring various dimensions of human existence, beginning with the thorny question of the relationship between the brain and the mind. In this connection, Büchner reviews the huge anthropometric, physiological, and biochemical literature correlating physical and chemical features of brain structure and function with notions of "intelligence" and mental health, repeating all of the purported evidence that supported belief in the mental inferiority of non-Europeans[49] and of the lower classes within Europe.[50]

It is in these sections on the physiological foundations of class and racial differences that Büchner most clearly separated himself from the more radical dialectical materialists. Like many liberal reformers, Büchner detested social injustices. He was a co-founder of the *Deutscher Bund für Bödenreform* (the German Confederation for Agrarian Reform), and he served as a liberal in local and regional governmental bodies for fifteen years. But he was not a believer in equality, nor did he believe in coercive social programs, such as those promoted by the communists and Social Democrats.[51]

Like the Parisian physiologists Cabanis and Bichat at the beginning of the century, Büchner admitted, for the most part, that he could not explain *how* mental phenomena arise from physical and chemical processes in the brain. "It is quite sufficient," he insisted, "to have proved by facts the necessary, indissoluble and normal connection between the brain and the mind."[52] Similarly, when he turned to discuss the special mental phenomenon of consciousness, he wrote: "How and in what way the atoms, nerve cells, or, to speak generally, matter, began to produce and bring forth sensation and consciousness, is quite unimportant for the purpose of our investigation; *it is sufficient to know that such is the case.*"[53] Yet he could not always rest satisfied with the convincing claim that mental events are correlated with physical ones, and in a chapter on "thought" he pressed on to claim that thought, or "psychical activity," was nothing other than "a particular manifestation of that great, universal, and simple natural force which sustains the eternal cycle of energies, revealing itself now as mechanical, now as electrical, now as mental force."[54] This notion that "psychic" energy was just another form of energy that had to be conserved unless transformed into another typical form seems to have been widely accepted among German materialists, and it played a major role in shaping Sigmund Freud's (1856–1939) theories of mental functioning.

Like the major works of the other scientific materialists, Büchner's *Force and Matter* was only indirectly political. Its major themes celebrated the ongoing triumphs of empirical science and claimed that the implications of that science were generally anti-authoritarian, anti-theological, and anti-idealist, but it stated no positions with respect to party politics.

Tactical Materialism, "Organic Physics," or Physiological Reductionism in the Service of the State

The aggressively anti-religious stance of the scientific materialists had little or no significant impact on the research practices of German scientists, but there was a more limited form of materialism that did. This form also spread into popular scientific literature, and it had a major impact on German governmental policies toward scientific research. Without seeking initially to explore the broad philosophical implications of their position, advocates of what came to be called "organic physics" focused on investigating physiological processes as if they were completely understandable in chemical and physical terms. As early as 1842, Emil Du Bois-Reymond (1818–96) wrote that he and a fellow student, Ernst Brücke (1819–92), in Johannes Müller's Berlin physiological laboratory, pledged to act as if the following statement were true: "No other forces than the common physical-chemical ones are active within the organism. In those cases which cannot at the time be explained by these forces, one has either to find the specific way or form of their action by means of the physical-mathematical method or to assume new forces equal in dignity to the chemical-physical forces inherent in matter, reducible to the forces of attraction and repulsion."[55]

Soon Du Bois-Reymond and Brücke were joined by another Müller *protégé*, Hermann von Helmholtz, and by Carl Ludwig (1816–95) from Marburg. Though it is probably not true that their views dominated German physiology in the second half of the century, it is certainly true that the research programs of the organic physicists constituted a serious and substantial segment of mainstream German biology.[56] It is even more true that the efforts of this group played a major role in generating the public and governmental support that made the German university system the center of experimental and clinical scientific research in the second half of the nineteenth century.

Like the scientific materialists, this group tended to be strongly opposed to *Naturphilosophie* and the idealism with which it was associated. Unlike the scientific materialists, however, they tended to be either silent about religious and political issues or openly orthodox and conservative. In part for this reason, and in part because they tended to be more philosophically sophisticated, these organic physicists or physiological reductionists sought to distance themselves from scientific materialists as much as from those influenced by *Naturphilosophie,* identifying both perspectives as fundamentally metaphysical rather than scientific. Thus, argued Helmholtz:

> There are two characteristics . . . which metaphysical systems have always possessed. In the first place man is always desirous of feeling himself to be a being of a higher order, far beyond the standard of the rest of nature; this wish is satisfied by the spiritualists [idealists]. On the other hand, he would like to believe that by his thought he was unrestrained lord of the world, and of course by his thinking with those conceptions, to the development of which he has attained; this is attempted to be satisfied by the materialists.
>
> But one who, like the physician, has actively to face natural forces which bring about weal or woe, is also under the obligation of seeking for a knowledge of the truth, and of the truth only; without considering whether what he finds is pleasant in one way or the other. His aim is one which is firmly settled; for him, the success of facts is alone decisive.[57]

There is something ironic about this passage as it relates to the scientific materialists, for in their public roles, the organic physicists became much more vocal apologists for the uses of science in manipulating the world than any of the scientific materialists had been. At its origins organic physics had relatively little to do with social applications, but over time both Du Bois-Reymond and Helmholtz became formulators of and spokespersons for Baden and Prussian scientific policy respectively, and they increasingly emphasized the central role of science in industrialization in order to promote state support for scientific research. By 1862, when he became rector of the University of Heidelberg, Helmholtz chose the importance of applied science as the theme of his inaugural address. "Knowledge is power," he insisted, "and no age has been in a better position to realize it than the present one. . . . Even the proudest and least cooperative absolutists states have had to acknowledge that the power of the state rests on its wealth, which depends on command over the forces of nature and their application to agriculture, industry, and transportation. . . . No nation which wants to remain independent and influential can fall behind in the task [of developing knowledge of the natural sciences and their technical application]."[58]

Nothing else that I can think of, however, approaches Du Bois-Reymond's *Kulturgeschichte und Naturwissenshaft* (*Natural Science and the History of Culture*) of 1878 in its praise of material progress and its linkage of that progress to natural science:

> Wherever something to be physically achieved remains withheld from man, there the calculus of his spirit presses forward with its magic Key. . . . What the wishing wand did by magic, geology freely performs: freely it brings forth water, salt, coal, petroleum. The number of metals continues to increase, and though as yet, chemistry has not found the philosopher's stone, tomorrow perhaps it will. For the present it vies with organic nature in the production of the useful and the pleasing. From the black, stinking heaps of refuse, which turn every city into a Baku, it borrows the colors by comparison with which the magnificence of tropical feathers turns pale. It prepares perfumes without sun or flowers. Has it not even solved Samson's riddle, how to make sweet things out of loathsome material?
>
> Gay-Lussac's preserving art has not merely wiped out the difference between the seasons on rich men's tables. The poison monger sees with angry despair his tricks unmasked. The scourges of smallpox, plague, scurvy are under control. Lister's bandage protects the wounds of soldiers from the entrance of deadly germs. Chloral spreads the wings of God's sleep over the soul in pain, indeed chloroform laughs, if we wish, at the biblical curse of womanhood.
>
> So were Bacon's prophetic words fulfilled: *Knowledge is power. All the peoples of Europe, of the Old and the New Worlds, travel along this road. . . . If there is one criterion which, for us, indicates the progress of humanity, it is . . . the level attained of power over nature.*[59]

These statements by Helmholtz and Du Bois-Reymond are consistent with the claim made by Timothy Lenoir that the organic physicists, in fact, played a major role in formulating a new ideology grounded in the satisfaction of material interests and that between the late 1850s and the 1870s, they "were among the leading spokesmen of the rising industrial bourgeoisie."[60]

There is a sense in which the reductionist presumption of the organic physicists was as much an imposition on *their* science as that of the scientific materialists was on *their* understanding of science. But generally speaking, the organic physicists imported their reductionist tactics for reasons linked to their research careers rather than for reasons related to broader social issues. Helmholtz offered a particularly candid discussion of how he became committed to organic physics in an autobiographical address that he gave when he was seventy years old. As a student in the *Gymnasium* where his father was a relatively poorly paid teacher of German literature, Helmholtz had become fascinated with optics, rational mechanics, and mathematics. But when it came time for him to go on to study at a university his father and family dissuaded him from following his inclination to study physics on the grounds that it offered very poor job prospects. Instead, he was convinced to train in medicine at state expense at the Friedrich Wilhelm Institute for the Education of Army Surgeons in Berlin. Medical training involved at least some scientific background, it virtually guaranteed employment, and it saved the family money.

In Berlin, Helmholtz worked in the institute library, where he began on his own to read mathematical physics through the classic papers of the Bernoulli's, D'Alembert, and so on. As a consequence of considering how various forces would have to be related to one another in order that perpetual motion might be possible, Helmholtz wrote his famous 1847 paper, "Über die Erhaltung der Kraft," on the conservation of energy, using the basic strategies of eighteenth-century rational mechanics. At the same time, he began to study physiology formally, and it soon became obvious to him that his knowledge of physics gave him a real competitive advantage among physiologists and medical students: "I ascribed my success in great measure to the circumstance that, possessing some geometrical capacity, and equipped with a knowledge of physics, I had, by good fortune, been thrown among medical men, where I found in physiology a virgin soil of great fertility; while, on the other hand, I was led by the consideration of vital processes to questions and points of view which were usually foreign to pure mathematicians and physicists."[61]

One of his first attempts to exploit his physical knowledge for physiological purposes involved the invention of the ophthalmoscope, an instrument for observing the retina of the eye, which depends for its functioning on an understanding of the optical properties of the eye. This instrument was so successful in generating recognition for Helmholtz that he quite self-consciously decided to "establish and maintain my reputation as an investigator," by applying physical principles to physiological problems. Helmholtz's commitment to organic physics was thus based principally on a pragmatic judgment that he could parlay his unusual combined background in physics and medicine into a successful scientific career.

Once he had established an assured position in this way, Helmholtz admitted, his commitment to a physicalist approach to living systems continued to intensify, at least in part for simple psychological reasons. "The ideas of an individual," he wrote, "which he himself has conceived, are of course more closely connected with his mental field of view than extraneous ones, and he feels more encouragement and satisfaction when he sees the latter more abundantly developed than the former. A kind of parental affection for such a mental child ultimately springs up, which leads him to care and to struggle for the furtherance of his mental offspring as he does for his real children."[62]

If Helmholtz's advocacy of organic physics seems by his own admission to have been driven to some degree by opportunism and psychological processes more than by any objective intellectual justification, that of Du Bois-Reymond, who was the chief organizer and entrepreneur among the organic physicists, was even less grounded in matters of intellectual principle, if one can believe the analysis of Timothy Lenoir.[63] Lenoir sees Du Bois-Reymond's rebellion against vitalist and teleological perspectives as a calculated move to achieve career advancement unjustified even by the kind of predisposition for physics that Helmholtz began with. Prior to 1842, Du Bois-Reymond had shown substantial interest in Hegel and Schelling, and he may have been drawn to Müller's laboratory precisely because of the latter's lingering tendencies toward teleology and vitalism.

Since Müller was deeply involved in university administration, he depended heavily on assistants in running the lab, and these assistants tended to be hypercritical of the younger students, who they sometimes recognized as intellectually superior and who they saw as potential competitors for any jobs that might be opening up. When Du Bois-Reymond arrived in Berlin, an extremely ambitious young man in 1840, he took up the tradition of embryological studies that was one of the early lines of research developed by Müller, and one that remained most closely linked to teleological issues. Carl Reichert, a senior graduate student who directed the laboratory's work in this area, was severely critical of Du Bois-Reymond's work, claiming that he did not "know how to think in the spirit of nature."[64] The dejected young man decided that if he was to succeed as a physiologist, he would have to move into a completely new field. Following another suggestion of Müller's, he took up a neurophysiological problem, the question of how certain electrical fishes manage to produce electrical discharges. In order to get support in learning the physics that he would need to take up his new subject, Du Bois-Reymond formed a scientific club, the forerunner of the Berlin Physical Society, along with Ernst Brücke and three other interested graduate students.

It turned out that Du Bois-Reymond was brilliant at developing instrumentation for detecting very small charges and currents in living tissues without being so intrusive that normal functioning was greatly distorted, so he was able to achieve significant new results on a number of electrophysiological problems. But now he and those who joined him, such as Brücke, faced a new problem. The dominant German traditions in physiology were embryological and morphological. How could the new upstarts convince others that their chemical and physical approaches were important enough to deserve support and to command academic appointments?

It was in this connection that Du Bois-Reymond's entrepreneurial, oratorical, and literary talents came in to play. And it was in connection with seeking patronage for his new discipline that Du Bois-Reymond focused on cultivating support for his kind of experimental science outside of the academic power structure, both among younger academics and among those outside of the university who were promoting industrialization and material progress. He did manage to convince the professor of physics at Berlin, Gustav Magnus, that physiological problems opened up a new and promising domain for physical investigation, and he recruited Magnus to become the single professor to join in establishing the Physical Society of Berlin in 1845. Moreover, he marshaled an additional fifty-three members, twenty-two of whom were reform-minded *Privatdozenten,* including Gustav Kirchoff and Hermann von Helmholtz, and the rest, military engineering officers (including Werner Siemens), instrument makers, and mechanics-ambitious men who were all interested in industrial entrepreneurship.

In 1858, partly because of his research successes, partly because of the support that members of the Physical Society were able to mobilize, and partly because of his growing reputation as a lecturer, Du Bois-Reymond succeeded Müller as professor of physiology at Berlin. Then from within that position he managed, with the help of Rudolph Virchow, to orchestrate a reform of the medical examination system, replacing the old examination on philosophy with a new examination on the physical sciences that focused on chemistry, physics, and experimental physiology. Thus the organic physics approach to physiology became a prerequisite for medical licensing, and enrollments grew.

Throughout the 1860s, wherever and whenever liberal attitudes prevailed, the organic physicists hitched their wagon to the cause of material progress and to the claim that scientific research was the engine that would drive increases in agricultural and industrial productivity, as well as in public health. Helmholtz, for example, provided leadership for academic reform in Baden, where state support for laboratory and clinical facilities grew rapidly through the 1860s. Du Bois-Reymond supported the anti-militarist policies advocated by the *Deutches Forscrittspartie* in Prussia between 1862 and 1866, when it seemed as if the advocates of a constitutional state committed to material and intellectual progress through science were ascendant. When it became clear to Du Bois-Reymond that Bismarck's monarchist-militarist policies would prevail, however, he remained true to his long-term principle: do anything to promote the discipline of organic physics. He became the chief academic spokesman for Bismarck's policies. Returning to the *volkish* rhetoric of Fichte, in major speeches during 1869 and in August of 1870, at the very beginning of the Franco-Prussian War, Du Bois-Reymond praised the Prussian military as the chief agent of that German unification that had long been sought, arguing in direct contradiction to his claims of two decades earlier, that it had been able to accomplish "what our railroads and telegraphs, our trade and industriousness, our laboratories and institutes of natural science could not achieve."[65]

On October 6, 1870, Otto von Bismarck wrote to Du Bois-Reymond expressing gratitude for his support and especially acknowledging the impact that his speeches had in convincing the English that the German attacks on France were "a great moral and national uprising against an unwanted enemy attack."[66] Within three months, Du Bois-Reymond had applied to the cultural minister of the new unified Germany to build the most expensive scientific research institute ever constructed. By the end of August, 1871, the first installment of funding had been approved.

None of this story is intended in any way to discount the real scientific achievements of the organic physicists or to suggest that they were not genuinely fascinated by the topics they explored and convinced that their physicalist program offered the most promising approach to studying them. Nor is it intended to deny that their promises that experimental science could provide a driving force for increasing agricultural production, industrial production, or military power contained a substantial grain of truth. It does, however, suggest that these arguments were less than completely disinterested and that they were often even self-consciously aimed primarily at increasing the status and resources of their own academic discipline. It is certainly true that the organic physicists managed to insinuate themselves within the structures of both academic and state power and that they used their positions to elevate the status of not just their particular approaches to science and scientific education, but of science and scientific education in general.

Dialectical Materialism, Marxism, and Scientific Socialism

The third form of nineteenth-century German materialism, which the Russian Marxist, G. V. Plekhanov (1856–1918), dubbed "dialectical materialism" in 1891, was less directly involved in the practices of mid-nineteenth-century natural scientists than either of the other two, although its proponents certainly understood it to be scientific and it did eventually come to play a role in directing the scientific activity of Soviet scientists in the twentieth century.[67] One could, however, make a strong case that it has played a substantially greater cultural role than either scientific materialism or the tactical materialism of the organic physicists. Through its incorporation into the corpus of Marxist thought, it became part of the theoretical backbone of almost all working-class political movements and radical revolutionary movements from the late nineteenth through the late twentieth centuries. Moreover, it played a key role in the formulation of much modern social scientific theory. This form of materialism, which its formulators, Friedrich Engels and Karl Marx, variously identified as "new," or "modern," materialism, grew up, like scientific materialism, as an extension of Ludwig Feuerbach's views. But instead of renouncing the Hegelian elements in Feuerbach, it actually amplified them and tried to link them to recent developments in the mathematical and natural sciences.

Neither Marx nor Engels had formal academic training as a scientist, but both to some degree, and Engels to a very great degree, educated themselves in mathematics and the natural sciences. In the second edition of his *Anti-Dühring* in 1885, Engels describes the commitments to dialectical logic, to materialism, and to a detailed understanding of contemporary science that went into the formulation of dialectical materialism: "Marx and I were pretty well the only people to rescue conscious dialectics from German idealist philosophy and apply it to the materialist conception of nature and history. But a knowledge of mathematics and natural science is essential to a conception of nature which is dialectical and at the same time materialist. Marx was well versed in mathematics, but we could only partially, intermittently, and sporadically keep up with the natural sciences. For this reason, when I retired from business and transferred my home to London, thus enabling myself to give the necessary time to it, I went through as complete as possible a 'moulting' as Liebig calls it, in mathematics and natural sciences, and spent the best part of eight years on it."[68]

In the division of labor between the two close collaborators, Marx and Engels, it was Engels who wrote most extensively on issues of dialectical materialist scientific method, especially as they pertained to the natural sciences. Marx concentrated on providing the social scientific foundations for the policies that he promoted for the various working-class revolutionary organizations that he and Engels provided leadership to over a nearly forty-year period. The two men worked closely, however, and not only did Marx read over and approve Engels's writings on scientific method as long as he lived, he implicitly and, less frequently, explicitly, incorporated dialectical materialist methods into his social analyses. Moreover, there is an extremely important sense in which the basic elements of what later became dialectical materialism informed all of Marx's writings, beginning with his dissertation, *On the Difference between the Democritean and Epicurean Philosophy of Nature,* which was completed in 1839. In that work Marx unambiguously committed himself to a materialist philosophy based on that of Epicurus, but informed by Hegelian logic.

It was extremely important to Marx, to Engels, and to subsequent nineteenth- and early-twentieth-century Marxists that dialectical materialism be understood to accurately reflect the best in contemporary scientific practice and that Marx's social investigations be understood to be rigorously scientific. This is so because it was on the basis of their claims to be providing objective, scientific assessments of the situation that Marx and Marxists were able to wrest leadership of the working-class movement from anarchist elements lead by Michael Bakunin, from charismatic figures such as Ferdinand LaSalle (1825–64), from established and venerated revolutionaries, such as Pierre Proudhon (1809–65), and from a host of other utopian visionaries who hoped to bring about liberal reforms and radical improvements in the living and working conditions of the laboring poor without the necessity of violent revolution.

Over and over again, Marx lashed out viciously at leaders of factions within the working-man's movement with whom he shared many basic goals, but whom he identified as promoters of policies that were unworkable and unrealistic because they were based on an inadequate understanding of the conditions that the proletariat faced. In an attack on Proudhon, whom Marx actually had substantial respect for, he articulated this position particularly clearly: "It is not enough to desire the collapse of these forms [the social forms that lead to the oppression of working people], one must know in obedience to what laws they came into being, in order to know how to act within the framework of these laws, since to act against them, whether deliberately or not, would be a futile and suicidal act and would, by creating chaos, defeat and demoralize the revolutionary class, and so prolong the existing agony."[69]

One consequence of Marx's tendency to attack virtually all of his fellow radicals, except for Engels, who showed no open need for independence, was that he eventually alienated nearly everyone with whom he worked. Bakunin, an early associate in Paris during the 1840s, ended up viewing Marx as "malicious, vain, quarrelsome, intolerant and autocratic . . . and insanely vindictive,"[70] and this seems to have been a widespread assessment. On the other hand, Marx and Engels were successful in using their claims to a uniquely objective perspective to eliminate opposition and to gradually unify proletarian movements around Marx's interpretation of history and his policies. Thus, when the Second International Workingman's Party was established in 1889, it was completely dominated by self-professed Marxists, although this did not mean that there wasn't internal conflict among variant interpretations of Marxism.

Like Positivism, with which it shared more than Marx or many of his successors would have been inclined to admit, Marxism has been understood differently by virtually every interpreter. And like Comte's, Marx's writings are often divided into earlier and later periods, with serious debate about how much continuity there was across the divide. In Marx's case, however, the earlier writings have a more humanistic cast, while the later are more scientific or scientistic, whereas the opposite was true for Comte. Furthermore, in Marx's case, there is little question that throughout most of the nineteenth and early twentieth centuries, it was the later, scientistic, Marx, as interpreted by Friedrich Engels, who was vastly better known and more important. In fact, the bulk of Marx's writings that predated the *Communist Manifesto* of 1848, were not published until the late 1920s and early 1930s. As a consequence, only in the twentieth century were the more humanistically oriented early writings widely disseminated and attended to. In what follows, though I will refer to some of the earlier writings to establish a background, my major focus will be on "scientific" Marxism as it appeared in *Capital*, as it was presented by Engels, and as it was understood by both the members of the First International and by the first generation of Marxist disciples, including Karl Kautsky (1854–1938), G. V. Plekhanov, and V. I. Lenin (1870–1924), who led the Second International and who guided the Russian Revolution.

Lenin was fond of saying that Marx's great accomplishment was in synthesizing elements from classical German philosophy, classical English political economy, and French socialism and revolutionary doctrines into a single coherent system.[71] This seems to be an accurate and almost universally acknowledged assessment, though different interpreters might assign the various elements differing levels of importance. There can be relatively little doubt that in terms of Marx's temporal development, French socialist elements preceded and continued to inform his reactions to German philosophy, to revolutionary rhetoric, and to political economy.

Marx was born at Trier in the Rhineland, May 5, 1818. His father, Heinrich (b. Herschel) had officially converted from Judaism to Lutheranism in 1817 in order to continue to practice law following the passage of Prussian anti-Jewish legislation in

1816. But Heinrich's religious sympathies seem to have been most consistent with the universalist and humanitarian emphases of the Saint-Simonian New Christianity, which spread in Germany during the early 1830s.[72] Karl's few extant *Gymnasium* exercises suggest that his own views were similarly shaped. In an essay on choosing a profession, for example, the young Marx emphasized "the good of mankind and our own perfection," and argued on the one hand that our choices are already limited by the social relationships that we find ourselves in and on the other that our choices should reflect the talents that we possess—all themes consonant with Saint-Simonian views.[73] More illuminating was his response to an examination question that asked him to "demonstrate, according to the Gospel of Saint John, XV, 1–14, the reason, the nature, the necessity, and the effects of the union of believers with Christ." Completely avoiding biblical references or the writings of any Christian theologian, Marx offered a historical discussion of the gradual development of human morality in which the Christian doctrine of unification in Christ plays an important and progressive but transient role on the way to a more universalist morality.[74] In a way, the most remarkable thing about this essay is that the examiner found its completely heterodox views worthy of a pass, though he did complain about the lack of specifically religious arguments.

Later in his career, Marx distanced himself from what he increasingly viewed as the sentimental moralizing tone of Saint-Simon and his own youthful essays. Indeed, in *The German Ideology,* which was written in 1846, after he had identified himself with communism, he insisted that "communists do not preach *morality* at all."[75] As he became more sophisticated with respect to political economy, he also gave up on the Saint-Simonian hope for an end to class conflict through inter-class cooperation and chided the Saint-Simonian socialists for their utopian views. Furthermore, Marx later viewed Saint-Simonian efforts to provide evidence for their views as dilettantish. Nonetheless, a number of key elements of subsequent Marxist doctrine almost certainly found their initial appropriation via Saint-Simon or his followers. The emphasis on an economic foundation upon which political and intellectual superstructures are constructed; the focus on class conflict as a mechanism for historical change; the identification of a series of great historical stages, including ancient, feudal, modern, and future, the first three of which find themselves undermined by changes in modes and relations of production that were nurtured within them; the argument that particular developments must be evaluated in terms of their roles in a particular developmental context—all of these are to be found in the Saint-Simonian corpus, and, if we are to believe Marx's 1859 preface to his *A Contribution to the Critique of Political Economy,* they continued to guide his studies. Thus, he wrote:

> The general conclusion at which I arrived, and which, once reached, became the guiding principle of all my studies can be summarized as follows. In the social production of their existence, men inevitably enter into definite relations, which are independent of their will, namely relations of production appropriate

> to a given stage in the development of their material forces of production. The totality of these relations of production constitutes the economic structure of society, the real foundation, on which arises a legal and political superstructure and to which correspond definite forms of social consciousness. . . . It is not the consciousness of men that determines their existence, but their social existence which determines their consciousness. At a certain stage of development, the material productive forces of society come into conflict with the existing relations of production or . . . with the property relations within the framework of which they have operated hitherto. From forms of development of the productive forces, these relations turn into their fetters. Then begins an era of social revolution. The changes in the economic foundation lead sooner or later to the transformation of the whole immense superstructure. . . . In broad outline, the Asiatic, ancient, feudal and modern bourgeois modes of production may be designated as epochs marking progress in the economic development of society. The bourgeois mode of production is the last antagonistic form of the social process of production—antagonistic not in the sense of individual antagonism but of an antagonism that emanates from the individual's social conditions of existence—but the productive forces developed within bourgeois society create also the material conditions for a solution to this antagonism. The prehistory of human society accordingly closes with this social formation.[76]

There is one further link between Saint-Simonianism and Marx's work that is tremendously important: that is the value that they both place on human self-actualization through labor. We saw in chapter 2 that Saint-Simon proclaimed that his highest goal was to encourage the creation of a society in which every individual would have the maximum opportunity to develop his or her own faculties through rewarding labor. Similarly, as we shall see below, one of the most persistent driving themes of Marx's work is not simply that workers are in some sense "robbed" of much of the exchange value of their labor, but more importantly, that capital exercises a "despotic" control over the laborer and turns work into a process that is intrinsically unrewarding, because not self-directed. This issue is articulated most clearly in *The German Ideology,* written around 1846, in a section in which Marx explains why the existence of a true community rather than a mere state is so important: "Only within the community has each individual the means of cultivating his gifts in all directions; hence personal freedom becomes possible only within the community. In the previous substitutes for the community, in the state, etc., personal freedom has existed only for the individuals who developed under the conditions of the ruling class, and only insofar as they were individuals of this class. The illusory community in which individuals have up till now combined always took on an independent existence in relation to them, and since it was the combination of one class over against another, it was at the same time for the oppressed class not only a completely illusory community, but a new fetter as well. In the real community the individuals obtain their freedom in and through their association."[77]

It is certainly true that all of the elements discussed above could have been found outside of the Saint-Simonian tradition, for the general sense that laws, politics, and intellectual life constitute a superstructure built upon the economic foundations of society had become commonplace during the second half of the eighteenth century both within the tradition of philosophical history flowing from Montesquieu, Turgot, Adam Smith et al., and within the French materialist tradition flowing from Helvetius.[78] The focus on the claim that labor should be self-fulfilling was present in the works of Smith and Ferguson and it was highlighted by the Ideologues, De Tracy, and Say, as well as by such Saint-Simonian contemporaries as Charles Fourier. Moreover, Marx's interpretation of historical change, with its emphasis on the notion that forces for change emerged from within any given social and economic structure, and with its insistence on the context dependency of value, was certainly reinforced by, if it did not originate in, his study of Hegel. Nonetheless, even though Marx seems to have been temperamentally unable to acknowledge his intellectual debts, Engels was inclined to emphasize the Saint-Simonian background of Marxist doctrine, writing in his *Anti-Dühring,* "In Saint-Simon we find the breadth of view of a genius, thanks to which almost all of the ideas of later socialism which are not strictly economic are contained in his works in embryo."[79]

Like Saint-Simon, Marx was also convinced that the detailed laws that are obeyed by society could only be discovered from the historical record. By sometime in the spring of 1850, he had decided to produce a detailed historical study of the processes by which the capitalist system of production and social relations had come into existence and by which it would surely self-destruct. Marx worked on *Capital* for nearly twenty years, interrupted by periods of journalistic and narrowly targeted polemical writing. By the time the third and final volume had been edited by Engels after Marx's death, it had reached nearly three thousand pages and constituted what Isaiah Berlin has characterized as "the most formidable, sustained and elaborate indictment ever delivered against an entire social order, against its rulers, its supporters, its ideologists, its willing and unwilling instruments, against all whose lives are bound up with its survival."[80]

Marx was obsessed with regard to the scientific validity of his analyses in *Capital,* so before we turn to its content, which involved a critical revision of classical political economy as well as a prediction of the imminent collapse of capitalism and the consequent triumph of the proletariat, we should consider its method, which brings us to the role of German philosophy and Hegelian dialectic. When Marx left the local *Gymnasium,* he entered the Law faculty at Bonn, largely because of parental pressure. After two years, he left for Berlin to study philosophy in spite of his father's qualms. Hegel had been dead for half a decade, and the young Hegelians, led by Bruno Bauer and David Strauss, were in the ascendant.

Marx was repelled by the speculative idealist side of Hegelianism, and he shared with Feuerbach an emphasis on the need to keep scientific theorizing closely controlled by an extensive experience of the phenomena being theorized about. "Those who dwell in intimate association with nature and its phenomena," he wrote at the beginning of his doctoral dissertation, "grow more and more able to formulate, as the foundations of their theories, principles such as admit of a wide and coherent development: while those whose devotion to abstract discussions has rendered unobservant of the facts are too ready to dogmatize on the basis of a few observations."[81] In part because the materialists of antiquity laid claim to being among the first to fuse empiricism with theoretical attempts to understand the hidden causes of phenomena, Marx chose to write his doctoral dissertation on the differential ethical implications of the systems of the two great classical materialists, Democritus and Epicurus. But Marx was also impressed by Hegelian dialectical logic and its emphasis on historical development: thus, his attempt to fuse materialism with dialectical principles.

In this fusion, one other element, probably derived from *naturphilosophisch* sources, is significant in Marx's own attempts to distinguish his materialism from the merely mechanical materialism associated with such figures as Descartes in the seventeenth century. Like Epicurus, Marx insists that motion is a quality inherent in matter, but he understands "motion" in a particularly expansive way: "Among the qualities inherent in matter, motion is the first and foremost, *not only in the form of mechanical and mathematical motion, but chiefly in the form of an impulse, a vital spirit, a tension—of a 'Qual,' to use a term of Jacob Bohme's—of matter. The primary forms of matter are the living, individualizing forces of being inherent in it and producing the distinctions between the species.*"[82] For Marx, then, as for the empiricist physiologists who had been inspired by *naturphilosophisch* ideas, material entities contained within themselves special forces that guided their temporal development. Nearly simultaneously, Friedrich Engels was studying *Naturphilosophie* and Hegelian philosophy on his own and coming to similar conclusions, although Engels remained much more openly sympathetic to the goals and methods of *naturphilosophisch* science than Marx. Indeed, in his first substantial exposition of dialectical materialism, *Anti-Dühring*, he wrote that, "the *Naturphilosophen* stand in the same relation to consciously dialectical natural science as the utopians [such as Saint-Simon] to modern communism."[83]

Given Engels's favorable understanding of the historical role of utopian socialism, though he viewed it as retrogressive since the rise of Marxist, "scientific" socialism, we can reasonably take this as a favorable comment on the historical role of *Naturphilosophie*. He was quick to admit that *Naturphilosophie* "contains a great deal of nonsense and fantasy." But, he continued, "that there was also in it much that was sensible and rational began to be perceived after the theory of evolution became widespread."[84] Especially important in stimulating dialectical understanding, for example, was the *naturphilosophisch* emphasis on polarities and the way in which opposites came together in such a way as to destroy one another in the process of creating a qualitatively new entity. Similarly, according to Engels, the *natuphilosophisch* insistence on transformations, metamorphoses, and the significance of change over time were all critical notions that, incorporated into dialectical materialism, helped to distinguish it from the crude and static "vulgar itinerant-preacher materialism of a Vogt and a Büchner."[85]

In both the *Anti-Dühring* of 1878 and *Dialectics of Nature,* which was written between 1873 and 1883 (though only published in 1925) Engels laid out what he viewed as the three central "laws" of dialectical logic articulated by Hegel, and purported to show how they could be "abstracted" from natural phenomena and human history. These laws are

> The law of the transformation of quantity into quality and vice versa;
> The law of the interpenetration of opposites; [and]
> The law of the negation of the negation.[86]

In *Anti-Dühring,* Engels purports to show for each of these laws how it is derivable from natural phenomena, how it is illustrated in some human or social phenomenon that is not related to Marxism, and then how it is utilized by Marx within the structure of *Capital.* For example, in discussing the law of the transformation of quantity into quality, Engels discusses phase transformations from ice to liquid water to steam, in which "merely quantitative changes of temperature bring about a qualitative change in the condition of the water." Then he discusses the several series of compounds of carbon, hydrogen, and oxygen, in which "we have a whole series of qualitatively different bodies formed by the simple quantitative addition of elements." Turning to human phenomena, Engels cites Napoleon's observation about combat between the French cavalry, which was well disciplined but lacked very highly skilled riders, and the Mamelukes, who lacked discipline, but were superb horsemen. According to Napoleon, whenever three Frenchmen met two Mamelukes, the Mamelukes always won, one hundred Mamelukes and one hundred Frenchmen usually fought to a draw, three hundred Frenchmen could usually beat three hundred Mamelukes, and a thousand Frenchmen could always defeat fifteen hundred Mamelukes. Thus, simply changing the scale of the confrontation between French and Egyptian cavalry changed the outcome of the confrontation. Finally, Engels pointed out that for Marx, a "minimum sum of exchange values was necessary to make possible its transformation into capital," and "the cooperation of a number of people . . . creates a new power, which is essentially different from the sum of [their] separate forces."[87]

The law of the negation of the negation is particularly clear in successive transformations of energy from one form into another or in the production of new life from the decay products of the old, and it is an absolutely essential characterization of the historical process by which capitalist production—itself the negation of petty bourgeois production—creates the conditions for its own destruction. In Marx's words, "Capitalist production begets, with the inexorability of a process of nature, its own negation. It is the negation of the negation."[88]

One might reasonably complain, as Dühring did, that the basic dialectical generalizations or laws are so vague and nonspecific as to be virtually useless for establishing any particular historical claim. But such a complaint would, according to Engels, suggest that the function of dialectical laws has been misunderstood. In characterizing the process by which capitalism will destroy itself as the negation of a negation, Engels argues, Marx does not seek to demonstrate the historical necessity of the process. "On the contrary," Engels insists, "it is only after he has proved from history that in fact the process has partially already occurred . . . he in addition characterizes it as a process which develops in accordance with a definite dialectical law."[89] That is, the assumption of dialectical laws does not play a role in *establishing* the particular claim regarding capitalism's self-destructive course. However, the fact that the claim happens to be formulatable as a particular case of a general dialectical law reinforces our assurance that it is correct.

One final issue should be addressed before moving to some of the central arguments of *Capital.* That issue involves the tension between the determinist implications of scientific knowledge and the notion of human agency and urgency that pervades Marx's writings. Such a tension is, of course, not unique to Marx and Marxism. It is implicit in every claim, materialist, Positivist, or otherwise, which sees scientific knowledge as a foundation for action in and manipulation of the world. If the phenomenal world is truly and absolutely deterministic, there can be no effective way for us to influence its course. But this tension seems highlighted in Marxism because advocacy and claims to objectivity are both so passionately promoted.

I have already insisted upon the extent to which Marx and Engels used the claim that their opponent's policies were not predicated on an accurate understanding of social conditions and dynamics as a bludgeon against enemies within the workingmen's movement. This position was closely linked to their insistence that historical developments are inevitably driven by impersonal forces associated with economic interests and class identities rather than, as in the Hegelian case, by the ideas of World-Historical individuals. At the same time, both Marx and Engels violently rejected the view, often later associated with Darwin, that consciously guided action in the world was impotent. In the eleventh of his "Theses on Feuerbach," Marx articulated his frustration with philosophers in general on this issue by writing: "The philosophers have *interpreted* the world in various ways; the point, however, is to *change* it."[90] And in the third of his theses, he specifically criticized materialist philosophy as it had developed through the writing of Feuerbach: "The materialist doctrine that men are the product of circumstances and that, therefore, changed men are products of other circumstances and changed upbringing forgets that circumstances are changed precisely by men. . . . The coincidence of the changing circumstances and of human activity can only be conceived and rationally understood as revolutionizing practice."[91] Similarly, in one of the few passages in which Engels shows anything but unqualified admiration for the application of Darwinian insights to human social development, he writes: "Darwin did not know what a bitter satire he wrote on mankind, and especially on his countrymen, when he showed that free competition, the struggle for existence, which the economists celebrate as the highest historical achievement, is the normal state of the *animal kingdom.* Only conscious organization of social production, in which production and distribution are carried on in a planned way, can lift mankind above the rest of the animal world as regards the social aspect, in the same way that production in general has done this in the specifically biological aspect."[92]

Just how conscious intervention to direct human affairs could be efficacious in a deterministic universe was never clearly spelled out within Marx's works. Indeed, the relationship between moral choice and scientific knowledge in Marx has been the focal point of many of the most contentious battles among Marxists, with Karl Kautsky representing an extreme technological or economic determinist version of Marxism in his *Materialist Conception of History* of 1927, and V. I. Lenin representing a much more voluntarist version in *What Is to Be Done,* published in 1902. But Marx's insistence that the truth of scientific knowledge could never be known except as a consequence of its transformative power in human hands highlighted his own unwavering central belief that it must be possible to combine knowledge with true human agency. Indeed, what drew Marx to Epicurean materialism in his dissertation was the fact that Epicurus constructed a version of materialism in which we humans could "free ourselves" from the fear both of the arbitrary will of the gods and of the uncontrollable character of natural phenomena. Marx's sense that human agency, always appropriately informed by scientific knowledge, was responsible for the transformation of the means of production as well as potentially capable of transforming social relations, remained central to his thought. It is also indispensable to any understanding of how his claims to scientific analyses could go along with the constant exhortations to action that characterize his works.

Marx's insistence upon the possibility, indeed, the psychological necessity, of some form of self-determination even in a scientifically understandable world predated and probably stimulated his gradual immersion in the communities of radical political activists during the 1840s. But it seems to have been at least reinforced, if not intensified, as he went into exile in Paris when the radical tone of his early work as a journalist led the Prussian government in April of 1843 to shut down the *Rheinische Zeitung.* In Paris, Marx interacted with most of the major European political agitators of the time; became a self-professed communist; formed the close association with Friedrich Engels that lasted for the rest of his life; read widely in political economy; began to formulate his project of writing a detailed analysis of capitalism; and formed the opinion that unless the various radical splinter groups, which often embraced nationalistic or ethnic agendas, could be brought together around some common vision and program, there could never be a successful proletarian revolution.

Expelled from Paris in 1845 at the behest of the Prussian government, Marx moved to Brussels, where he began to lecture and write for the *Deutsche Brüsseler Zeitung* in order to prepare working men for their role in what he viewed to be the imminent proletarian uprising. On a trip to England with Engels, Marx came into contact with the London-based German Workers Educational Association, which had recently joined a loose affiliation of local revolutionary societies to form the Communist League. Marx undertook organizing activities on behalf of the growing Communist League, and in 1847 the League commissioned Marx and Engels to produce a general statement of its goals and unifying beliefs. This attempt to provide a unifying vision and program for working class-revolution, *The Communist Manifesto,* was published just weeks before the first outbreak of Europe-wide revolutionary activity began in Paris early in 1848.

Though the stirring rhetoric of *The Communist Manifesto* has undoubtedly made it the most frequently read, relished, and reviled revolutionary tract in the Western world, it had little impact on the abortive, largely bourgeois-led, and liberally aimed revolutions of 1848. Marx's subsequent analyses of the revolutionary activity in France and in Germany during 1848 and 1849 confirmed him in his opinion that no "revolution" short of an international and probably violent uprising of the entire proletariat could be successful. Moreover, it convinced him of how ill informed and unprepared the proletariat was; thus it reinforced his desire to provide that scientific analysis of capitalism that could fully prepare people for the coming struggle.

This analysis appeared in *Capital,* of which Marx completed the first volume in 1867. Engels brought out the second and third volumes posthumously in 1885 and 1894. *Capital* provided most of the basic Marxian economic and sociological theorizing known during the nineteenth century. Many key Marxian concepts, such as alienation, utilized in *Capital,* had been explored more extensively in either *The German Ideology,* written during 1845 and 1846, but not published until 1926, or in the *Economic and Philosophical Manuscripts of 1844,* which were not published until 1932; but nineteenth- and early twentieth-century Marxism had to do without the insights provided by these works.

The basic strategy of *Capital,* which Marx had been previewing since the publication of *The Communist Manifesto,* was to offer an analysis of the capitalist system of production and capitalist social order that showed how it produced its own negation, which Marx believed would constitute a classless society of cooperative producers: the proletariat, free at last. But in a curious letter to Engels written in December of 1867 and aimed at discussing the most effective strategy for planting favorable reviews to stimulate interest in and sales of *Capital,* Marx suggested that the analytic structure be decoupled from the possibly disturbing conclusions, laying emphasis on the former, and downplaying the latter. In his sample review, Marx wrote: "We must distinguish between the author's positive contributions . . . and the tendentious conclusions which he draws. The former constitute a direct enrichment of science: he has produced a new analysis of real economic conditions using a materialist approach. . . . Example 1) the analysis of money; 2) the 'natural' development of cooperation, the division of labor, the system of machinery and the corresponding social relations and social ties. . . . On the other hand, the author's subjective tendency, i.e., the way he imagines or presents the final result of the present movement, of the real social process, has actually nothing to do with the analysis itself—he was perhaps morally obliged and committed to this standpoint by virtue of his political affiliation and his past."[93]

In proffering this reading of *Capital,* Marx actually anticipated the way in which the work has been most often read for over a century. While orthodox Marxists and vehement anti-Marxists alike have always understood the work as a promise of the ultimate success of the proletarian revolution, a wide range of thinkers from almost every imaginable intermediate political perspective have managed to separate elements of Marx's discussion of capitalism and of the general relationships between economic life and other aspects of culture from the overarching political aim of the text and to incorporate his concepts and/or insights as central elements in virtually all of the social sciences in the twentieth century.

It is beyond the scope of this study to follow Marx's analyses in all of their often bewildering detail, but discussion of a few key concepts can suggest something about the ways in which Marx's work departed from the approach of other nineteenth-century political economists, something about how thoroughly he was able to integrate concepts and analytic strategies from the various traditions from which he drew, and something about the new directions that he promoted in the social sciences.

In connection with virtually every concept that he used, Marx began by appropriating some traditionally defined notion or set of notions; then he transformed them so that they took on different meanings. This is particularly clear in his discussions of "alienation," which is used almost exclusively in its Marxian sense in modern sociological literature. The concept of alienation had been developed by Fichte and Hegel and it was used extensively by Feuerbach to indicate a self-initiated process by which the spirit or mind projected a part of itself onto something external and in some sense, opposed to itself. It was thus through a process of self-alienation that man created the gods as powers outside of himself, and therefore beyond his control.

In *Capital,* Marx retained the notion that alienation involves both a separation from the self and a loss of control. But he placed the cause of that separation and loss, not in some mental act, but in a historical process. This process gradually transformed a primitive system of production in which members of a community directly exchange services to meet one another's needs, into the capitalist system of production in which workers are forced in isolation to create surplus-value for their employers. Moreover, into this process, he integrated the discussions initiated by Montesquieu regarding the tendency of commercial competition to undermine personal ties within communities,[94] as well as the discussions of Ferguson and Smith regarding the dehumanizing consequences of the division of labor pushed to its extremes in factory production.[95]

In the first stages of the process, the exchange of use-values whose meaning in the lives of community members is well understood, is transformed through the division of labor into an exchange of commodities, whose use is divorced from their production.[96] As the historical process unfolds, driven by the division of labor, social forces are created that increasingly seem beyond the control of individuals: "The social power, i.e., the multiplied productive force, which arises through the co-operation of different individuals as it is determined by the division of labor, appears to come about naturally, not as their united power, but as an alien force existing outside of them, of the origin and goal of which they are ignorant, which they thus cannot control."[97]

By the time full-fledged capitalism has emerged, virtually every aspect of human experience has been divorced from and turned against the individual laborer: "Within the capitalist system all methods for raising the social productivity of labor are put into effect at the cost of the individual worker; . . . all means for the development of production undergo a dialectical inversion so that they become means of domination and exploitation of the producers; they distort the worker into a fragment of a man, they degrade him to the level of an appendage of a machine, they destroy the actual content of his labor by turning it into a torment; they *alienate* from him the intellectual potentialities of the labor process in the same proportion as science is incorporated in it as an independent power; they deform the conditions under which he works, subject him during the labor process to a despotism the more hateful for its meanness; they transform his lifetime into working time."[98] Ultimately, within capitalism, then, humans become alienated from the products of their labor and from the processes of labor, for neither of these any longer relates directly to the satisfaction of human needs or to the accomplishment of human purposes. Virtually all subsequent discussions of alienation as a significant feature of advanced industrial societies draw from Marx's analysis.

For the most part, Marx and Engels accepted the rough scheme of Montesquieu as temporalized by the Scottish philosophical historians in describing the historical movement of dominant modes of production and social structures from hunting to pastoral to agricultural to commercial, though these categories were correlated with the Saint-Simonian categories of Primitive communitarian, Ancient tribal, Feudal, and Industrial (Capitalist). Moreover, Marx also accepted the existence of a separate static Asiatic system of production and society based on irrigation agriculture and the communal ownership of property.[99] But overriding these categories was an implicit distinction between social structures dominated by communitarian notions—in which production was primarily for immediate use, in which relations among members were primarily personal and cooperative rather than impersonal and competitive, and in which the division of labor was limited by the existence of constricted markets—and societies dominated by commodity production, in which social relationships tended to be impersonal and competitive and in which the division of labor was constantly intensifying.

For Marx, though the division of labor, with its concomitant drive toward commodity production and the use of money, had been constantly negating elements of community throughout the entire historical process, it was in the transition from feudal society to capitalist society that the most dramatic shift occurred. In 1887, Ferdinand Tönnies formalized this set of distinctions and made it central to modern sociology and political theory by identifying premodern social structures with the term *Gemeinschaft* (community) and modern social structures with the term *Gesellschaft* (society).

One of the central tests of the extent to which a particular society had moved away from existence as a community, according to Marx, was the extent to which a sense of common interests and purposes had been displaced by well-developed "class" distinctions and class consciousness. Once again, Marx took a term that had been gaining currency within the eighteenth-century tradition of philosophical history as a generic synonym for "estate," "order," "rank," or indeed, any group of persons bound together by some set of common interests or attributes, and he gave it a meaning that is more restricted in its extent but much richer in its connotations. Starting from the general insistence that the most fundamental and meaningful processes and relationships in society are derived directly from the character of its economy, Marx insisted that the most important groupings of people must depend upon the economic roles that exist within society.

In communitarian societies there are no interests that divide any group of persons from any other. Though some division of labor may occur based on gender or on the differential distribution of talents, all members of the community stand in a relationship of equality to one another as producers of goods or services that are consumed directly by their fellow community members and as consumers of the goods and services of others. But in all societies beyond the stage of primitive communism, divisions based on economic function emerge, and "classes" are formed. Furthermore, these classes inevitably have different interests, and the conflict that grows out of the struggle of each class to promote its own interests provides the driving force for social change. Marx's view of the role of classes and class conflict are most directly and simply stated at the beginning of the first major section of *The Communist Manifesto:*

> The history of all hitherto existing society is the history of class struggles. Freeman and slave, patrician and plebeian, lord and serf, guild-master and journey-

> man, in a word, oppressor and oppressed, stood in constant opposition to one another, carried on an uninterrupted, now hidden, now open fight, a fight that each time ended, either in a revolutionary reconstitution of society at large, or in the common ruin of the contending classes. . . . The modern bourgeois society that has sprouted from the ruins of feudal society has not done away with class antagonisms. It has but established new classes, new conditions of oppression, new forms of struggle in place of the old ones. . . . Our epoch, the epoch of the bourgeoisie, possesses, however, this distinctive feature: It has simplified the class antagonisms. Society is more and more splitting up into two hostile camps, into two great classes directly facing each other—bourgeoisie and proletariat.[100]

In *Capital,* Marx explores in excruciating detail the many mechanisms through which the capitalist system of production can concentrate capital and drive members of intermediate classes down into the proletariat, so that large-scale capitalists and members of the proletariat eventually face one another in a stark confrontation. Moreover, he rehearses the way in which the increasing use of machinery in modern factories enables capitalists to extract surplus-value from workers in ever more efficient ways, increasing their control over the workers and driving the share of the products of their own labor retained by the proletariat ever downward.

He discusses the processes by which large-scale capital begins to drive workers into unions out of a sense of self preservation: by bringing many workers together into factories, pushing to make the working day longer, making wages lower, and appropriating to itself all of the value of the increased productivity that it generates. Advanced capital thus creates the conditions under which the proletariat gradually becomes aware of its existence as a class and of the interests shared by working persons everywhere. Finally, Marx details the way in which the fundamental driving principle of capitalist production, the tendency to extract maximum surplus value from labor by increasing the production of commodities without regard for their use-values, initially leads to increased material plenty and to the continuing accumulation of capital, but eventually to overproduction, the creation of ever-intensifying economic crises that necessitate ever-intensifying attempts to exploit labor and, presumably, eventually to a revolution of the proletariat.

One might reasonably ask why neither the capitalists nor the proletariat can see what is happening and take steps to avoid the impending collapse. Some "revisionist" Marxists, such as Eduard Bernstein (1850–1932), actually argued that they might, and in his *Evolutionary Socialism* of 1899, Bernstein projected the possibility of a peaceful creation of the ultimate socialist society as a consequence of changes in capitalist economic developments. But for most Marxists, such a possibility was precluded by the fact that every class, other than the proletariat, was blocked from an objective understanding of its own circumstances by the existence of a "false consciousness," or "ideology."

Like alienation, ideology was a term appropriated and transformed by Marx.

Among the French *Idéologues*, the term signified any set of largely unexamined presuppositions shared by a significant number of people regarding salient features of the social world in which they lived. Such presuppositions might accurately reflect an independently existing external reality or not, and the groups that shared ideologies might be constituted in any number of ways, based on the sharing of interests that might be founded in common membership in religious communities, occupational roles, legal status, political groupings, wealth, and so on. The only certain thing that one could say is that every person lived his or her life from within some constellation of presuppositions, or *ideology*, for quite literally, people live in a world of ideas.

Marx does not disagree with the claim that people base their actions more on what they believe to be the case than on what is "really" the case. But he refocuses attention on the material factors that go into the creation of the "consciousness" of any person or class in the first place. As we saw above, in the *German Ideology*, Marx insisted that, "it is not the consciousness of men that determines their social existence but their social existence that determines their consciousness." But, for Marx, the social existence lived by any person is just that of his or her own class, so it would seem that the consciousness, the world of ideas and beliefs, of any human must be limited by the experiences and interests of the class to which he or she belongs. There is, however, an important caveat to this notion. Until a class develops its own consciousness, because the means for inculcating ideas and values in any society are controlled by the ruling class of that society, "the ruling ideas of each age have ever been the ideas of its ruling class."[101] As a consequence, the dominant *ideology* in any time or place reflects the immediate interests of the ruling class, rather than anything approaching an objective understanding of conditions.

At the time of the publication of *The Communist Manifesto*, argued Marx, the proletariat had not yet developed its own consciousness. Those who spoke on behalf of working men up to that point had been members of the bourgeoisie, and because "the bourgeoisie naturally conceives the world in which it is supreme to be the best,"[102] the "bourgeois socialism" that preceded Marxism was naturally formulated to perpetuate bourgeois values and a social structure that left the bourgeoisie comfortably in control. Thus, for example, we get the Saint-Simonian society managed by bankers and entrepreneurs.

Especially through the works of Karl Mannheim (1893–1947) in the 1920s and early 1930s, the Marxian suggestion that no understanding of social relations can be free of values and political implications—that every conceptualization is somehow ideologically loaded—has come to represent the all but universal view among social theorists. Indeed, it has become the foundation of a field of sociology called the "sociology of knowledge."[103] Many of those who emphasize the inescapable ideological content of social theorizing continue to argue that economically based class considerations constitute the ultimate foundations of ideologies. Moreover they insist that because ideologies inevitably reflect the interests of the dominant class at any time, they must be conservative in their very nature. In the late twentieth century there was a growing sense among those who were not orthodox Marxists that group identities based on gender, nationality, ethnicity, religion, and so on, are vastly more important in shaping the life experiences, and thus the ideological orientations, of individuals than Marx and Engels had allowed.

Even without considering the possible noneconomic elements that might enter into ideologies, Marx's emphasis on the inevitable presence of ideological elements in social theories created an important set of paradoxes and problems for Marxists and non-Marxists alike. If Marx was correct, how could nineteenth-century Marxist theorists, almost all of whom came from bourgeois family backgrounds, claim that *they* were capable of objectivity and value neutrality in apparent violation of their own principles? One might imagine that in a future "classless" society there would no longer be recognizable ideologies because the salient features of the experience of all human beings would be identical. Then an "objective"—that is, a universally agreed upon—understanding of society would be virtually assured. But the nineteenth century remained a century dominated by class antagonisms, and even Marx and Engels were members of the bourgeoisie. Even if Marxism was capable of expressing the true interests of the proletariat, those interests were still "class" interests in the nineteenth century and could not be understood as the foundation for a value-free and politically neutral science of society. On the other hand, if Marx was correct in a strong sense, no other group could claim to have a nonideological scientific perspective from which to condemn the Marxist perspective as illegitimate because it was informed by particular class interests.

At a very fundamental level, Marxism and Positivism faced precisely the same intellectual problem with respect to the incorporation of socio-cultural presuppositions into scientific theory, although the problem was vastly more visible in Marxism because of its heightened emphasis on the connection of theory with practice and because of the Marxist restrictive insistence that these socio-cultural presuppositions must inevitably be grounded in economic class considerations. Furthermore, the Positivist and Marxist responses to their respective problems were remarkably similar. In each case a relativist and progressivist vision of understanding emerged in connection with the simultaneous claims that: 1) at any stage even the best knowledge available must inevitably contain at least implicit, and sometimes explicit, residues of the cultural presuppositions of its creators; 2) a method, or philosophy of science, can be developed that will allow one to minimize, but not necessarily eliminate, both the number and the impact of these cultural assumptions; and 3) the ultimate test of the philosophy of science or method adopted is the extent to which the theories created allow one to predict the outcomes of events and to transform the world through the knowledge produced.

It was in connection with claim number 2, that Positivism and Marxism diverged. Positivism focused on the need for scientific thinkers to distance themselves from their own hopes and fears in creating scientific theories, to give up the hope of discovering the necessary causes of phenomena, and to avoid all hypotheses not capable of direct observational testing. On the other hand, Marxist dialectical materialism drew from the German tradition from Kant through the *Naturphilosophen* and Hegel in insisting that passionate engagement with one's subject of study was more likely to help than hinder knowledge production, that the search for causes was a central feature of any science, and that hypotheses should be restricted not by their direct accessibility to empirical test so much as by their consistency with rational criteria (in this case, dialectical logic).[104]

There is no doubt that Marxism was a scientistic movement. That is, it openly sought to extend methods derived from mathematics and the natural sciences to deal with social phenomena. With respect to Marxist claims that Marxism is scientific, however, we come back to an issue raised at the beginning of chapter 4. If we insist on applying positivistic criteria by which to judge such a claim, Marxism undoubtedly fails. If we open up our criteria to include those that had characterized German *Naturphilosophie* in the early part of the century, it is probably a reasonable claim. Once again, however, even if we were to agree that Marxism was and is indeed scientific, that would not justify the most important inference that Marx, Engels, and subsequent Marxists have wished to draw: that it was therefore also correct in all of its claims.

CHAPTER 6

Early Victorian Culture: Public Science and Political Science

The first half of the nineteenth century was a period of tremendous conceptual and institutional growth and change in the natural sciences in Britain. During the early part of the century, there was perhaps no single science in which British scientists were the best in the world. Leadership in analytic mechanics, for example, remained in Paris. And world leadership in organic chemistry probably migrated to Germany in the 1820s. But the richness and extent of British scientific activity across the board was probably as great as, if not greater than, that of any other country by mid century. The British government provided almost no financial support for any scientific work or institution. With the singular exception of the Astronomer Royal at Greenwich, whose activities were held to be essential for enabling the precise navigation upon which British shipping depended, no British scientist received a government salary primarily for engaging in scientific research in 1830. The government did support the expenses—and sometimes the salaries—of physician-naturalists on naval expeditions, because their major functions were to provide medical care and to inventory the regions visited for resources that might be exploited. Otherwise, there was almost no public funding for science in England and little beyond local support for the (very important) universities in Scotland.

In spite of the low levels of state support, however, British investigators were very active in almost every natural scientific discipline. John Dalton (1766–1844) at Manchester, Humphrey Davy (1778–1829) in London, and Thomas Graham (1805–69) in Edinburgh were certainly among the outstanding chemists of the age. William Smith (1769–1839), James David Forbes (1809–68), and Charles Lyell (1797–1875) helped British geology surpass that of the French by the 1830s. William Rowan Hamilton (1805–65), George Boole (1815–64), and George Green (1793–1841) began to challenge Continental authors in pure and applied mathematics. Thomas Young (1773–1829), Michael Faraday (1791–1867) and James Prescott Joule (1818–89) were among the preeminent experimental physicists of the time. John Herschel (1792–1871) and George Biddell Airy (1801–92) were outstanding astronomers, and Charles Bell (1774–1842), and Richard Owen (1804–92) rivaled George Cuvier and Karl von Baer as anatomists.

The Royal Society of London, which had become relatively moribund during the late eighteenth century, remained a favorite target of scientific reformers until mid century,[1] but specialized scientific societies modeled on the 1788 Linnaean Society began to proliferate with the founding of the Geological Society of London in 1807, the Royal Astronomical Society in 1820, the Zoological Society of London in 1826, and the Statistical Society in 1834. University College, London, the first secular scientific university in England, was established in 1826. And in 1831 British scientific reformers banded together to establish the British Association for the Advancement of Science, their answer to Oken's *Versammlung deutscher Naturfoscher und Artz.*[2]

The stories of conceptual and institutional development within the early Victorian scientific community have been often and well told elsewhere, so we will virtually ignore them.[3] Our initial concern in this chapter will rather be with exploring how scientific knowledge, practices, and attitudes were extended beyond the gentlemanly British community of self-styled natural scientists into the various segments of society at large so that we can understand how and why scientistic movements had substantial appeal to multiple British publics.

Britain had shared the eighteenth-century vogue for scientific lectures and entertainments with the rest of Europe. But a more serious interest in scientific knowledge and attitudes had already begun to spread among agricultural and industrial entrepreneurs in eighteenth-century Britain, providing one of the critical conditions for and stimuli to the revolutions in food production and manufacturing that constituted the first industrial revolution.[4] As a consequence, among progressive landowners and merchant manufacturers there was a widespread interest in science that led to the establishment of numerous provincial philosophical societies, many of which persisted through the nineteenth century.[5] During the first three decades of the nineteenth century, these philosophical or literary and philosophical societies—which usually sponsored lectures, established a scientific library and a laboratory of some sort, and required the presentation of member's papers on their own investigations—continued to appear.[6]

Both Anglican and dissenting clergy in England were also much more involved both as consumers and practitioners of science than their counterparts on the Continent because Anglicanism had developed a much greater dependence on natural theology than either Catholicism or Continental Protestantism,[7] and many dissenters had their intellectual roots within Broad Church Anglicanism.[8] Going into the nineteenth century, then, an interest in science was unusually widespread within a British cultural elite that included not only clergy, lawyers, physicians, and urban aristocrats, but also rural gentlemen and urban merchant/manufacturers.[9]

But in early Victorian Britain, science spread far beyond the artisans, educated upper middle class, and aristocratic boundaries that for the most part contained it on the Continent, involving many elements of the middle class and the working class. A substantial fraction of the most outstanding British scientists during the first half of the nineteenth century, including William Smith, John Dalton, Humphrey Davy, Michael Faraday, James Prescott Joule, Joseph Hooker (1817–1911), and Alfred Russell Wallace (1823–1913), for example, never attended, much less graduated from, a university; nor were any of them born into the aristocracy. Indeed, Dalton, Smith, Faraday, and Wallace were all from working-class backgrounds.

It is also the case that decades before men like Karl Vogt and Ludwig Büchner found ready audiences in the German middle class for their works of popular science, Jane Marcet's *Conversations on Chemistry* (1806), Harriet Martineau's *Illustrations of Political Economy* (1834), and Mary Somerville's *Connection of the Physical Sciences* (1834), as well as the many volumes published in the Library of Useful Knowledge series by Henry Brougham's Society for the Diffusion of Useful Knowledge and in Dyonisus Lardner's competing Cabinet Cyclopedia series, were finding very large markets in Britain.[10] Moreover, dozens of popular science journals—including the *London Journal of Arts and Sciences,* the *Penny Mechanic,* the *English Mechanic,* the *Mechanic's Oracle,* the *New London Mechanic's Register,* the *Naturalist,* the *Magazine of Natural History,* and the *Geological and Natural History Repository*—began to appear in London beginning in the 1820s.[11]

On the other hand, the professional class of camerally trained civil service officers in Germany were vocal and active promoters of scientific education and the application of science to productive activities, and the civil service of France was increasingly constituted of graduates of the *École polytechnique* and the *écoles de sante* at the beginning of the nineteenth century. Professional civil servants in Britain, primarily graduates of Oxford and Cambridge, however, rarely had any significant scientific education, and they generally opposed the application of scientific expertise and the expansion of scientific education.[12] Political pressures to expand the roles of science in British public life came instead primarily from elected politicians—usually Scottish educated—or from voluntary organizations and informal networks of scientists and educational reformers.

It will be a primary aim of this chapter to explore some of the ways in which scientific knowledge, attitudes, and practices came to expand their place in the cultures of the middle class and the labor aristocracy of Britain during the first half of the nineteenth century, as well as the ways in which working-class science occasionally differed from "elite" science.[13] In this process, we will consider the close links between science and both orthodox Christianity and dissident religious movements, and we will explore some of the developing relationships between science and political life.

The Legacy of the "Academy of Physics": Science for the Literati and for the Masses

In early 1797, a group of about a dozen students at the University of Edinburgh, almost all of whom had attended both the aging Joseph Black's (1728–99) lectures on chemistry and at least one of Dugald Stewart's (1753–1828) moral philosophy classes, formed a new student society. This society, the "Academy of Physics," was promoted by its founder, the eighteen-year-old Henry Brougham (1778–1868), as one that would be devoted to the study of "the *Newtonian Philosophy*, comprehending every subject to which induction and reasoning can be applied: 1. Mathematics, pure and mixed, 2. Physics, 3. Mind, 4. History."[14] Though the society lasted for only about three years, its leading members went on to play an important role in bringing their kind of inductive, "Newtonian" science to an ever-wider British public.

Brougham became a lawyer and politician and went on to hold the office of Lord Chancellor of England during the period leading up to the passage of the Reform Bill of 1832. He also published avocationally in physical optics over a period of more than fifty years. But for present purposes, he is important primarily as one of the most vigorous promoters of the secular University of London, as the leader of the movement that established the Society for the Diffusion of Useful Knowledge and its literary offshoots, the *Library of Useful Knowledge* and the *Penny Magazine*, and as the author of more than eighty popular scientific articles for the *Edinburgh Review*.

Second only to Brougham among the academy's members in long-term importance for the spread of scientific knowledge and attitudes was George Birkbeck (1776–1841), a Quaker medical student who subsequently left his friends in the academy to become Professor of Natural Philosophy at Anderson's Institution in Glasgow, where he initiated a special free natural philosophy class for local mechanics. This class, which was continued by Birkbeck's successors, stimulated the founding of the Edinburgh School of the Arts in 1821 and the Glasgow and London Mechanics' Institutes in 1823. Birkbeck, who had moved to London after leaving Glasgow, became the first president of the London Mechanics' Institute and chief spokesman for the entire Mechanics' Institute movement, which involved more than 120,000 members in 700 different local Institutes in Britain and Ireland, as well as many more in America, New Zealand, and Australia at its peak popularity in the 1850s.[15]

Third in importance within the group was probably Francis Jeffrey (1773–1850). Slightly older than the rest of the group at twenty-four, Jeffrey had attended Edinburgh High School before heading to Glasgow to study law with John Millar and then going to Oxford, which he detested, for a two-year period. Returning to Edinburgh, he studied law, and in 1794 he was admitted to the bar. Though committed to law as his profession, Jeffrey devoted substantial time to avocational interests in the sciences, especially chemistry, on which he made presentations before the academy. It was at Jeffrey's home that the *Edinburgh Review* was first projected, and between 1803 and 1829, he was the editor whose shrewd policies made the *Edinburgh Review* so successful that its circulation exceeded that of the London *Times*, reaching thirteen thousand copies per issue by 1814 and maintaining that level in spite of the appearance of numerous imitators and competitors.[16] Even while serving as editor, Jeffrey continued to practice law, leaving his editorship when he became Dean of the Faculty of Advocates at Edinburgh in 1829.

Among its other founding members, the Academy of Physics numbered the twelve-year-old Leonard Horner (1785–1864), who subsequently established the Edinburgh School of Arts and became president of the Geological Society of London. In 1826, Horner became the first warden of University College London and thus the first scientist to head a British university.[17] In addition, there was Leonard's brother, Francis (1778–1817), who, at age nineteen had already translated Leonard Euler's *Elements of Algebra* into English, and who, along with Sydney Smith (1771–1845), Francis Jeffrey, and Brougham, was one of the founders of the *Edinburgh Review* in 1802. Like Brougham, Francis Horner became a member of Parliament, and he was an outstanding political economist. Finally, there was Thomas Brown (1778–1820), who succeeded Dugald Stewart as Professor of Moral Philosophy at Edinburgh, and who also contributed to the *Edinburgh Review.*

In 1798 the founders were joined by Sidney Smith, an Oxford graduate who was in Edinburgh as companion and tutor to a member of the English landed gentry. While his student took classes, Smith studied with both Stewart and Black, preached in the Charlotte Street Chapel, and hung out with the young members of the Academy of Physics. It was Smith who proposed establishing a new periodical that would reflect their shared views on scientific and social issues. He served as editor of the first number of the *Edinburgh Review,* which appeared in October of 1802, before he departed for London, leaving Francis Jeffrey in charge. Later, Smith became a distinguished clergyman and a famous humorist, but he continued his connection with his Scottish friends by contributing over one hundred reviews, many on geography and travel literature, to the *Edinburgh Review.*

No other group of seven personally connected persons played a greater role in promoting the spread of scientific knowledge and attitudes within all segments of the British population than this group. On the one hand, the *Edinburgh Review* promoted interest in scientific method, natural philosophy, chemistry, geology, and political economy among the intellectual elite by giving the sciences a central place in *the* model British highbrow intellectual journal of the century.[18] At the other end of the social spectrum, the Mechanics' Institute movement and the cheap literature movement promoted by the Society for the Diffusion of Useful Knowledge attempted to bring scientific knowledge and education to members of the working classes.

The vision of how science and scientific knowledge related to all classes in society promoted by the members of the Academy of Physics reflected their Scottish education, incorporating the views of Black and of Stewart in particular.[19] This vision was undoubtedly appropriated selectively and was modified to varying degrees by the groups of patrons and working-class audiences with whom they worked to extend scientific education, but it provides a useful starting point from which to note divergences. The emphasis of the group was principally secular, and their shared expectations regarding human progress focused on progress in knowledge and material well-being rather than on progress toward salvation in an otherworldly sense. Furthermore, they were all advocates of the kind of improving mentality that they attributed to Francis Bacon and that viewed scientific knowledge as the engine driving continuing material progress in society.

The academicians' shared understanding of the relationship between philosophers and artisans was one that had been promoted by Joseph Black and by his teacher, William Cullen (1710–90). It assumed that the philosopher, or scientist, was to be the producer of knowledge that the artisan then applied. Communication between the two groups was critical, for without awareness of what the philosopher had learned, there was no systematic way for the artisan to improve his materials or practices. Black and Cullen thus both sought to teach in a way that made chemical knowledge accessible to the broadest possible audience.[20] This orientation goes a long way to explain why Black's lectures drew up to six hundred auditors at a time and in an institution in which the next most populous classes, the introductory moral philosophy courses, seldom drew one hundred.

Though communication between scientists and artisans was critical, there was no question among members of the Academy of Physics that there was a clear division of labor between the artisan and the scientist, nor was there any question that the proper flow of information was from the scientist to the artisan, rather than vice versa. Black had made this crystal clear every year in the first lecture of his chemistry course: "Chemistry is not an art, but a science. The artisan is he who puts in practice what the philosopher conceives. . . . But he who studies deeply and thinks on rational methods is a philosopher. . . . The study of the mechanical arts has no doubt done a good deal of service to this science *but these arts are more obliged to chemistry than it to them.*"[21]

The view that science education was a one-way, top-down, process was one that was perpetuated by the members of the academy in all of their subsequent promotional activities. It should not be terribly surprising that the condescension of this view produced resistance among the workers who were the supposed recipients of the wisdom of scientific experts and that it sometimes led to attempts to develop autonomous, worker-directed, science and science education.

One reason for the presumption that workers cannot be innovative in their own right was initially articulated by Adam Smith (1723–90), and Smith's argument will lead us to a second and nonvocational reason for extending scientific education to the working poor. When Smith had discussed the growing division of labor, which makes higher productivity and economic progress possible, he had argued that there is a very significant negative impact on the workers when the division becomes extreme: "The man whose life is spent performing a few simple operations, of which the effects too are, perhaps, always the same, or very nearly the same, has no occasion to exert his understanding, or to exercise his invention in finding out the expedients for removing difficulties which never occur. He naturally loses, therefore, the habit of such exertion, and generally becomes as stupid and ignorant as it is possible for a person to become."[22] The lack of stimulation of the mental faculties that goes with work under an extreme division of labor thus leaves the worker's abilities to innovate and improve the processes he is involved with "obliterated and extinguished."[23] But it does something even worse to the worker from the standpoint of moral philosophy: "The torpor of his mind renders him, not only incapable of relishing or bearing a part in any rational conversation, but of conceiving any generous, noble, or tender sentiment, and consequently of forming any just judgement concerning many even of the ordinary duties of life."[24]

Adam Ferguson (1723–1816) and his student Dugald Stewart were, if anything, even more concerned than Smith had been about how the division of labor produces a narrowness of vision, not just for workers, but for employers as well; for the employers become almost equally narrowly and illiberally focused on making a profit.[25] Virtually the only suggestion that Ferguson and Stewart could make to mitigate the profoundly negative consequences of specialization in the workforce was to engage all classes in a "liberal" education that would occupy and stretch their intellects. Indeed, Stewart insisted that the primary aim of all education was not to convey information or to develop specialized vocational competence, but rather, "to cultivate *all* the various principles of our nature, both speculative and active, in such a manner as to bring them to the greatest perfection of which they are susceptible."[26] Only someone who had been exposed to a range of experiences and perspectives could be rendered "happy as an individual, and an agreeable, a respectable, and a useful member of society."[27]

The Mechanics' Institute Movement

It became a chief goal of Stewart's students, Birkbeck, Brougham, Leonard Horner, and their friends, to implement programs for the liberal education of working men to match their mentor's vision. Thus, when Birkbeck initiated his first free course for mechanics in Glasgow, he chose to offer lectures on "the mechanical properties of solid and fluid bodies," because that was what the workers he talked with wanted to hear about. But he did not expect that the knowledge they gained would be of much immediate use to the auditors in their jobs. Instead, he said that he was convinced that "pleasure would be communicated to the mechanic in the exercise of his art, and that mental vacancy which follows a cessation from bodily toil, would often be agreeably occupied, by a few systematic ideas, upon which, at his leisure, he might meditate. It must be acknowledged too, that greater satisfaction in the execution of machinery must be experienced, when the uses to which it must be applied, and the principles on which it operates, are well understood, than where the manual part alone is known, the artist remaining entirely ignorant of everything besides."[28] In this way he indicated that the intellectual stimulation to be obtained from the course was much more important from his perspective than the technical content itself. Similarly, though the curriculum of the Edinburgh School of Arts focused almost exclusively on mathematics and the natural sciences in its early years, its goal was not to teach trades but rather to teach "the various branches of science . . . so that they [i.e., the mechanics] may the better comprehend the reason for each individual operation that passes through their hands."[29] It expressly avoided excessively vocational subjects such as veterinary medicine.[30]

There are many ways in which the Mechanics' Institutes and the complementary Society for the Diffusion of Useful Knowledge failed to achieve what various constituencies hoped they might. To the extent that the initiators sought to engage the mass of unskilled factory workers, who most needed intellectual stimulation according to Smith's and Stewart's views, for example, their projects were largely unsuccessful because they presumed basic levels of literacy and commitments of time that very few unskilled workers were able to offer.[31] Instead, it was largely workers in skilled crafts and trades, clerical workers, shopkeepers, and a sprinkling of gentlemen and ladies who flocked to the Mechanics' Institute courses and purchased the publications of the Society for the Diffusion of Useful Knowledge (hereafter, the "SDUK").[32] But those scholars who have interpreted the tendency for Mechanics' Institutes to increasingly offer literary and more broadly "philosophical" courses and events as they matured as a sign of failure are discounting one of the most important goals of the founders of the movement. For them, the demand for less immediately practical experiences among artisans could be viewed as a sign that the Institutes were succeeding as sources of liberal education.[33]

The initial goals of the Stewart-inspired promoters of the Mechanics' Institute and cheap literature movements were intended to be liberating to members of the lower classes. From the very beginning of the Mechanics' Institute movement, however, there were workers who saw the motives of the middle-class promoters as more repressive than liberating. And there were unquestionably many among local patrons of the movement whose interests were more in procuring a technically capable and socially malleable workforce than in enhancing the quality of life or enhancing the capacity for active citizenship of the working class.[34]

In London, before Birkbeck gained control of the Mechanics' Institute, Thomas Hodgskin, the associate editor of the working class *Mechanics' Magazine*, had argued very strongly that the Institute should be completely mechanics controlled and supported. "Men had better be without education," he argued, "than be educated by their rulers; for then education is but the mere breaking in of the steer to the yoke; the mere discipline of a hunting dog, which, by dint of severity, is made to forego the strongest impulse of his nature, and instead of devouring his prey, to hasten with it to the feet of his master."[35] Those who sought financial support from wealthy patrons, and who therefore emphasized technical education and social control in their promotional materials, carried the day in London. But the constitution of the London Institute, which was widely emulated elsewhere, ostensibly ensured control by the working-class membership by insisting that two-thirds of the Managing Committee should be "working men."[36] At Glasgow, no person who was not a student could serve on the General Committee that governed the institution; but that committee too sought outside patronage.[37]

Generally, the managing committees of Mechanics' Institutes, even when packed with workers, were unwilling seriously to offend patrons. So for the most part, Mechanics' Institute courses and libraries were shaped to steer clear of politically sensitive and religiously sensitive topics. Thus the Institutes served, on the face of it, to support the existing social order, as Hodgskin had feared.

A few Mechanics' Institutes, including those at Keighley, Halifax, and Huddersfield, did manage to avoid obligations to wealthy patrons.[38] Those were more likely to be linked to radical politics and to include newspapers and political materials in their libraries; but they also tended to be very short lived. At Bradford, for example, 1825 saw the attempt by a group of politically and religiously radical "operatives" to establish a Mechanics' Institute controlled by the mechanics themselves. This project failed within a matter of months, but in 1832, a second, long-lived, Mechanics' Institute, emerged. This institute, supported by local dissenting ministers and industrialists, repeatedly emphasized in its reports that it would "preserve rank in society, keep anarchy at bay, and provide the common ground on which different classes and sects could meet without either sacrifice of principle or danger of collision."[39]

One should be a little careful about assuming that because the public pronouncements and the course and library contents of most Mechanics' Institutes were apolitical and socially conservative, the overall impact of the movement was so. Margaret Jacob has made a very compelling argument that eighteenth-century Freemasonry was apolitical in much the same way with regard to the formal content of topics of discussion, and so on. But she has insisted that the process of constitutional self-government in which members of different social classes intermingled in nontraditional ways made the societies of Freemasons a critical source of experience in political action that could be and was transferred into the public domain during the French Revolution and after.[40] Similarly, it may well be the case that the experience of hundreds of thousands of voting members of

Mechanics' Institutes and the more significant and complex experiences of tens of thousands of members of their governing committees created experiences and expectations of self-governance that contributed to greater political activism among the middle and working classes in Victorian Britain.

A secondary consideration may be equally important. As time went on, individual lectures and short courses of lectures tended to replace long courses of lectures for a variety of reasons, including both time and fiscal constraints. Moreover, unpaid lectures by Institute members, often mechanics with some special interest, gradually supplanted paid lectures by professional itinerant lecturers. Thus, by the early 1850s, members were making approximately two-thirds of all presentations, providing important opportunities for members of the working class in public speaking and organizing the presentation of material for mixed audiences.[41]

When we turn to the content of the lectures, classes, and library acquisitions of the Mechanics' Institutes, it is clear that in spite of some commonalities, there were huge regional differences as well as important changes over time. Basic literacy and numeracy were much more extensive in Scotland than in England, so auditors in the scientific and mathematical courses in Glasgow, Edinburgh, Aberdeen, Stirling, and so on, were more likely to be able to benefit from them than working-class students in England, where basic education was less often available. As a consequence, the originator's emphasis on scientific subjects was initially greater in the Scottish Institutes, and it persisted for a longer time there.

In the industrial cities of northern England, on the other hand, lectures and classes were never as heavily focused on the sciences, and the ratio of scientific to nonscientific topics declined relatively fast. In the period between 1835 and 1839, for example, 235 out of 394 lectures (60 percent) at the Manchester Mechanics' Institute were on the physical sciences. During the period between 1845 and 1849, that ratio had dropped to 88 out of 229 (38 percent). At the competing Manchester Athenaeum, which drew a slightly more upper-middle-class audience, the figures were 173 of 352 (49 percent) for the period from 1835 to 1841 and 57 out of 394 (15 percent) for the period from 1842–49.[42]

Because one of the most critical reasons for the relative lack of support for a continuing scientific emphasis in the Institutes of the industrial cities was the deficient preparation of students, many of these urban Institutes set up classes in basic reading, writing, arithmetic, and commercial subjects, as well as "mutual instruction" courses in applied science given by artisans for apprentices. Such courses gradually drew more students than the "regular" scientific curriculum.[43] At Manchester, Leeds, and at Liverpool, the Mechanics' Institutes even sponsored day schools. The school at Liverpool eventually became one the finest academic high schools in Britain. It was certainly the earliest to offer laboratory instruction in chemistry and the physical sciences in 1835.[44]

The Content and Character of Science in the Mechanics' Institutes

Turning from the mix of scientific and nonscientific courses and lectures in Mechanics' Institutes to the character of those that *were* scientific, we can reasonably ask how the science taught to mechanics was related to that available to students in universities or promulgated to the intellectual elite in such journals as the *Edinburgh Review.* At least in the early years, when curricula were most easily shaped by the middle-class promoters and patrons, scientific courses focused almost exclusively on topics from natural philosophy and chemistry, in part because these were the topics closest to the natural scientific interests of the members of the Academy of Physics, and in part because they were topics that seemed most directly related to the work experiences of the primary intended audience of mechanics (defined as those who worked with their hands).

Political economy, which drew much attention in the elite-targeted *Edinburgh Review,* was almost completely absent from the curricula and lectures of the early Mechanics' Institutes because of the fear that it might inflame political passions and lead to disruptive action both inside and beyond the Institutes. When it *was* proposed for inclusion, it was almost inevitably by those followers of Malthus who saw population growth among the poor as the great cause of the economic ills of capitalist society. Thus, Thomas Chalmers (1780–1847), whom we shall meet later in a different context, argued for its inclusion on the grounds that it would demonstrate that "the infuriated operatives, instead of looking to capitalists as the cause of their distress, should look to one another."[45]

Natural history and geology were also all but absent from the Mechanics' Institute courses in the early years, although they accounted for a large fraction of university-level science enrollments and for a substantial fraction of the scientific articles in the *Edinburgh Review.* In these cases the explanation seems to be almost entirely linked to the early utilitarian ideology of the Institutes, for there is good reason to believe that natural history and geology were every bit as interesting to artisans as the more exact sciences[46] and no reason to believe that they were seen as socially threatening, at least prior to the late 1840s. Indeed, both geology and natural history were heavily represented among the SDUK publications.[47] As time went on and lecture topics and courses diverged from the original pattern in the Mechanics' Institutes in response to members' demands, both natural history and geology tended to appear more frequently in the curricula.

Some authors have argued that regardless of topic, the Mechanics' Institutes taught scientific knowledge to working-class students in ways that differed substantially from those by which universities conveyed it to the budding young members of the intellectual elite. Steven Shapin and Barry Barnes have characterized those differences in a particularly clear way: "The central notion, shared by very many of the projectors of Mechanics' Institutes, was precisely this: to show 'how things really are in nature,' rather than to stress, or in some cases even to allude to, the provisional nature of scientific knowledge. The world of workers' science was a world of facts and laws, not a world of theories, so identified. Where Brougham and Horner might orient themselves in a body of scientific knowledge which was partly hypothetical, wholly provisional, and recognized as theoretically informed, the scientific knowledge presented to mechanics was to have none of those characteristics. It was hard, factual, solid and enduring: in no way tentative or revisable."[48]

In addition, Shapin and Barnes have claimed that these differences were intended to enforce social control in a relatively subtle, though ultimately unsuccessful way.[49]

This characterization of the differences between "high" and "low" scientific cultures exaggerates the differences significantly. Black's chemistry courses and those given by his successor, Thomas Charles Hope (1766–1844), to Edinburgh University students were little less inductivist, experimental, and anti-speculative than were those offered in the Mechanics' Institutes, though it is true that there were more speculative chemists, such as Dalton and Davy, working in England. Secondly, though it is also true that John Playfair (1748–1819), for example, spent a substantial amount of time talking about theories and hypotheses in his natural philosophy courses, his major emphasis was on the induction of general facts from particular facts, and the philosopher most frequently appealed to in his methodological discussions was Francis Bacon, who also received accolades in the Mechanics' Institutes.[50]

In the contexts of their moral philosophy classes, Scottish university students would have had more extended discussions of the relativity and revisability of scientific theories,[51] and in their articles for the *Edinburgh Review,* Scottish scientists in particular were inclined to explore complex issues in the philosophy of science, including the legitimacy of certain kinds of hypothetical argumentation. But within their science courses themselves, Scottish university students were exposed to very nearly the same demonstration-experiment based, inductivist-oriented pedagogy that was used in the Mechanics' Institutes. Ian Inkster has also argued that the early Mechanics' Institute courses drew heavily on the professional corps of independent lecturers who purveyed scientific knowledge to the upper middle classes through the Literary and Philosophical Society network, and there is no evidence that they changed their course content or pedagogy significantly to suit different audiences.[52]

Many of the middle-class promoters of working-class scientific knowledge certainly did hope that such knowledge would promote social stability. The *Edinburgh Review* article, "Thoughts on Popular Education," from 1825, undoubtedly spoke for many in arguing that "knowledge of science for the lower classes, like wealth for the rich, gives them a direct interest in the peace and good order of the community, and renders them solicitous to avoid whatever may disturb it."[53] It is, however, not at all clear that the more inductive and less theoretical features of the science taught to working-class people in the Mechanics' Institutes had anything to do with conscious attempts at social control. Indeed, many of those same features were adopted as characteristics of worker-initiated science and used to attack the pretentiousness of academic science and to promote a radically democratic understanding of the scientific enterprise that would have been anathema to many of the patrons of the Mechanics' Institutes.[54] The constant emphasis on experiments as a source of scientific knowledge that pervaded the pedagogy of almost all Mechanics' Institute lecturers even seemed to some working men to offer an advantage to the working man in scientific discovery: "The experimentalist has to put up forges, or furnish laboratories, at great trouble and expense; but the smelter, the blacksmith, the founder, the glass blower, and a hundred other mechanics and operatives, have all this apparatus daily before them, and therefore, without any trouble, might sound the depths and scan the heights of knowledge. Nothing would be required but a little observation."[55]

Science in the *Edinburgh Review*

When we turn to science as it was presented within the pages of the *Edinburgh Review,* we do find a substantial difference from that found in the Mechanics' Institutes. As Shapin and Barnes suggest, *this* science was portrayed as much more theory dependent and provisional. Thus, in his favorable review of Playfair's *Illustrations of the Huttonian Theory of the Earth* in 1803, Brougham wrote: "It cannot be denied . . . that observations accumulate but slowly when unassisted by the influence of system. The observer never proceeds with more ardour than when he theorizes; and every effort to verify or disprove particular speculations necessarily leads to the evolution of new facts and to the extension of the limits of real knowledge. Hence it seems to be the business of real philosophy rather to point out the imperfections, to detect the errors, to restrain the presumptuousness of the theorist, than to extinguish altogether a spirit, which, however incomplete and insufficient may be the materials on which it has to work, must at least facilitate generalization, and render the approach to truth less tedious."[56]

Similarly, in his review of Auguste Comte's *Cours de Philosophie positive* in 1838, David Brewster (1781–1868) chose to emphasize the use of heuristic hypotheses that did not meet the Positivist's criteria of acceptability, which had focused on the observability of any posited entity: "If the mind rests on any hypothesis as disclosing the real cause of the phenomena which it explains, and thus paralyzes all our efforts in searching for any other, and perhaps the true cause; then science has not in this case performed her proper functions. But if the undulatory, or any other hypothesis, is adopted, and used only as a temporary auxiliary—as a bond of cement which unites a number of isolated facts, or even as a fertile principle which may indicate new phenomena—science cannot be considered as having overstepped its limits."[57]

Contributors to the *Edinburgh Review* frequently disagreed regarding the utility and appropriateness of particular hypotheses and theories. Brougham, for example, never accepted the wave theory of light or the notion of a luminiferous ether that the wave theory seemed to imply. David Brewster, on the other hand, could accept both as provisional. But all unquestionably did communicate their understanding of scientific knowledge as "partly hypothetical, wholly provisional, and recognized as theoretically informed."[58]

As a general literary periodical rather than a strictly scientific journal, the *Edinburgh Review* naturally covered a much wider range of subjects than those of interest to the early Mechanics' Institutes. Among those topics usually referred to as scientific or philosophical, the most important for the *Edinburgh Review* was undoubtedly political economy, which was downplayed in the Mechanics' Institutes. According to the historian of economic thought, Frank W. Fetter, "For the better part of a half century, it was the closest approach that Great Britain had to an economic journal. It reviewed practically all the significant economic literature as it appeared, and it discussed the great economic controversies of the day. . . . Its rivals laid less emphasis on economic subjects than did the *Edinburgh.* Their economic articles were less numerous, and they did not have the tone of authority or the influence on public Opinion."[59]

Science and "Orthodox" Religion in England: 1802–58

The distinctive relationships between science and both religious developments and governmental operations that characterized early Victorian Britain were associated in large part with the special character of Oxford and Cambridge universities and the roles of "Oxbridge" graduates in public life. By the beginning of the nineteenth century, most Continental and Scottish universities included professional schools of law and medicine as well as theology. Furthermore, almost all teaching, even in the pre-professional liberal arts curriculum, was done by a specialized professorate. Finally, at least a substantial number of students in most of those universities, and virtually all students in such specialized institutions as the *École polytechnique* and the various schools of mines cropping up throughout Europe, had well-focused career goals in mind, for which specific training was sought. Indeed in Scotland, especially, it was atypical for a student to complete a degree.[60] Students matriculated, took the courses they wanted to serve their own purposes, and then left.

Oxford and Cambridge, on the other hand, had initially been developed in the Middle Ages almost exclusively to provide an educated clergy for the Catholic Church. When the Anglican Church split off from the Roman Catholic Church in the sixteenth century, Oxford and Cambridge remained primarily religious institutions, though over time they became increasingly important in providing a general moral and liberal education, not vocational training, for "gentlemen." Legal education in England was accomplished primarily through apprenticeship arrangements at the Inns of Court in London, and entry into the profession did not presuppose a university liberal arts background, though some aspiring lawyers did get a university degree before they went to London.

Similarly, most English physicians and surgeons trained through apprenticeship throughout the eighteenth century (though membership in the elite Royal College of Physicians generally presupposed an Oxbridge background until 1835). The growing number of middle-class students in the early nineteenth century who sought university training in medicine and surgery went north to study at Edinburgh, which probably provided the best medical education in the world by the end of the eighteenth century, or, when the University of London was established, they either studied there or in one of a number of private medical colleges that sprang up.

A small number of professorships in special subjects had been endowed at Cambridge and Oxford by the beginning of the nineteenth century, and these included some in mathematics and the sciences; there were nine at Cambridge by 1800.[61] The terms of professorships, however, seldom required residence in Oxford or Cambridge. Usually they stipulated that the professor give a few lectures each year, and during the early nineteenth century it was not unusual for professorial lectures to have no student auditors at all. In other cases, there was no formal educational requirement. Thus, Charles Babbage, who held the Lucasian Professorship of Mathematics at Cambridge from 1828 to 1839, never gave a single lecture. A few science professors took their teaching obligations seriously. At the beginning of the century, for example, the popular E. D. Clarke (1769–1822) drew two hundred students to his descriptive mineralogy lectures, but his successors, John Henslow (1796–1861), William Whewell (1794–1866), and William Miller (1801–80) drew far fewer students, and in the 1830s and 1840s, as Britain entered into an economic depression, attendance at all science lectures dwindled to near nothing.[62]

Throughout the first half of the nineteenth century, tutors (who were fellows of the residential colleges) did most of the teaching at Oxford and Cambridge. Moreover, a standard practice was for tutors to coach their students in all subjects in the liberal arts curriculum. There was only one universal requirement to be a fellow of one of the Oxbridge colleges and thus to be eligible to be a tutor: one had to take orders in the Anglican Church. As a consequence, all teachers at the two English universities were Anglican clergy, usually chosen not for any special competence they had in a single academic subject, but rather for their moral uprightness and their breadth of knowledge across the curriculum, which focused heavily on classical languages and literature and moral philosophy, with some mathematics and a very small amount of natural science.

These features of Oxbridge education had two important and, to some extent, contradictory, consequences for the place of science in English upper- and upper-middle-class culture. On the one hand, they meant that most products of Oxbridge education had little exposure to mathematical and scientific learning and that the exposure they did have was usually at the hands of tutors whose interest in these subjects was marginal. The ambitious students in this majority category took the classics honors examinations, high placement in which was often a key to entrance into government service or service in the East India Company, and they often retained an antipathy to science and scientists throughout their careers.[63]

On the other hand, the strong liberal or Broad Church Anglican emphasis on natural theology that had developed during the seventeenth and eighteenth centuries culminated in William Paley's (1743–1805) *Natural Theology, or Evidences of the Existence and Attributes of the Deity Collected from the Appearances of Nature* (London, 1802), a book that was required reading for divinity students and was among the favorite readings of the young Charles Darwin. Paley's work remained central in the study of Divinity at Oxbridge through the first three quarters of the nineteenth century, and it continued to promote some interest in the natural sciences within the clergy for at least that long. Indeed, during most of the first half of the nineteenth century there were small but intellectually vital communities of scientifically active clergy connected with both Oxford and Cambridge (of the approximately two hundred fellows actually teaching at Oxford and Cambridge at any given time during the 1830s, perhaps fifteen, or 7.5 percent, published works in the natural sciences).[64] These men not only stimulated many of their brightest students to become interested in the natural sciences, they also encouraged those who were going into the church to sustain an interest in natural theology.

At Oxford, William Buckland (1784–1856), who became reader in mineralogy in 1813, was the most distinguished of the group. His published works, including *Reliquiae Diluvianae* (1823) and *Geology and Mineralogy Considered with Reference to Natural Theology* (1836), focused on finding geological evidence to support the biblical flood story, and his students included Charles Lyell, probably the most distinguished geologist of the century. Baden Powell (1796–1860), Savilian Professor of Geometry from 1827 to 1860, did substantial work on radiant heat in the infrared region and on the dispersion of light and published *The Connection between Natural and Divine Truth* in 1838. Charles Daubeny (1795–1867), one of the early promoters of the British Association for the Advancement of Science, was also among the most vocal early supporters of Dalton's atomic theory. In 1822 Daubeny became professor of chemistry at Oxford; in 1834 he added the professorship in botany; and in 1840 he also added the chair in rural economy, holding all three chairs until his death in 1867.

At Cambridge in the 1830s the group of scientifically oriented fellows was both larger and generally more distinguished. It included the mathematician and computing innovator, Charles Babbage (1792–1871); the astronomers George Biddell Airy and John Herschel; the chemist James Cumming (1777–1861); the geologist Adam Sedgwick; and the botanist, John Henslow, who was Darwin's closest mentor. The universally acknowledged leader of the Cambridge scientists, and the man who coined the word "scientist," was William Whewell. Author of *Astronomy and Physics Considered with Respect to Natural Theology* (1833), Whewell was almost the ideal Oxbridge fellow because of his command of virtually all aspects of the curriculum. His more than thirty monographs covered topics in mathematics, mechanics and dynamics, mineralogy, architecture, moral philosophy, German literature, political economy, history and philosophy of science, as well as natural theology.

Not all Oxbridge scientific clergy spoke with a single voice on any issue. Nonetheless, their common simultaneous commitments to Anglican Christianity, broadly conceived, and to the pursuit of science, as well as their constant polemicizing regarding the mutual support given to each by the other, ensured that at least within the English intellectual and clerical elite there was no widespread feeling that science and Christianity were incompatible or conflicting until after mid century.

Though institutional contexts were quite different in Scottish universities than at Oxford and Cambridge, two special sets of considerations ensured that students at Aberdeen and Edinburgh in particular, were likely to be convinced that scientific activities and both moderate and evangelical Presbyterian religion were mutually supportive. By the early nineteenth century almost none of the natural philosophy, mathematics, chemistry, or natural history professors in Scotland were also clergy. In fact, a few, such as John Leslie (1766–1832), professor of mathematics from 1805 to 1819 and then of natural philosophy at Edinburgh from 1819 to 1833, were openly irreligious.[65] On the other hand, the moral philosophy curriculum, which was the centerpiece of Scottish undergraduate education, was conceived and taught during the first third of the century by a series of powerful intellectuals and popular teachers who collectively go under the title of "Common Sense Philosophers." And it was a central aim of Common Sense Philosophy to demonstrate the compatibility between science, especially Newtonian science, and Christianity.[66]

At Aberdeen, in the moral philosophy classes of Thomas Reid (1710–96), of James Beattie (1735–1803), and of his son and his grandson, as well as at Edinburgh in the moral philosophy classes of Dugald Stewart from 1785 to 1810 and of Thomas Brown from 1810 to 1820 and in the Logic classes of Sir William Hamilton (1788–1856) from 1836–56, students studied what we now call the philosophy of science and natural theology as well as other topics.[67] Many of these students, including the natural philosophers David Brewster, James Clerk Maxwell (1831–79), Balfour Stewart (1828–87), and Peter Guthrie Tait (1831–1901); the neurophysiologist, Charles Bell; and the chemist William Prout (1785–1850), went on to become outstanding both as scientists and as promoters of Christianity.

Maxwell, for example, in introducing his natural philosophy lectures at Aberdeen in the 1850s, specially emphasized their value for students intending to go on to study theology: "Those who intend to pursue the study of Theology will also find the benefit of careful and reverent study of the order of Creation. They will learn that though the world we live in, being made by God, displays his power and his goodness even to the careless observer, yet that it conceals far more than it displays, and yields its deepest meanings only to patient thought."[68] Stewart and Tait produced one of the most intriguing natural theological texts of the second half of the century in their *The Unseen Universe, or Physical Speculations on a Future State* (London, 1885). Both Bell and Prout wrote natural theological works sponsored by the estate of the Earl of Bridgewater.[69]

Almost equally important in ensuring that religion and science would continue to be seen as intimately linked within Scottish intellectual life through most of the nineteenth century was the appointment of Thomas Chalmers to the professorship of Divinity at Edinburgh from 1828 to 1843. Born in 1780, Chalmers had studied Divinity at St. Andrews and had been licensed to preach by the time he was nineteen. But he also had serious interests in mathematics and natural philosophy, so he came to Edinburgh to study mathematics with John Playfair, natural philosophy with John Robison (1739–1805), and chemistry with Charles Hope. After a brief stint as assistant to the professor of mathematics at St. Andrews and as a private mathematics teacher, Chalmers became minister of the Tron Church in Glasgow. While there, he preached a series of sermons on the relationship of astronomy to Christian Revelation. These became so popular that when he published them as *A Series of Discourses on the Christian Revelation, Viewed in Connection with the Modern Astronomy* in 1817, they sold twenty thousand copies in the first year.[70]

His experiences in a poor Glasgow parish interested Chalmers in social reform and led him to develop interests in political economy, so when he returned to St. Andrews as professor of moral philosophy, he added political economy to the moral and religious topics covered. He came to Edinburgh in 1828. In 1834 he published his *On the Power, Wisdom, and Goodness of God, as Manifested in the Adaptation of External Nature to the Moral and Intellectual Constitution of Man,* and over the next decade he turned the school of Divinity at Edinburgh in new, and increasingly evangelical, directions. The school's curriculum emphasized natural theology, but the school also incorporated political economy into its work on practical pastoral education, and it initiated a chair in modern biblical criticism. As a consequence of his leadership of an evangelical movement within the Scottish Presbyterian Church, Chalmers was ejected from the church in 1843, becoming Moderator of the Scottish Free Church and resigning from his position in the university. His curriculum reforms, however, continued to inform the education of Scottish clergy. As at Oxbridge, then, during the early nineteenth century, both the scientific and clerical elites of Scotland were being educated to believe that their activities were mutually supportive, and the movement back and forth from scientific to religious activities was quite common in individual careers.

While it may be true that many scientists throughout Britain, like many in the confederation of German states, were largely indifferent to religious issues, when an English or Scottish scientist spoke out on a religious issue during the period between 1830 and 1860, it was almost certain to have been in defense of some variant of organized Christianity, whereas on the Continent the probability was high that it would have been in opposition.

The Role of the Bridgewater Treatises

In discussing the way in which science and religion tended to be seen as mutually supportive within both English and Scottish institutions of higher education during the early nineteenth century, I mentioned William Buckland's *Geology and Mineralogy Considered with Reference to Natural Theology,* William Whewell's *Astronomy and General Physics Considered with Reference to Natural Theology,* and Thomas Chalmers's *On the Power, Wisdom, and Goodness of God, as Manifested in the Adaptation of External Nature to the Moral and Intellectual Constitution of Man.* These three works of natural theology represent a diversity of scientific subject matters as well as a substantial range of theological positions, extending from Buckland's biblical focus, through Whewell's Broad Church probabalism, to Chalmers's evangelical emphasis on the human condition. Yet all three works had been commissioned by Davies Gilbert, president of the Royal Society of London, along with the Archbishop of Canterbury and the Bishop of London, as treatises on "the power, wisdom and goodness of God, as manifested in the creation," under the terms of the will of the eighth Earl of Bridgewater, Francis H. Egerton (1736–1803).

The eight commissioned Bridgewater Treatises plus the unsolicited *Ninth Bridgewater Treatise: A Fragment,* produced by Charles Babbage in 1837, constituted one of the most highly visible vehicles by which natural theology extended its reach beyond the universities and the clergy in Britain to spread the view of harmonious cooperation among the sciences and religion, and they served to promote what Robert M. Young has called a "common intellectual context," at least within the middle- and upper-class reading public in Britain.[71]

Davies chose his authors from across a broad spectrum of scientific as well as theological positions—even selecting the anti-Newtonian theologian, William Kirby, to cover the religious implications of the habits and instincts of animals. Moreover, he gave the authors little direction regarding the audience toward which they should aim their works. As a consequence, some clearly aimed at an extremely well-educated audience and supposed that readers could read Greek and knew classical literature. Others used their treatises in part to report new scientific findings of interest primarily to their scientific peers. Still others tried to be accessible to a much wider readership. In spite of all these divergences, however, all of them sought, and in varying degrees succeeded, in Whewell's words, "to lead the friends of religion to look with confidence and pleasure on the progress of the . . . sciences, by showing how admirably every advance in our knowledge of the universe harmonizes with the belief of a most wise and good God."[72]

Initially, the authors and Davies expected relatively small sales to a discriminating and wealthy audience of aristocrats and gentlemen, and the prices of the volumes (they averaged almost one pound) were high enough to ensure that they could not be purchased by working-class readers, whose weekly income was likely to be under 30 shillings per week.[73] But demand turned out to be vastly greater than expected. More than sixty thousand of the high quality editions were sold over a fifteen-year period before the *Treatises* were brought out in cheap editions in the early 1850s, making them available to nearly anyone.[74]

By general agreement, none of the treatises was particularly original with respect to its theological arguments. Their collective importance lay rather in the fact that they tended to explicate contemporary science in a religiously safe way. One consequence of this feature of the Bridgewater Treatises is that they were often used as texts in the scientific education of the working classes, either in Mechanics' Institute courses or through extracts published in such popular periodicals as the *Youth's Magazine and Evangelical Miscellany,* the High Church *Saturday Magazine,* or the liberal *Penny Magazine.*[75]

Science and Religious Heterodoxy in Early Victorian Britain

Science and mainstream variants of Christianity were widely advertised as mutually supportive within the religious center of early Victorian society, even though within that broad center of the religious spectrum there were fairly significant disagreements. From the perspective of this work, however, there were equally interesting relationships between science and the religious views of those who stepped well beyond the limits of orthodox Christianity into Free Thought, Secularism, Robert Owen's (1771–1858) Universal Community Society of Rational Religionists, Unitarianism, or Comte's Religion of Humanity.

Abandonment of traditional Christianity increased both within the working class and among middle-class intellectuals throughout the nineteenth century, even though middle-class church attendance rose in general.[76] It has been widely held that different explanations account for lower-class rejection of Christian orthodoxy and for the rebellion of the intellectuals. Most working class religious non-attendees were probably disinterested in religion altogether, but an important segment was actively hostile to traditional religion because it had shown itself all but totally uninterested in improving the lot of the working poor. The Secularist preacher, Robert Cooper (?-1868), spoke the nearly universal opinion of this group when he wrote: "As a *class,* willingly or unwillingly on their part, [the Christian clergy] are the stumbling block in the way of every effort to enlighten and emancipate mankind. Talk of *social* reform, and they exclaim that poverty is a *divine* ordinance; that God made both *poor* and rich, and that the people must, therefore, 'be content in the situation in which Divine Providence has placed them.' Talk of *political* reform, and they remind you that it is our duty, by command of the inspired word of heaven, to submit 'to the powers that be.'"[77]

The traditional account of the causes of growing unbelief among early Victorian intellectuals is quite different. It was characterized and only to a slight extent caricatured by Howard Murphy as follows: "The prevailing impression seems to be (a) that, because Lyell and Darwin had showed that neither the origin of the earth nor the origin of man as described in Genesis can be reconciled with the findings of science, therefore, thinking people became atheists or agnostics; and (b) that, because a number of German Scholars had shown that neither the Old nor the New Testament can be taken at face value, therefore honest men had no recourse but to abandon Christianity altogether."[78]

Almost nothing about this interpretation seems correct. First, the retreat from orthodox Christianity accelerated among Victorian intellectuals during the 1830s and 1840s, before Darwin's work was known and before the works of German higher criticism had been translated into English. No one that I know of read Lyell's work as a challenge to Christianity. Moreover, Mary Anne Evans, better known as George Eliot, did the first English translations of both David Strauss's *Life of Jesus* (in 1846), and Ludwig Feuerbach's *The Essence of Christianity* (in 1854), as a *consequence* of her turn away from evangelical Christianity, rather than as a prelude to it.[79]

When the biographical details of those middle- and upper-class intellectuals who turned away from Christianity before the 1860s are explored, many do show a common pattern. A religious crisis initially occurred when the person recognized a fundamental conflict between some religious dogma, frequently the doctrine of election or that of vicarious atonement, and a sense of justice or compassion grounded in the assumption that the life of humans on earth both can and should be improved through human intelligence and effort. That is, they turned away for much the same reason that members of the lower classes did, and they were quite likely to be strong supporters of working-class causes, though frequently in a condescending way. Following their religious crisis these individuals usually turned away from orthodox Christianity, but not from religion, and they then sought, after the fact, to appropriate evidence from contemporary scientific developments to support their new position and to attack the old.[80]

Those who have read chapter 5 will recognize this pattern as that followed by David Strauss and Ludwig Feuerbach in Germany as well. Strauss, for example, had argued in his *Life of Jesus* of 1835 that taking the Gospel stories of Jesus as factual distorted the ethical message of Christianity. Subsequently, in his *Doctrine of Faith* of 1840, he began to appeal to scientific evidence to challenge the Genesis account of the origins of humanity. Similarly, in his *Thoughts on Death and Immortality* of 1830, Feuerbach focused on the way in which the traditional Christian doctrine of life after death undermined people's will to help themselves and one another in this world. Only later, and especially in his *Essence of Christianity* of 1839, did he begin to style himself as a "natural philosopher in the domain of the mind" and to use what he understood as the methods of the sciences to undermine traditional theology in favor of a purely humanistic religion. Science thus did come to play an important role in anti-Christian apologetics in nineteenth-century Germany and Britain, much as natural philosophy had played an important role in Christian anti-Pagan apologetics in the early years of Christianity.[81] But interest in science, by itself, rarely if ever seems to have initiated religious crises. More frequently religious crises led to a search for alternative sources of value, which often ended in some form of scientific position, either secular or religious.

When individuals moved away from traditional Christianity in nineteenth-century Britain, a whole series of alternative religions, including Swedenborgianism, Unitarianism, the Society of Rational Religionists, Secularism, and the Comtean Religion of Humanity were available to offer institutional homes for those who still felt that they needed some organized religion. All of these groups attacked Christian theology on the grounds that it "fixed men's thoughts upon a visionary future, rather than on the world around them,"[82] and all of them to some degree viewed scientific knowledge as the foundation of a more socially responsible religion. The Rational Religionists even designated their meeting places as "Halls of Science" rather than as chapels, churches, or meeting halls.

The Religion of Humanity (Again)

Of these alternative religions, the Religion of Humanity was one of the most significant, not so much because of its numbers (the subscribers to the sacerdotal fund of the main London church never exceeded 137 and there were never more than six or seven English congregations outside of London), but because its members and friends had a disproportionate impact on Victorian culture through their lectures and literary productions.[83] The influence of the Religion of Humanity, or the Positivist Church, peaked in England near the very end of the century; but I treat it here because it emerged before 1860 and because it remained essentially unchanged throughout the remainder of the century in spite of subsequent developments in the sciences.

In chapter 3 we saw that Comte believed that Christianity had played a critical role in promoting a more universal morality than earlier civic religions had offered, but he believed that Christianity had failed to evolve to meet the religious needs of the industrial age. Furthermore, he believed as his mentor Saint-Simon had argued, that left unregulated, the egoistic behavior encouraged by the political economists must eventually lead to a despotism of the wealthy. There was a critical need for a new religion to offer an altruistically based morality suitable to the needs of the modern age. The fundamental goal of Comte's Religion of Humanity was thus to promote and strengthen the altruistic, social, or sympathetic passions relative to the animalistic or egoistic ones and to do so by redirecting human feeling, thought, and action.

On the emotional level, Comtean worship emphasized the subordination of every individual human being to a greater unity, the Universe, as well as the connection between each individual and all of humanity, past, present, and future. Moreover, it dealt with the critical issue of death through the notion of "subjective immortality," or ongoing life in the memory of others. It also built altruistic feeling through a new form of "subjective fetishism"—the worship of objects such as the busts of famous scientists, whose sacred character was self-consciously understood to be subjective rather than objective. Finally, it used artistic productions, especially literary ones, to heighten altruistic emotions by creating models of behavior for people to emulate.

On the intellectual level, theology was succeeded by the natural and social sciences, which replaced superstition with the knowledge of when one must become resigned to endure the inevitable and how one might act to change that which is capable of change. Finally, at the level of action, the Religion of Humanity sought to convince those with wealth to use that wealth for the benefit of others. It also aimed to convince those in positions of power and authority to subordinate political action to moral considerations. Additionally, as the material improvements made possible by the exploitation of natural knowledge allowed humans to meet their material needs with less and less labor, it proposed to redirect human activity increasingly away from unsatisfying work into artistically creative forms of self expression. Virtually all of these goals were ideally suited to respond to the hopes of those middle-class intellectuals who had faced crises of Christian faith because of a perceived conflict between Christian doctrine and social justice, but who needed some kind of religious support and who could not identify with the extreme egalitarianism or the potential violence associated with most working-class movements.

A handful of professional scientists, including Alexander Williamson (1824–1904), professor of practical chemistry at the University of London; George Allman (1812–98), professor of natural history at the University of Edinburgh; and Alexander Bain (1818–1903), psychologist and professor of logic at the University of Aberdeen, went through brief periods of fascination with the Religion of Humanity early in their careers; but each split away in reaction against Comte's authoritarianism. On the other hand, though literary figures John Stuart Mill and Henry Lewes both rebelled against many of the details of Comte's proposals for worship and ritual and against Comte's authoritarianism, they tended to be sympathetic enough to many Positivist positions, including the general goals of the Religion of Humanity, that they continued to offer qualified support. Mill wrote in 1848 to John Nicol that Comte had provided the grounds for believing that "the *culte de l'humanité* is capable of fully supplying the place of a religion, or rather, (to say the truth) of *being* a religion—and this he has done, notwithstanding the ridiculousness with which everybody must feel in his premature attempts to define in detail the *practices* of this *culte*."[84] Similarly, in his diary for January 4, 1854, Mill wrote that, "there is no worthy office of a religion which this system of cultivation does not seem adequate to fulfill."[85] What seems to have held Mill back from any public pronouncements favoring the Religion of Humanity, at least until the 1860s, was his feeling that openly acknowledging his atheism would severely damage his own reputation.[86]

Lewes, already a Unitarian when he was introduced to Comte's works through Mill, never formally renounced his Unitarianism, but he became an even more avid promoter of Positivism than his mentor. Lewes, like Mill, complained about the details of Comte's religious plans, but he and his unofficial wife, George Eliot, attended lectures and worship services at the Chapel Hill Positivist "church" from its formal establishment in 1859, until Lewes's death in 1878.

Both Mill and Lewes had been raised under religiously heterodox circumstances. Their initial attraction to Comte came primarily through *The Positive Philosophy*. And though sympathetic to the goals of the Religion of Humanity, neither became an enthusiast on its behalf. The cases of Richard Congreve, who was to become the leader of the Religion of Humanity in England, and of George Eliot, whose novels did vastly more to promote its principles than all of Congreve's lectures and didactic writings, were quite different. Congreve (1818–99) was educated at Rugby before attending Oxford, where he became a fellow of Wadham College in 1844 and a tutor in 1849, after a brief stint as master of Rugby School. He was very well liked as a tutor and seemed to be "on the high road to the highest preferment in the Anglican Church," in 1851.[87] He had been introduced to Comte through Mill's *System of Logic* (1843) and Lewes's *Biographical History of Philosophy* (1845–46), and he had even met Comte on a visit to Paris in 1849. At that time he did not see Comte as a religious leader. In 1852, however, Congreve seems to have suffered a severe crisis of faith focused around a sense that Anglicanism in particular and Christianity in general did not adequately promote benevolence and philanthropy.

At this point Congreve again visited Comte, becoming attracted to Positivism as a religious system. He read and eventually translated the *Catechism of Positive Religion*. Then, in 1854 he resigned his position at Oxford to marry. Over the next two years he became increasingly committed to the Religion of Humanity in all of its details and rituals. In 1857 he formally resigned his orders in the Anglican Church, and in the same year, with Comte's personal encouragement, he wrote the anti-imperialist tract, *India*, which drew several future Indian Civil Service officers to Positivism. Finally, in January of 1859, Congreve began the first services of the Religion of Humanity in his own home and began identifying himself as "*un Ministre de la Religion de l'Humanité.*" In order to meet Comte's conditions to become a priest of the Religion of Humanity, Congreve began to study medicine, becoming a member of the Royal College of Physicians in 1866.[88] Although Congreve was apparently anything but a dynamic and charismatic speaker, his lectures did attract up to seventy auditors and a number of strong supporters, including Frederick Harrison (1831–1923), lawyer and Trades Union supporter, who would eventually provide the vigorous leadership that spread interest in the Religion of Humanity to the provinces.[89]

In spite of the long-term failure of the Religion of Humanity to attract a large following, it almost certainly did have a significant impact in promoting the growing importance of scientific authority in Victorian Britain through the literary productions of its members and its friends. Almost all of the novels of George Eliot and George Gissing (1857–1903), and to a lesser extent, those of Thomas Hardy (1840–1928) and George Meredith (1828–1909), promoted both specifically Positivist doctrines and a general submission to scientific expertise, and the thousands of articles published by Congreve, Harrison, Eliot, Lewes, and others in the literary periodicals of the nineteenth century kept the Positivist enthusiasm for science and its meliorist message before the public.

Owenism, the Cooperative Movement, and Early British Socialism

During the period between the 1820s and the 1850s, the Mechanics' Institute movement and the Society for the Diffusion of Useful Knowledge were engaged in promoting working-class scientific knowledge at least in part as a means to encourage the development of productive working-class citizens who were respectful of both the extant stratification of society and orthodox religion. At the same time there was another, radical, working-class movement that was deeply political and often openly hostile to religious orthodoxy. That movement was guided by a particular version of social science promulgated by Robert Owen and was taught in the Halls of Science that constituted its meeting places during the 1840s.

In the opening decades of the nineteenth century, British working-class radical politics was shaped by the rationalist, republican, and largely anti-clerical literature of the French *philosophes* and of Thomas Paine (1737–1809) on the one hand, and by social crises brought on by early industrialization on the other. In this second category, the technological unemployment of skilled workers, primarily in the woolen industry, led during 1811 and 1812 to the famous Luddite rebellion in which groups of unemployed workers broke into a number of mills and destroyed the machinery that was displacing them. The cotton mills, of course, had shown a virtually omnivorous appetite for unskilled labor from the late 1780s until 1813, creating a scandal through the employment of huge numbers of children, some as young as six and seven years of age, under terrible conditions. Thus, one of the key ongoing issues of working-class radicals was the improvement of working conditions, especially for children. A particularly severe crisis came in 1813–14, after the defeat of Napoleon, when the market for military uniforms collapsed at the same time that huge numbers of soldiers came home to seek jobs in an industry in a deep slump.

This situation produced a triple set of interrelated problems: one, how to cope with the basic needs of masses of unemployed or underemployed people; two, how to keep wage rates from dropping below subsistence level in the face of an oversupply of labor; and three, how to pull the economy out of the depression. In this environment Robert Owen developed and promulgated a set of social doctrines that were grounded in associationist psychological doctrines.[90]

Owen was born in 1771 in Central Wales to a saddler. A precocious child, he was acting as a teacher's assistant in a local school by the time he was seven. By age ten, he informs us in his autobiography, he had decided that "there must be something fundamentally wrong in all religions," because they seemed more to increase the amount of hatred and misery in the world rather than the amount of happiness.[91] In 1781 Owen left home for good. After a few weeks of traveling, he became a shop assistant to a high-class linen merchant at Stamford in Lincolnshire. A few years later he headed for Manchester, a major center of commercial and industrial growth. With a hundred pounds that he managed to borrow from a brother, Owen became a partner in a business that made spinning machinery. When the business failed in 1791 as a consequence of the incompetence of his partner, he took three of the machines and set up his own small spinning factory.

Within a year Owen had used the success of his tiny factory to secure a job managing a five-hundred-worker factory. Soon, Owen had been elected a member of the Manchester Literary and Philosophical Society, where his interests focused on factory management and rational education for the improvement of humanity, that is, on topics associated with associationist psychology and political economy. Then in 1798, at age twenty-seven, Owen, with financial backing from two relatively progressive factory owners, purchased "the land, village, and mills of New Lanark."[92] What Owen proposed to do was to use his control of the mill environment to demonstrate that productivity could be radically increased through education and the creation of community sentiment.

From the beginning, New Lanark was a financial success, but Owen had trouble convincing his partners that his expenditures on high-quality worker housing, on education, and on community buildings were justifiable. So Owen arranged for another group of backers, including Jeremy Bentham, who were more interested in the social experiment than in maximizing short-term profits, to buy out his initial partners. Eventually the autocratic Owen quarreled with these partners as well, but in the meantime New Lanark was a rousing success, both as a commercial enterprise and as a social experiment. Visitors came from all over the world (more than twenty thousand in the ten years from 1809–19 signed the visitors' book) to observe what was being done.

The following extract from the report of a delegation from the Leeds Poor Law Guardians gives a sense of what they found:

> In the education of the children the thing that is most remarkable is the general spirit of kindliness and affection which is shown towards them, and the entire absence of everything which is likely to give them bad habits, with the presence of whatever is calculated to inspire them with good ones; the consequence is that they appear like one well-regulated family, united together by ties of the closest affection. We heard no quarrels from the youngest to the eldest; and so strongly impressed are they with the conviction that their interest and duty are the same, and that to be happy themselves it is necessary to make those happy by whom they are surrounded, that they had no strife but in the offices of kindness. . . . We could not avoid the expression of a wish that the orphan children in our workhouses had the same advantage of moral and religious instruction, and the same prospects of being made happy themselves and useful to the families in which they are placed.[93]

After about a dozen years as managing partner at New Lanark, Owen, whose credibility was high because of his success as a businessman, began to carry his doctrines to broader audiences through lectures, through the publication of numerous pamphlets and articles in periodicals, and through direct attempts to promote legislation to improve working conditions and to protect children from destructively exploitative practices. In *A New View of Society; or Essays on the Principle of the Formation of the Human Character,* published in 1813, Owen announced the simple foundational assumptions and principles that had guided his experiments at New Lanark and that would continue to inform all of his subsequent theories and plans for implementation.

He argued that humans were morally plastic and that it was their experiences that ultimately produced their moral character. As a consequence, Owen insisted that "any general character, from the best to the worst, from the most ignorant to the most enlightened, may be given to any community, even to the world at large, by the application of proper means; which means are to a great extent at the command and under the control of those who have influence in the affairs of men."[94] It was not the case that humans were without natural instincts. Quite the contrary. Human behavior was subject to natural laws and it was particularly important in Owen's view that systems of law and morality be revised to take cognizance of the "Science of Man" that discovered these laws, for it was the failure to do so that had led to the creation of positively harmful religions and laws in the past.[95] In addition, Owen argued that humanity is continuously passing from ignorance toward ever greater intelligence, and in that process man [sic] is discovering that "his individual happiness can be increased and extended only in proportion as he actively endeavours to increase and extend the happiness of all around him."[96]

In this formulation, Owenist doctrines look very much like Bentham's Utilitarianism, but there was a huge practical difference in implementation. Bentham had focused on what he called "sanctions" against anti-social behaviors in order to coerce people into acting for the general good rather than in their own private interests.[97] Owen, on the other hand, focused exclusively on what later behaviorists would call the positive reinforcement of altruistic behaviors and on the elimination of social conditions that were destructive of the health and well-being of people.

At the beginning of 1815, Owen launched a campaign to reform the conditions of factory life by proposing an Act of Parliament that would do the following four things: 1) limit the regular working day to twelve hours; 2) raise the minimum age for children to work in factories to ten and limit their hours of employment to six per day until age twelve; 3) require that no child be employed in a factory until after he or she had learned to read, to write, to do simple arithmetic, and (for girls) to sew; and 4) establish a system of inspectors to monitor factories to ensure compliance.[98] By the time the first Factory Act was passed in 1819, it had been gutted by the elimination of the third and fourth provisions; its provisions were restricted to the cotton industry; it lowered the age for hiring children to nine; and it allowed children to work up to twelve hours per day; but Owen had initiated the first factory reform legislation.[99]

While Owen was in London promoting factory legislation, the major post-war slump in textile production hit, and the government established a Committee for the Relief of the Manufacturing Poor to try to suggest some constructive response to the situation. Owen was invited to testify and then to draw up a plan based on his testimony. This plan, first presented in 1817 as a "Report to the Committee for the Relief of the Manufacturing Poor," began by identifying the crisis as a consequence of technologically driven overproduction and unemployment exacerbated by low wages, which left workers unable to purchase any reasonable fraction of what they could produce.[100]

Given the cause of the crisis, Owen argued that there were in principle three different possible national responses. First, there was the Luddite response: "The use of mechanism must be greatly diminished." But, argued Owen, this response could not work because if any nation unilaterally reduced the use of machinery and thus increased its cost of production of goods, other nations would continue to drive production costs down, leading to the economic destruction of the nation with the more primitive system of production. Second, there was the Malthusian response: "Millions of human beings must be starved, to permit its [i.e., mechanization's] existence to the present extent." Such a solution was unthinkably barbaric in Owen's view. This left the third possibility as the only one that could be seriously contemplated: "Advantageous occupation must be found for the poor and unemployed working classes, to whose labor mechanism must be rendered subservient, instead of being applied, as at present, to supersede it."[101]

In order to accomplish this goal, Owen proposed the establishment of Cooperative villages of between five hundred and fifteen hundred persons that would incorporate all of the educational and community-building features that had been tested at New Lanark as means of eliminating vice and creating able and intelligent workers. These villages would feature common kitchens and mess halls to both cut the cost of food preparation and ensure its quality. They would develop cooperative stores to cut costs by purchasing in volume and by eliminating the profits of middlemen. They would encourage the direct bartering of services among community members. They would incorporate market-linked manufacturing enterprises that would provide both work and income. And they would use labor-intensive agricultural methods to remain relatively self-sufficient while providing work for nearly every person when demand for manufactured goods dropped off.[102]

Like Auguste Comte, Owen often became obsessed with details of his schemes, so he included in his proposals minute recommendations concerning the architectural features of the proposed communities, the agricultural methods to be used, and so forth. This inclusion of detail did, however, allow Owen to provide cost estimates for his proposals. He estimated the complete capital cost, including land, for a village of twelve hundred persons to be £96,000, or £80 per person. This cost, he admitted, might seem high until one realizes that the capital costs could easily be repaid at normal interest out of the profits of the community ventures, so investment in such villages would not be a drain on the treasury but would rather offer an opportunity for the profitable use of capital. The net result would also be to all but eliminate public expenditures on the poor.[103]

Much to Owen's surprise, virtually no one among the political and intellectual elite took him seriously. Radical leaders attacked his plans as overly simplistic, paternalistic, and insensitive to the emotional realities of human life. Almost no one was willing to risk capital on Owen's untried proposals, and the Tory government response to lower-class unrest was repressive rather than helpful.[104] Only a few full-scale Cooperative villages were attempted in the British Isles—one at Orbiston, Scotland (1825–27), one at Ralahine, Ireland (1831–33), and one at Queenwood in Hampshire (1839–45). Owen himself was convinced by a group of American sympathizers to try to establish one at New Harmony, in Indiana (1825–27). In every case, practical problems led to rapid disintegration of the project.[105] But while there were few complete Owenite villages, the Owen-inspired notion that working class cooperation in the production, exchange, and purchasing of goods and services could benefit the working class spread like wildfire.

Beginning with the establishment of the avowedly Owenite London journeyman printer's "Co-operative and Economical Society," under the leadership of George Mudie in 1821, and the "Owenian, Friendly, or Practical Society of Edinburgh" a few months later, the cooperative movement grew by 1832 into a formally organized network of nearly five hundred cooperative producer and consumer societies, trading with each other and joining in establishing "labor exchanges" in which goods valued by the labor time invested in them were exchanged.[106] The British cooperative movement of the 1830s in turn spawned or influenced two major worker's institutions. On the one hand, out of the consumer-cooperative side it led through the Rochdale experiments of the 1840s to the international cooperative movement that persists at the beginning of the twenty-first century, especially in the developing world.[107] On the other hand, through the producer cooperatives and labor exchanges, the movement played a significant role in the pre-Chartist days of the Trades Union movement in Britain.

Local trade unions, often limited to the workers in a single shop, predate the cooperative movement by decades. But the repeal of the Combinations Act, which had made most union activity illegal, in 1824, revitalized union activity. Among the most important organizers was John Dougherty, an Owenite, who first organized the Grand General Union of Spinners in 1829 and then the National Association for the Protection of Labor in 1830.

Owen himself became involved in union activities in 1833, first, leading an ambitious and ultimately unsuccessful reorganization of the Operative Builders Union into the Grand National Guild of Builders, and then proposing the establishment of the Grand National Consolidated Trades Union, which would cover all workers in Great Britain and would aim "to establish for the productive classes a complete domination over the fruits of their own industry."[108]

The year 1834 saw the collapse of the Consolidated and the Owenite phase of the Trades Union movement. In November of that year Owen started a new working-class journal, *The New Moral World.* In the first issue of the journal Owen signaled his disillusionment with the union movement and his intention to rededicate his efforts to create a new organization whose goals would be "to form an entirely new state of society through the reorganization of production and distribution, for reorganizing character and providing beneficial government for all."[109] Through a succession of organizations, beginning with the Association of All Classes of All Nations to Form a New Moral World (1835), and including the Universal Community Society of Rational Religionists (founded in 1839), Owen continued to promote his plans to establish cooperative communities.

Increasingly, however, instead of seeking capital from large investors to establish communities, local Owenite groups began to establish assembly halls, or Halls of Science, where working men and women could come to be educated in the principles of cooperation and in the sciences of man and society that underlay them. The initial expectation was that members would then purchase small shares in the communities that would eventually be founded. In connection with this new trend toward worker ownership in both the Halls of Science and the proposed communities, many of the members of these Owenite societies began to call themselves socialists. In effect, the Owenite socialist Halls of Science came to offer a complement to the Mechanics' Institutes, offering adult classes, and in some cases sponsoring day schools for children.[110]

The Owenite movement was never as large as the Mechanics' Institute movement, peaking in the early 1840s with somewhere near twenty-five thousand members in seventy-seven local branches. But according to a study done by the Society for the Diffusion of Useful Knowledge, even though membership in the Owenite groups was smaller, the attendance rate was much higher.[111] Education at New Lanark and in the Owenite day schools emphasized the natural sciences, but the lectures in the Halls of Science, which reached far greater numbers, never exhibited a major interest in those topics dealt with within the Mechanics' Institutes. Instead, they focused on what we now call the social sciences, including political economy, and they promoted both a concern with social and economic issues and a commitment to act on that concern.

Political Economy and Political Reform: ca. 1800–1860

The working classes remained politically disenfranchised in Britain throughout the first half of the nineteenth century in spite of the Great Reform Bill of 1832, which increased the pool of eligible voters from roughly the wealthiest two percent of the adult male population to about 14 percent. But the period from 1800 to the 1860s did see a number of major governmental reform initiatives, most of which involved the application of scientific expertise to social problems. Virtually all of these reforms were initiated and/or promoted by members of two closely interlinked networks of intellectuals. One of these networks was localized in London, and was centered on Jeremy Bentham, David Ricardo (1772–1823), John Stuart Mill, and Edwin Chadwick. It was associated, at least initially, with Jeremy Bentham's Utilitarian doctrines. With rare exceptions, the members of this group identified themselves with the Radical party in Parliament. The other network centered on Henry Brougham and his *Edinburgh Review* friends. It was associated primarily with the moral philosophy of Dugald Stewart, and all members identified themselves as Whigs.

One thing that unified the members of these two groups, in spite of their very different perspectives on such topics as religion and the importance of historical knowledge relative to psychologically based knowledge in guiding political policies, was a common sense that the traditional ruling class of Great Britain, with its disdain for direct involvement in commercial activities and its classically based education, was unprepared to cope with the problems of an increasingly industrializing and urbanizing nation. They were little less paternalistic than those they sought to replace, however, for with few exceptions, they viewed themselves as a new aristocracy of talent who had turned away from the idleness, waste, snobbery, and concern with appearances that they associated with the old aristocracy, in order to become what Thomas Carlyle called a "service elite," dedicated to the well-being of the modernizing nation.[112]

They all agreed with Stewart that social scientific knowledge must play a major role in national policy making because modern societies were becoming so different from those of the past that unanalyzed historical precedent was inadequate as a guide.[113] They disagreed, however, about precisely how the social sciences were to be constituted and legitimated and about which social sciences were most important in providing guidance on particular issues. Unlike the traditional ruling class, they were also very much concerned with molding and mobilizing public opinion, for they believed that public support could make or break the success of any policy. As a consequence, they were tireless polemicists, using the burgeoning periodical press as a critical resource for spreading their views. Not just through the *Edinburgh Review,* but also through the *Times,* the *Quarterly Review,* the *Westminster Review,* the *Cornhill Magazine,* the *Bee Hive,* the *Positivist Review,* and later, through the *Saturday Review,* the *London Review,* and the *Fortnightly Review,* among others, they made their views known. Moreover, they virtually all joined with Brougham and Robert Owen in establishing the National Association for the Promotion of Social Science in 1857.

On the issue of informing and mobilizing public opinion regarding the foundations for public policy, they were joined by several superb popularizers of scientific ideas. The most important of these were Jane Marcet (1769–1858), who followed her immensely popular *Conversations on Chemistry* (1806) with *Conversations on Political Economy: In Which the Elements of that Science Are Familiarly Explained,* which appeared first in 1817, and Harriett Martineau, the translator of Comte's *Positive Philosophy,* who produced a series of nine short novels, each illustrating a principle of political economy, between 1831 and 1833 and who then turned to write two more volumes on *Poor Laws and Paupers Illustrated* in the following year.[114]

Finally, though the philosophical radicals of the south had a distinct preference for more democratic and individually oriented governmental forms, the philosophical Whigs of the north preferred the balancing of class interests represented in the British "constitution" since 1688. They all agreed with Stewart, however, that political improvements could best be effected, "not by delineating plans of new constitutions, but by enlightening the policy of actual legislators."

For that reason, political economy, the science that investigated "those universal principles of justice and expediency, which ought, under every form of government, to regulate the social order,"[115] played the central role in promoting their practical political agendas.

There is little doubt that the most important British political economist of the first half century in terms of his long-term impact on the development of economic analysis was David Ricardo. His *On the Principles of Political Economy and Taxation* of 1817, began the trend toward casting political economy in the form of a hypothetico-deductive science. Similarly, there is very little doubt that the most important early teacher of political economy in terms of his short-term impact on policy was Dugald Stewart. Stewart began teaching a separate political economy course in 1799 and was the teacher of James Mill as well as of at least five members of Parliament and all of the early *Edinburgh Review* authors. Stewart's importance as a teacher was approached only by James McCulloch, the chief interpreter of Ricardo to those who could not penetrate his precise but extremely awkward prose. McCulloch wrote for the *Edinburgh Review,* taught private courses on political economy in London, and ended up as the first professor of political economy at the University of London in 1827.[116]

The most important innovation in political economy in terms of its impact on early nineteenth-century politics, however, was provided by Thomas Malthus (1766–1834), a Cambridge-trained mathematician and clergyman who published the first edition of *An Essay on the Principle of Population, as It Affects the Future Improvement of Society. With remarks on the Speculations of Mr. Godwin, M. Condorcet, and Other Writers,* in 1798, with subsequent, substantially revised editions in 1803, 1806, 1807, 1817, and 1826. The first edition of Malthus's *Essay on Population* was written as a counter to the optimistic views of his father, an admirer of Condorcet and William Godwin, both of whom saw the French Revolution as ushering in a period of continuing human progress and well-being. To oppose this view, Malthus argued that the combination of humanity's two greatest needs, the need for food and the need to procreate, must always operate together to produce misery and suffering because of a simple pair of mathematical relationships: "Population, when unchecked, increase[s] in a geometrical ratio, and subsistence for man in an arithmetical ratio. . . . Taking the population of the world at any number, a thousand millions, for instance, the human species would increase in the ratio of—1, 2, 4, 8, 16, 32, 64, 128, 256, 512, &c. and subsistence as—1, 2, 3, 4, 5, 6, 7, 8, 9, 10, &c. [Assuming that population doubles in a twenty-five year period,] in two centuries and a quarter, the population would be to the means of subsistence as 512 to 10: in three centuries as 4096 to 13, and in two thousand years the difference would be almost incalculable."[117] Furthermore, according to Malthus in the first edition of the *Essay on Population,* nothing short of the checks of famine, disease, and war could limit population to that which could be carried by the food-producing capacity. Even these checks would produce an oscillation of population about the line of food production that must include periods of extreme misery.[118] To some extent, Malthus was willing to admit that inequalities of wealth and power exacerbate the misery of the poor; but, he argued that "though the rich by unfair combinations contribute frequently to prolong a season of distress among the poor, yet no possible form of society could prevent the almost constant action of misery upon a great part of mankind, if in a state of inequality, and upon all, if all were equal. . . . The superior power of population cannot be checked without producing misery or vice."[119] This had to be true because Malthus was convinced that no voluntary restraint on procreation could be strong enough to keep a population below the carrying capacity of its relevant food-producing region.

Criticisms of the first edition of *On Population* led Malthus into a several-year period of travel and data collection throughout Europe and to a major change of perspective. He now came to believe that moral restraint could possibly, under favorable circumstances, provide a sufficient check on a population to avoid disaster. Thus, in the second edition of *On Population* (1803), he incorporated extended discussions of voluntary checks on population. But for most short-term purposes relating to public policy, this new perspective had little significance, for neither Malthus nor most of his readers believed that the poor of England were presently capable of adequate restraint. Malthus remained convinced that his basic claims regarding the geometrical increase of unchecked population and the linear increase of foodstuffs were correct, so even if some moral restraint were shown, it would just delay the time before population reached carrying capacity.

Up to the time of Malthus, political economists had been united in treating population primarily as a source of labor and thus of wealth. Indeed, from William Petty on, human beings were treated almost exclusively as a form of stock or capital by political economists. Adam Smith had pointed out that in any industry, the wages paid to labor will be greater when there is a labor shortage and lower when there is a labor glut, but it was Malthus who first began to explore the implications for an entire economy of overpopulation and excess labor supply and to suggest policy implications.

At the beginning of the nineteenth century, English procedures for caring for the aged, infirm, and unemployed or underemployed poor were still based on the Elizabethan Poor Law of 1601. According to this law, the nation accepted a responsibility to alleviate the suffering of the poor, and it established a system of poor relief in which the Anglican Church of each parish was charged with administering aid to the poor while local justices of the peace established a special tax rate to cover the costs of poor relief.

In 1722, parishes were allowed to establish workhouses and to refuse relief to any who would not enter. Some parishes contracted out care of the poor to workhouse management companies, which tended to exploit the system and abuse the poor whom they were intended to help; so during the later part of the eighteenth century evangelical reformers had sought modification of the system to make it more responsive to the needs of the poor. The Gilbert Act of 1782 removed the requirement of entering workhouses and established a form of "outdoor relief." After 1795 this outdoor relief often took the form of subsidized wage rates for the working or "industrious" poor when food prices rose beyond the ability of standard wages to meet family needs.[120] Additional laws sought to encourage the establishment of voluntary Friendly Societies among workers by relieving them of traditional tax liabilities, to ease the conditions of residence that had to be met to qualify for aid, to guarantee aid to the families of military personnel, to provide low-cost housing, and to provide national standards for the aid granted to those eligible.[121]

Through most of the eighteenth century, the total percentage of the population on poor relief was around 3 percent, with a high in 1776 of 3.8 percent. But as a result of increasing rural unemployment, that number steadily grew after 1780; and during the post-Napoleonic industrial depression it soared to 13.2 percent. At the same time the per capita poor rates rose from 5 shillings in 1784 to 13 shillings, 3 pence in 1818.[122] During the early nineteenth century, then, there was a dual pressure to do something to reform the poor laws. On the one hand, there was a growing number of persons in severe distress in spite of the fact that they were usually quite willing to work. On the other hand, rate payers were complaining because poor rates had nearly tripled over a period of thirty years.

Into this situation Malthus's analysis appeared, suggesting that virtually all poor relief must be self-defeating. His attacks on supplemented wages and on the provision of housing were particularly compelling. With respect to augmented wages, he argued that by increasing wages without increasing the quantity of food available, one would just drive up the price of food to everyone, and since the wealthy would still be able to outspend the poor, the poor would be no better off.[123] But this is only the beginning of the problem, for if the poor could expect to be supported by the state, they would marry and procreate at a faster rate. As a consequence, Malthus wrote, the poor laws "may be said therefore in some measure to create the poor which they maintain, and as the provisions of the country must, in consequence of the increased population, be distributed to every man in smaller proportions, it is evident that the labor of those who are not supported by parish assistance will purchase a smaller quantity of provisions than before and consequently more of them must be driven to ask for support."[124]

Virtually every political economist of the first half of the nineteenth century found Malthus's arguments regarding poor relief compelling. Political economists took the lead in opposing the extension of relief and instead, pushed to abolish it completely for all able-bodied individuals. Even Radicals such as David Ricardo and Whigs such as Brougham led Parliamentary attempts increasingly to restrict poor relief until the New Poor Law of 1834, which had been drafted largely by the Benthamite Edwin Chadwick, finally revoked all outdoor relief for the unemployed or underemployed.[125] A few years later, when famine conditions in Ireland demanded emergency consideration, in spite of the fact that even Malthus had suggested that short-term relief that did not create expectations of continuation might be appropriate, Parliament, responding to the impassioned speeches of most of its twenty-eight political economist members from all parties, refused to grant aid to the able bodied.[126] The new guiding principle, articulated by Brougham in the debate on Irish poor relief, was that no legislation should be passed that "proceeded on the assumption of the doctrine that it is part of the duty of any government whatever to feed the people of any country, or to provide food, wages, or labor for the people."[127]

While it is true that rate-payer unrest over the high cost of poor relief might eventually have produced some other rationale for renouncing the traditional obligation of governments to help their citizens during periods of economic hardship, there is no doubt that virtually all explicit arguments in favor of poor-law reform during the early nineteenth century were drawn from political economy. Nor is there any question that the philosophical Radicals and Whigs played a leading role in promoting tougher attitudes toward poor relief.

If political economy seems to have promoted regressive policies in connection with the Poor Laws, it did quite the opposite in connection with monopolistic practices in general and in connection with the so-called Corn Laws in particular. At the beginning of the nineteenth century, the British economy was still dominated by protective and monopolistic practices. The East India Company, for example, had an exclusive right to carry on British trade with India; the Navigation Acts required that all trade with British colonies be carried out in British ships; and British farmers were virtually guaranteed premium prices for their grain by the existence of extremely high import duties on imported grains ("corn" in nineteenth-century terminology). Of all the protective practices, the protection of grain prices seemed most egregious to many Britons because it ensured high food prices and exacerbated the problems associated with the poor. But up until the reform of 1832, Parliament was dominated by landed interests; and in spite of the fact that Radical and new Whig leaders such as David Ricardo and Francis Horner spoke vigorously against these effective price supports, they were not only continued but they were made increasingly restrictive through the Corn Laws of 1815.

In 1822 Ricardo sought to introduce an amendment that would substantially reduce import duties on wheat, but it was defeated in the House of Commons by a vote of 217 to 25. Four years later, three members of the House of Lords, all of whom had either attended James McCulloch's (1789–1864) lectures on political economy in London or taken private lessons from McCulloch, began to attack the Corn Laws, arguing that they encouraged the inefficient production of grains on unsuitable land. By 1833, the reformed Parliament came closer to amending the Corn Laws, defeating a move to reduce duties by a vote of 74–47 in the House of Commons. Robert Peel (1788–1850) led the opposition, claiming that the repeal of the Corn Laws would create an undesirable "revolution in the relations of the different classes of society."[128] The philosophical Whig, Robert Torrens (1780–1864), expressed the disgust of his scientifically inclined colleagues: "Those who were endeavoring to keep up the system of restriction and monopoly were contending against an irresistible power—they were contending against the power of knowledge, and struggling with the omnipotence of truth."[129]

Once again in 1842, the free market coterie, led in part by David Ricardo's nephew John Lewis Ricardo, tried to repeal all corn duties, with the *Edinburgh Review* contributor and chancellor of the exchequer from 1835 to 1839, Lord Monteagle (1790–1866), claiming in debate that "there can be no permanent rest, no quiet, no safety with respect to this subject, until the trade in corn shall be as free as air, as far as protection is concerned."[130] Once again, however, they were defeated. In 1846, when Peel became converted to a free-trade position, the Corn Laws were finally abolished, with Brougham and the Marques of Landsdown, another of Dugald Stewart's students, leading the charge in the House of Lords.

In all of the debates leading up to the repeal of the Corn Laws, the principle arguments in favor were explicitly drawn from political economy and most often they were delivered by the philosophical Whigs from Edinburgh and the philosophical Radicals from London. The same is true of the subsequent debates leading to the abolition of the Navigation Acts in 1849 and of the earlier debates that ended the exclusive trade privileges of the East India Company in 1833.

With respect to two other major early-nineteenth-century policy issues, the reform of Parliament and the regulation of working conditions in factories and mines, the scientifically oriented political groups surrounding Bentham and Brougham played major roles. At the beginning of the nineteenth century, the 638 members of the House of Commons were elected according to arrangements that had been made in the sixteenth and seventeenth centuries. Most of the relatively new towns, including Manchester and Birmingham, which were major population centers, had no representatives, while many representatives came from small or deserted villages whose votes were effectively controlled or purchased by a single powerful individual. (The secret ballot was not established until 1872, so pressure could easily be brought against voters who did not agree with those they were dependent on.)

The reform sponsored by Brougham, while it did only a modest amount to expand the franchise to members of the middle class, abolished most of the boroughs with fewer than sixty voters; assigned those members to the more heavily populated counties, towns, and cities; and established uniform voting requirements. These changes made it possible to contest far more elections and for candidates with a serious interest in government to be elected. One consequence was an immediate increase in the number of members of Parliament (MPs) with significant knowledge of political economy from twenty-six in the

Parliament elected in June 1831 to forty-three MPs in the Parliament elected in January 1833.[131] A second consequence was a rapid increase in the percentage of MPs who actually attended debates and voted.

On the issue of the regulation of working conditions, the overwhelming support of free-market conditions by scientifically oriented politicians led to paradoxical consequences. On the one hand, both Benthamite and Broughamite factions were leaders in repealing the laws against the combination of workers (unionization) that had been promulgated in 1799 and 1800 in the panic after the French Revolution. McCulloch's January 1824 *Edinburgh Review* article against the Combination Acts had a powerful impact on public attitudes, leading to the repeal of restrictions on worker combination later in 1824, although a flurry of strikes during the subsequent year led to a partial reversal of this liberalization.[132] On the other hand, politicians committed to political economy were loath to interfere with the autonomy of employers in order to protect workers.

Though Francis Horner spoke in favor of restricting the employment of children under ten in factories in 1815, when the Owen-supported bill to forbid factory employment for children under nine and to limit the hours of children's work to twelve per day was proposed, no Benthamite or Broughamite spoke in favor of the bill, though it did pass. When debate over the conditions of labor in factories resumed in the 1830s, after the evangelical Tory, Michael Sadler (1780–1835), introduced a bill to reduce the hours that could be worked by anyone under eighteen to ten in a day, the Royal Commission appointed to study the problem and recommend policies was dominated by Benthamites.

Sadler's ten-hour bill was clearly an attempt to regulate adult hours as well; for assembly-line work was organized so that minors and adults worked together. Chadwick and his colleagues were deeply opposed to the regulation of adult working conditions, but favored the restriction of hours and improvement of other conditions, including ventilation and drainage, for children. Thus, they proposed a compromise that would reduce the hours for those under age thirteen to eight. This compromise, they pointed out, would allow two children's shifts per sixteen-hour shift for adults. The Benthamites also recommended the establishment of mandatory schooling for children working in the factories and the establishment of factory inspectors to ensure compliance with the law.

With some significant changes, including the abandonment of the educational plan, the increase of daily hours to nine hours for children under age thirteen, and the addition of a bitterly opposed twelve-hour limitation on the hours of work for "young persons," age fourteen to seventeen, the committee's recommendations were passed. The inspectorate was too small to ensure substantial compliance, and the law sustained the principle of noninterference with the conditions of adult laborers; but the Factories Act of 1833 was the first major piece of social legislation passed primarily on the recommendation of expert advice, involving national regulation of industry, and containing a procedure for supervision.[133]

After another report of the Royal Commission on Children's Employment exposed shocking conditions in mines and factories in 1842, a few philosophical Whigs, such as the McCulloch-trained Lord Howick (1764–1845), insisted that in some cases other considerations had to override those of political economy. Thus in a speech to the House of Lords he argued: "You altogether misapply the maxim of leaving industry to itself when you use it as an argument against regulations, the object of which is, not to increase the productive power of the country, or to take the fruits of a man's labor from himself and give it to another, but, on the contrary to guard the laborer himself and the community from evils against which the mere pursuit of wealth affords us no security. The welfare, both moral and physical of the great body of the people, I conceive to be the true concern of the Government. . . . He must indeed have a low and mean idea of our nature, who thinks that mere wealth is all in all to a nation, and who does not see that in the too eager pursuit of wealth, a nation like an individual, may neglect what is of infinitely higher importance."[134] But Howick was in a distinct minority. Brougham, for example, reluctantly supported a bill to limit the use of children in collieries, but he absolutely insisted that he would not support any legislation to regulate the conditions of adult laborers.[135] In 1844, another bill to limit the hours of work for women and children to ten hours per day was defeated without support from either the Benthamite or Broughamite group, and the final factories bill of the first half-century, in 1847, once again failed to reduce the working day to ten hours. Sir James Graham (1792–1861) expressed the common view of the philosophical Whigs and Radicals: "The question is, shall you, by indirect legislation restrain industrious men from working twelve hours a day for the purpose of earning their livelihood, though they are willing to undergo the fatigue? . . . [the bill is] a departure—a flagrant departure—from the strict rules of political economy—a science which in some quarters of this House appears to be treated almost with contempt, but it is a science which I have always considered as tending towards the benefit and general happiness of the nation; and I doubt if any legislation will be found safe if you depart from the great rules of that important science."[136]

I argued in discussing science and religious heterodoxy in Victorian Britain that rejection of traditional Christianity was often associated with a feeling that churches were doing too little to ameliorate the plight of the poor and that one response of middle-class intellectuals was to turn to science as a source of value and support. Through the first half of the nineteenth century, however, it became increasingly clear to many that political economy, the science most directly related to governmental policies regarding the plight of the poor, was no more helpful. Indeed, political economy was frequently used, even by those sympathetic to the conditions of the poor such as Sir James Graham, to justify inaction on issues that might improve conditions for the working poor. Rather than abandoning their commitment to science, many of those who viewed science as a source of human well being either sought to focus on the limits of applicability of political economy or tried to find alternative scientific grounds for directing social action. It is to the foundation for these alternatives that we turn in chapters 7 and 8, though as we shall see in chapters 9 and 10, the new approaches were, at best, ambiguous with respect to their socially progressive policy implications.

PART III

Evolutionary Perspectives

CHAPTER 7

The Rise of Evolutionary Perspectives: 1820–59

By the second decade of the nineteenth century, historical perspectives had begun to replace the hyper-rationalism of sensationalist psychology as the foundation of French radical thought at the hands of the Saint-Simonians. Moreover, a nationalistically grounded historicism had provided the central vehicle for German anti-universalist sentiment among conservatives and liberals alike from the beginning of the century. In Britain, on the other hand, if any perspectives may be said to have dominated liberal and radical understandings of social issues between 1820 and 1860, those were the relatively ahistorical Utilitarianism and classical political economy. After 1860, however, historical, or evolutionary social theories, often misleadingly known as "Social Darwinisms," became increasingly dominant as the theoretical underpinnings for radical and liberal ideologies in Britain, spreading to both reinforce and modify historicist trends on the Continent and in the United States.

In this chapter, we focus first on the initial rise of evolutionary social theories in Britain, exploring the reasons for and character of the historical turn in social thought. Then we turn to several theories of biological evolution, or transformism, and the way they interacted with ongoing themes in social and political discourse to influence public perceptions regarding such issues as gender, mental illness, morality, and race. Finally, we briefly consider the evolutionary social theory of Herbert Spencer, one of the pervasive scientistic social theories in mid-nineteenth-century Britain and America. This background relating to pre-Darwinian evolutionary ideas is crucial to an understanding of the character of those later and more pervasive evolutionary social theories that are often identified as Social Darwinist theories; for well into the twentieth century many evolutionary social theories emphasized themes that predated Darwin's work and stood in opposition to some of his most important ideas.

In the next chapter, we turn to Charles Darwin's formulation of evolutionary theories in biology and anthropology; for it was Darwin who made the doctrine of evolution by natural selection respectable within the community of professional biologists. Moreover, it was the attention to transformation with descent, or evolution, stimulated by Darwin's works—especially *The Origin of Species* of 1859 and *The Descent of Man and Selection in Relation to Sex* of 1873—that promoted attempts to use evolutionary ideas to address a huge range of psychological and social issues that seemed of major importance in the mid nineteenth century. Evolutionary perspectives were applied to "the woman question," which involved deciding whether women should have the same educations as men, whether they should enjoy the right to vote, and whether they were more fragile emotionally. They provided a foundation for discussions of racial differences and the appropriateness of slavery and colonialism; and they provided a new way to understand mental illness and the extent to which mental illness was or was not curable.

In chapter 9, we will turn to explore the spread of post-Darwinian theories of social evolution in Britain, the United States, and on the Continent, considering both national differences in evolutionary social theories and attempts by groups with differing political agendas to appropriate the growing authority of evolutionary biology. These theories would form the basic justification for a wide range of militarist and imperialist, as well as socialist and pacifist ideologies during the late nineteenth century and much of the twentieth.

In Charles Dickens's novel, *Hard Times*, published in 1854, the Utilitarian schoolmaster, Mr. Gradgrind, and his two sons, Adam Smith and Malthus, are portrayed repellently because a lack of compassion, emotion, and capacity for joy and spontaneity seem to follow from their extreme emphasis on reason. Dickens's portrayal is symptomatic of a deep distrust of and disgust with both Utilitarianism and political economy that seems to have accelerated during the 1840s among liberals and radicals, at least in part as a consequence of their roles in opposing humanitarian directions in Poor Law and factory legislation.

In its Smithian form, political economy had seemed to promise well-being for all if a nation followed the path of noninterference in economic affairs. In the form promulgated by Malthus, Ricardo, and their followers, however, while political economy offered a way to increase total wealth, it seemed to offer no effective way to avoid poverty and misery in the numerically superior working class. To be fair to Ricardo, John Stuart Mill, and virtually all other reasonably sophisticated political economists, they had pointed out that in assuming that economic actors operate only from a foundation of self-interest, they were making an admittedly counterfactual assumption that allowed them to predict and possibly influence some, but not all, behaviors. Mill, perhaps the most outstanding political economist of the generation after Ricardo, wrote, "Political Economy considers mankind as occupied solely in acquiring and consuming wealth, and aims at showing what is the course of action into which mankind, living in a state of society, would be impelled if that motive . . . were absolute ruler of all their actions. . . . Not because any political economist was ever so absurd as to suppose that mankind are really thus constituted, but because this is the mode in which science must necessarily proceed."[1] In fact, Mill's exposure to Saint-

Simonian ideas had opened him to the possibility that the commitments to free competition, private property, and the inviolability of inheritance that dominated classical political economy were all challengeable.

For Mill, as for other political economists, such considerations did not delegitimize the enterprise. They simply called attention to ways in which its conclusions might have to be qualified. But British advocates of applied political economy in the press and in Parliament had a difficult time keeping such qualifications in mind. So after mid century, increasing numbers of thoughtful persons complained with the wealthy philanthropist and social scientist, Charles Booth (1840–1916), that "the *a priori* reasoning of political economy, orthodox and unorthodox alike, fails from want of reality. At its base are a series of assumptions very imperfectly connected with the observed facts of life."[2]

The psychological underpinnings of Utilitarianism came under serious criticism simultaneously. Few, if any, were inclined to deny the basic Utilitarian claim that humans are pleasure-seeking and pain-avoiding creatures. On the other hand, it became increasingly difficult to sustain the Benthamite notion that human nature is sufficiently constant over time and place that one could identify the universal causes of pain or pleasure and use that knowledge to establish effective sanctions to direct behavior. Accumulating historical and ethnographic evidence suggested that what motivates people is very often culturally conditioned.[3] John Austin (1790–1859), a Utilitarian jurist and friend of John Stuart Mill, expressed the growing frustration among those who had once hoped to legislate for a universal human nature, writing that both history and daily experience seemed increasingly to demonstrate "the extraordinary pliability of human nature."[4]

Among the philosophically aware, this problem was exacerbated by the fact that the sensationalist psychological foundations for Utilitarianism, which came from Locke and Hartley, were under severe attack by Kantian and German idealist philosophies. The history of philosophy thus came less to seem to offer a triumphant story of how British empiricism gradually achieved a correct understanding of the human mind than to provide a catalog of mutually conflicting epistemological and ontological positions among which there were no adequate criteria for making a conclusive choice. At the same time, as we saw in the previous chapter, traditional religious institutions were under attack because they also seemed unable to offer compelling moral guidance or intellectual coherence.

As a consequence of the discrediting of both traditional religion and the newer guides of a priori political economy and Utilitarian theory, many British intellectuals in the mid nineteenth century found themselves searching for some secure foundation for their beliefs and actions. Moreover, this floundering seems to have been almost equally severe for traditionalists, such as Matthew Arnold (1822–88), and for progressive thinkers, such as Harriet Martineau. In his *Culture and Anarchy* of 1869, Arnold wrote: "In all directions our habitual courses of action seem to be losing their efficaciousness, credit and control, both with others, and even with ourselves; everywhere we see the beginnings of confusion, and we want a clue to some sound order and authority."[5] Similarly, in her 1853 introduction to Comte's *Positive Philosophy,* Martineau wrote: "The supreme dread of everyone who cares for the good of nation or race is that men should be adrift for want of anchorage for their convictions. I believe that no one questions that a very large part of our people are now so adrift."[6]

In this situation, British thought took a historical turn under the simultaneous influences of Comtean Positivism, German historicism, and biological theories of evolution. G. H. Lewes, recently converted to Positivism, offered his view of the situation in an 1844 article on "The State of Historical Studies in France," in the *British and Foreign Review:* "[Ours is] an age of universal anarchy of thought, with a strong desire for organization; an age, succeeding one of destruction, anxious to reconstruct—anxious but as yet impotent. The desire of belief is strong; convictions are wanting; there is neither spiritual nor moral union. In this plight we may hope for the future, but can *cling* only to the past: that alone is secure, well grounded. The past must be the basis of certainty and the materials for speculation."[7] Particularly appealing to liberals and radicals initially were progressive philosophies of history, such as those of Comte and of Herbert Spencer (1820–1903), that offered what amounted to secular theodicies. It was true that serious evils such as poverty and suffering were and must be present in contemporary society, as Malthusian political economy demonstrated. But that suffering could now be understood and accepted as part of a process that would lead to an ever better world. Spencer, probably the most influential British social theorist of the second half of the nineteenth century, wrote in his *Social Statics* of 1851: "The well-being of existing humanity and the unfolding of it into this ultimate perfection are both secured by that same beneficent, though severe discipline to which the animate creation at large is subject: a discipline which is pitiless in the working out of good: a felicity-pursuing law which never swerves for the avoidance of partial and temporary suffering. The poverty of the incapable, the distresses that come upon the imprudent, the starvation of the idle, and those shoulderings aside of the weak by the strong, which leave so many in 'shallows and miseries,' are the decrees of a large, far-seeing benevolence."[8] In Spencer's case, a powerful progressive philosophy of history was even capable of reinvigorating elements of Utilitarianism and political economy, but in ways that acknowledged a much greater range of socially conditioned human desires and behaviors than had been the case in their early-century versions. Over time, Spencer, who was a deeply committed pacifist and advocate of democracy, became disillusioned with the imperialist and militaristic directions of European societies. His initial optimism and providentialism gave way to increasingly pessimistic and agnostic perspectives. Especially in America, however, his followers never lost their initial sense that whatever suffering may be a necessary part of the working out of social laws, the progress that it entailed was well worth it.

Mill's *System of Logic* and the Selective Appropriation of Comte's Historical Emphasis

In the sixth book of his *System of Logic, Ratiocinative and Inductive,* which John Stuart Mill devoted to "The Logic of the Moral Sciences," Mill gave an account of the social sciences that helped his readers put in perspective the apparent recent failures of applied political economy and Utilitarianism. This account linked those perceived failures to reasons for turning toward historical studies for understanding many social phenomena. For Mill, as for Comte, human social phenomena are so complex and are subject to the operation of so many causes that the attempt to induce particular causal mechanisms directly from social phenomena is doomed to failure, even though we may remain confident that such phenomena are subject to causal laws. The best one can hope to accomplish inductively is the identification of certain general descriptive or empirical laws.

Comte had gone on to argue that, in principle, because whatever causal factors are at work may interact with one another strongly, even if we could know most or all of the relevant causes, there would be no way to deduce phenomena from the effects of the several causes acting individually. Consequently, there is no point in trying to develop separate social sciences, such as political economy, which presume to be able to deduce the consequences of one particular set of human causes; that is, each individual's search to maximize his or her own wealth. This is so because in any particular social situation, each cause must interact with so many others that its effect must be literally incalculable.

While Mill agreed that the practical difficulties of developing and using separate social sciences were tremendous—thus the failures of applied political economy to suggest how to cope with problems of the impoverishment and exploitation of the working class—he insisted that, in principle, a complete social science could be constructed out of the aggregation of a sufficient number of specific and admittedly incomplete ones. Thus, he wrote, "the effect produced, in social phenomena, by any complex set of circumstances, amounts precisely to the sum of the effects of the circumstances taken singly."[9] Furthermore, he insisted, contra-Comte, that all social laws should ultimately be reducible to laws about individual human beings: "The laws of the phenomena of society are, and can be, nothing but the laws of the actions and passions of human beings united together in the social state. Men, however, in a state of society are still men; their actions and passions are obedient to the laws of individual human nature.... [They] have no properties but those which are derived from, and may be resolved into, the laws of the nature of individual man. *In social phenomena the Composition of Causes is the universal law.*"[10]

Mill's insistence on the related principles of what many have called "methodological individualism" and "the composition of causes," rather than holism, dominated British thought through the middle decades of the nineteenth century, distinguishing it from much Continental thought. The chief consequences were that it left individual human psychology as the cornerstone of social understandings in Britain, even among those who sought to develop sociology. Moreover, it promoted the exploration, not just of a comprehensive sociology, but also of presumably isolated social domains, including law, politics, religion, national character, economics, and so on.

At the same time that he insisted that in principle, social causes should be understood as operating on individuals, Mill admitted that in practice, the difficulty of actually calculating the effects of many individual causes acting on large numbers of persons, each of which has a set of psychological characteristics determined by a whole set of prior causes, is intractable. This being the case, he advocated also developing Comte's sociological methods, with their historical emphasis. Following Comte, Mill distinguished between Social Statics and Social Dynamics. The first of these would seek to understand the relationships among various features of the "state of society" at any given time; with the understanding that there must be certain "uniformities of co-existence" which regulate the different aspects of societies to form a coherent pattern. That is, society forms a system in which the various components, or social institutions, interact to shape and constrain one another. Social dynamics, on the other hand, would seek to uncover the laws "which regulate the succession between one state of society and another."[11]

Of these two kinds of sociological inquiry, the dynamical (or historical), was the more fundamental for Mill, as it was for Comte: "The mutual correlation between the different elements of each state of society is . . . a derivative law, resulting from the laws which regulate the succession between one state of society and another; for the proximate cause of every state of society is the society immediately preceding it. The fundamental problem, therefore, of the social science [i.e., sociology] is to find the laws according to which any state of society produces the state which succeeds it and takes its place."[12] The method for approaching this problem "consists in attempting by a study and analysis of the general facts of history, to discover the law[s] of progress; which law[s], once ascertained, must . . . enable us to predict future events, just as after a few terms of an infinite series in algebra we are able to detect the principle of regularity in their formation, and to predict the rest of the series to any number of terms we please."[13]

Historical study, guided by the presumption that it will reveal some principle of progress, came to guide much British social thought in connection with the growing interest in Positivism. At the same time, as we saw in chapters 4 and 5, both German idealist and dialectical materialist philosophies thrust temporal development into the center of intellectual life, reinforcing the interest in history among otherwise divergent British scholars, including William Whewell and Spencer, both of whom had become fascinated with German cultural developments.

Progressive historical understandings tended to focus on the differences between how people function at different stages of historical development. One consequence of this view was that it could be appropriated, even by people who advocated radical democracy for Englishmen, to justify treating non-Europeans in paternalistic ways that would have been viewed as intolerable when applied to civilized Europeans. For example, John Stuart Mill argued in his famous essay, *On Liberty,* that "despotism is a legitimate mode of government in dealing with barbarians, provided the end be their improvement."[14] Many others insisted that the study of anthropology, that is, the differences in human characteristics at different stages in social development, should play a key role in the exercise of imperial power and that it "must be consulted for the future help and guidance in the government of alien races."[15] As a consequence, even before biological evolution became linked to racist justifications of imperialist policies, historical investigations were already being used extensively to justify notions of European superiority to and responsibility for those in earlier stages of social progress. One of the great ironies, of course, is that the appeal to history that authorized the European domination of others less advanced than themselves was, as we have seen, often a consequence of Europeans' awareness that they were unable to discover adequate rational or religious grounds to guide their own social and political lives.

Varieties of Biological Evolution in the Early Nineteenth Century: Degenerative and Progressive

By the early 1870s, most philosophical approaches to history with any widespread public impact in Europe had become associated with theories of biological evolution. Darwin's theory of evolution by natural selection, with its self-conscious rejection of teleology, eventually triumphed within the Anglophone biological community in the 1920s; but long before that it generated, or at least promoted, powerful ideological positions. Both "degenerationist" theories of biological transformation and "progressive" transformist theories vied for adherents among students of natural history before Darwin and Alfred Russell Wallace emphasized the concept of natural selection. Progressive evolutionary theories remained not only alive but dominant among biologists in America, France, and the Soviet Union well into the twentieth century. And both progressive and degenerative evolutionary theories continued to have substantial impact on social thought and policies in Europe and America, even where and when natural selection dominated biological thought.

In order to provide a background for understanding the appropriation of biological models of evolution for social purposes, we will very briefly consider the growth of evolutionary theory in biology, concentrating attention on those features that had the greatest cultural significance or that produced the major distinctions between degenerative, progressive, and nonteleological versions. For our purposes, any biological theory will be said to be evolutionary if it acknowledges gradual changes in species over time and accounts for such changes in a naturalistic and noncataclysmic fashion. Whether such a theory posits one or more than one "original" organism is irrelevant. Furthermore, how it accounts for the initial organism(s) is also irrelevant. I use the term "evolution" because most modern students of nineteenth-century biological thought do so. It should, however, be noted that the term evolution was seldom used until the twentieth century and it appears nowhere in the most famous book on evolutionary theory, Charles Darwin's *The Origin of Species by Means of Natural Selection, or the Preservation of Favored Races in the Struggle for Life* (1859), though Darwin does use the term "evolved" to describe forms that now exist. Darwin's most common expression was "descent with modification," though he also used "transformation." Charles Lyell tended to use "successive development," while others, such as Robert Chambers (1802–71), were inclined to use "development" or simply to talk about a "law of progress" and occasionally about "degeneration" or "reversion."

The most comprehensive and influential eighteenth-century evolutionary theory was offered by Georges-Louis Leclerc, Comte de Buffon (1707–88), a man of wide-ranging interests whose work encompassed Newtonian mathematics and natural philosophy as well as geology, anthropology, and natural history. In 1739, Buffon became superintendent of the *Jardin du Roi* in Paris and began to work on a comprehensive survey of natural history that appeared as *Histoire Naturelle, générale et particulière,* in fifteen regular installments plus seven supplemental volumes between 1749 and 1789. Though he seems to have begun his major work believing that "species" are artificial human categories, Buffon soon came to consider them as groups of individuals "belonging to a distinct population maintained by reproduction."[16] In the early 1750s he openly asked whether it was possible that some closely related species, such as horses and donkeys, were both descendants of a common ancestor, suggesting that the structures that determined the characteristics of species were fixed in such a way that that was impossible.

By the time he had published volume 14 in 1766, however, Buffon had changed his mind. Now, probably because he was becoming increasingly interested in the differences among human races as well as in the geological evidence for changes in animal populations, Buffon suggested that it was most likely that all members of a Linnaean genus came from a single original interbreeding population. The structure of members of this population was determined by an "internal mold" that shaped the organism much as an external mold shaped a plaster statue or a wax candle. When different groups from the original species migrated to different regions, they faced different physical environments; and these environments produced changes in the progeny because different nutriments had to be utilized in the reproductive processes.[17] That is, new materials were shaped by the internal mold so that the finished product was slightly different, and new species were formed. Changes were limited, however, because regardless of the different materials available to the body in different places, the internal molds (which now defined the genuses) did not change.

When Buffon discussed the creation of new species out of a single ancestral form, he spoke of a "degeneration" from the initial form, understanding this term in a definite evaluative way. He thus came under fire from Thomas Jefferson because he insisted that New World cats had degenerated from their Old World ancestors, as was evidenced by their inferior size. Buffon was not an open advocate of the design argument, but his notion of degeneration implicitly seemed to assume that the internal mold that produced the ancestral form of current species worked best with the materials available at the time and place of its origin, so that both time and distance produced inferior organisms. Though Buffon was inclined to believe that only material forces operated in creating all of the forms of life produced in nature, the presumed stability of the interior molds that shaped every species allowed for a compatibility between this theory and natural theological arguments that posited a guiding designer of the whole process.

In the 1778 supplemental volume, *Les èpoques de la nature,* Buffon incorporated his understanding of changes in species into an important new theory of earth history. Probably to allay religious fears, he retained the traditional Christian notion that there were seven distinctive periods in the creation of the earth as we know it. But he offered a new, experimentally based, time scale, much longer than that accepted by Biblical Chronologists such as Archbishop Usher, who had determined that the earth was created in 4004 B.C. Moreover, the key events of Buffon's seven epochs had little to do with the biblical days.

According to Buffon, the earth was produced initially when a molten drop of material, largely iron, splashed out of the sun when it was struck by a comet. During a second epoch, the earth cooled and produced a solid crust that buckled as it shrank, producing geological features and allowing for the venting of trapped gasses including water vapor. In the third epoch water condensed out of the atmosphere; and as cooling continued, conditions became possible for the formation of organic chemicals that spontaneously organized themselves into the first life-forms.[18] During the next epoch, as cooling continued, these life-forms migrated toward the equator, eventually dying out as conditions became too cold for their persistence. During the fifth epoch, a new set of organic materials and organisms, the ancestors of those species that exist today, were spontaneously generated; and in the next epoch these ancestral species migrated into new environments, changing in limited ways and getting smaller as the earth continued to cool. Now, in the final epoch, climatic changes have become so slow that the rate of change of species has become so small that they seem to be stable.

Experimenting with the cooling rates of iron balls of differing diameters from nearly molten to near current atmospheric temperatures, and extrapolating to a ferrous sphere the size of the earth, Buffon was able to estimate the age of the earth at approximately one hundred thousand years, very nearly twenty times the age allowed by Biblical Chronologists. Thus, each of his epochs was thousands of years.

Note that in using experiments done on small ferrous balls in the present to estimate the age of the earth, Buffon was implicitly using an extension of an argument that he had offered in the first volume of his *Natural History,* which had appeared nearly thirty years earlier. Borrowing from the first of Isaac Newton's "Rules of Right Reasoning in Natural Philosophy" from book 3 of the *Principia,* Buffon wrote: "To give consistency to our ideas, we must take the earth as it is, examine its different parts with minuteness, and, by induction, judge of the future [and the past] from what at present exists. We ought not to be affected by causes which seldom act, and whose action is always sudden and violent. These have no place in the ordinary course of nature. But operations uniformly repeated, motions which succeed one another without interruption, are the causes which alone ought to be the foundation of our reasoning."[19] This basic argument, which insisted that natural processes acted uniformly over time and that it is illegitimate to posit the existence of past processes that one does not experience in the present, became in the hands of the Scottish geologists James Hutton and Charles Lyell, the foundational principle of what came to be called "uniformitarian" geology.

Buffon's theory accounted for many of the gross features of both the fossil record and the present distribution and stability of species, including the apparent extinction of many early life-forms and the fact that the ensemble of early fossils did not contain the "higher" forms, including all mammals. But it failed dismally to account for a feature of the fossil record that had become increasingly clear by the beginning of the nineteenth century. While his theory suggested that the progeny of any ancestral form should become smaller and less complex, evidence increasingly suggested that on balance, earlier species were smaller and less complex than their successors.

Among the theories intended to account for this apparently "progressive" character of the transformation of species, far and away the most important was a theory developed between 1802 and 1809 by Buffon's former protégé at the *Jardin du Roy,* Jean Baptiste Lamarck. Lamarck, whose *Philosophie zoologique* appeared in 1809, had few immediate and close followers for several reasons that we will mention, but variations on his progressive theory of transformism were widely influential in both France and England between 1820 and 1860 and then again on the Continent and in America toward the end of the century.

Like Buffon, Lamarck argued that life originated spontaneously out of nonliving materials. But his study of invertebrates, begun when he was named to the chair of invertebrate zoology at the *Muséum d'hisoire naturelle* in 1794, convinced him that only the simplest organism could arise spontaneously. He speculated that this happened when an electric current passed through a gel, producing a tubular structure that could serve as a gut. This process occurred many times in the course of geological history, and each time it produced the same simple organism rather than the multiplicity of forms that Buffon had postulated. From the repeated creations of this one simple form, all other life-forms arose through natural processes.

Here, at the very beginning of his theory, Lamarck lost most competent French scientists; for the chemical concepts that underlay his particular theory of spontaneous generation were idiosyncratic and incompatible with the almost uniformly accepted chemistry of Lavoisier. Moreover, his insistence that only a single form could emerge from nonliving matter and that such a form was the ancestral form of all living things placed him in opposition to Georges Cuvier, holder of the more prestigious chair of vertebrate zoology at the Muséum, and one of the most respected scientists in France.

Cuvier had proposed an earth history with a series of catastrophic events in violation of the uniformitarian geological assumptions of Buffon and Lamarck. Furthermore, he was able to offer an argument to justify his approach that many geologists and biologists found compelling at the beginning of the nineteenth century. Much of the fossil evidence appealed to by Lamarck was drawn from the Paris Basin, which Cuvier had studied intensively for many years. If Lamarck was correct and organisms changed only gradually over time, then one should be able to trace these gradual changes in the fossils contained in successive layers of rock, assuming that successive layers of rock represented successive periods of time, with the oldest on the bottom and the most recent on the top. But Cuvier found no such evidence. Successive layers of rock did contain fossils that became generally more complex as one approached the present; but they did not do so continuously. A layer of freshwater fossils might be succeeded by a layer with land animals and then by a layer with saltwater organisms. Moreover, the breaks seemed sudden. As a consequence, Cuvier argued that relatively stable periods on the earth's surface were punctuated by dramatic and discontinuous events, including floods and freezes. The Paris Basin, for example, had occasionally been inundated by flash floods, which then receded rapidly. According to Cuvier: "It is . . . extremely important to notice that these repeated inroads and retreats were by no means gradual. On the Contrary, the majority of the cataclysms that produced them were sudden. This is particularly easy to demonstrate for the last one which by a double movement first engulfed and then exposed our present continents, or at least a part of the ground which forms them. It also left in the Northern countries the bodies of great quadrupeds, encased in ice and preserved with their skin, hair, and flesh down to our own times. If they had not been frozen as soon as killed, putrefaction would have decomposed the carcasses. . . . This development was sudden, not gradual, and what is so clearly demonstrable for the last catastrophe is not less so of those that preceded it."[20]

Cuvier's "catastrophist" geology abandoned Newton's rules of reasoning, but it seemed to many to be much more closely tied to the geological "facts," or evidence, and it seemed to justify Cuvier's additional claim that after each catastrophe there was a new creation of a new group of species. Furthermore, it allowed such religiously oriented geologists as William Buckland and Adam Sedgwick to argue that catastrophism was consistent with the biblical flood story. Thus, in 1823, Buckland published his *Reliquiae Diluvianae: or, Observations on the Organic Remains Contained in Caves, Fissures, and Diluvial Gravel, and on other Geological Phenomena Attesting to the Action of a Universal Deluge.*

The geological arguments of William Smith carried out during the 1790s offered a possible way around Cuvier's discontinuous evidence; for Smith argued that in any given place the sequence of strata was incomplete because some layers were scoured away by the action of moving water or winds. The fossil record at any particular place was thus discontinuous; but one could construct a continuous geological column by pooling evidence from many locations. For those who were wedded to a strongly empiricist notion of science and who were also influenced by the biblical flood story, the direct evidence seemed to favor catastrophism. For those more wedded to Newton's rules of reasoning, uniformitarianism, bolstered by Smith's interpretation of stratigraphic evidence seemed most plausible, and Lamarckian ideas had substantial appeal. Thus, at the beginning of his three-volume *Principles of Geology*, Charles Lyell explained why he preferred uniformitarian geology in spite of some of the puzzling phenomena put forward by Cuvier: "When difficulties arise in interpreting the monuments of the past, I deem it more consistent with philosophical caution to refer them to our present ignorance of all the existing agents, or all their possible effects in an indefinite lapse of time, than to causes formerly in operation, but which have ceased to act."[21]

Once Lamarck had accounted for the origin of life in the simplest of organisms, he drew a key notion from his *Idéologue* friend Pierre Cabanis, who had argued that new and completely unanticipated properties emerge in connection with highly organized material entities. For Lamarck, the property of progressive self-organization emerged with life. That is, from the beginning, living beings had achieved a level of complexity sufficient that they contained within themselves a capacity not only to reproduce themselves, but also to generate increasingly complex life-forms with ever more specialized organs. Moreover, since this capacity was a determinate thing, all other things being equal, successive life-forms constituted a single lineage that would be repeated in all of its details each time life was again produced spontaneously. Currently existing species were supposed to constitute a single hierarchical scale of life-forms, the most complex and specialized of which represent the descendants of the earliest spontaneous emergence of life, while the most simple and homogeneous represent the descendants of the most recent spontaneous emergence. While it was thus not true for Lamarck that all current life forms were the progeny of a single original organism, it was true that they were products of a series of parallel progressive developmental paths.

While this aspect of Lamarck's theory constituted a direct challenge to most Christian accounts of creation, it held tremendous appeal for materialist-leaning

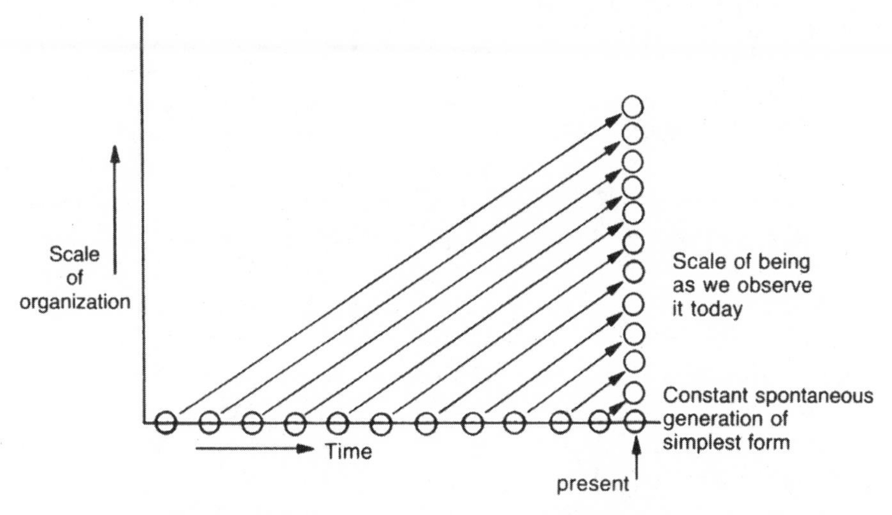

Fig. 7.1. Lamarck's Theory of Organic Progression. From Peter J. Bowler, *Evolution: The History of an Idea,* rev. ed. (Berkeley: University of California Press, 1989, p. 85).

comparative anatomists such as Geoffroy Saint-Hilaire and his followers, who saw a unity of composition in all organisms, in contrast with Cuvier's multiplicity of basic patterns. Beginning in the 1820s, Saint-Hilaire and his close follower Étienne Serres offered a "recapitulation" theory based on Lamarck's transformism that began to account for the fact that, while it was sometimes difficult to see the relationships between many structures in adult organisms, it was often very easy to see them in the embryonic stages. That is, they argued that in their development, the embryos of higher organisms pass through the stages at which the embryos of lower organisms stop. Human embryos, for example, pass through, or recapitulate, a gilled stage that fish embryos never pass beyond.[22]

Through Saint-Hilaire, Lamarckism then entered the teaching of anatomy at secular medical schools in France and England, becoming a standard feature of "radical" medical education from the 1820s through the early 1840s. It is likely that the young Charles Darwin first became aware of evolutionary theories through the teaching of the medical lecturer Robert Grant in Edinburgh in the late 1820s, though they apparently made little impression on him at the time. For better or worse, Lamarck's theories were linked closely in British medical education both to Saint-Hilaire's notion that there was a single hierarchical series of organisms of increasing complexity and specialization and to theories of recapitulation. So, in the early 1840s, when Richard Owen managed to severely challenge both the presumption of a single series and that of recapitulation, Lamarckian transformism was generally rejected in England, even though, as we shall see, Lamarck's theory logically entailed neither the existence of a single progressive series nor embryological recapitulation.

It was obvious to Lamarck, as it was to many others, that there was in fact, *not* a single path of transformations that passed linearly through fish, birds, and mammals as those features of his theory discussed so far would suggest. Instead, he argued, the path from simple, homogeneous, organisms toward complex, heterogeneous ones could be deflected by changes in the environment faced by organisms at any stage. As one developmental sequence passed through reptiles, for example, environmental conditions favored the production of birds rather than mammals, while in another case, the environment favored the production of amphibious mammals. The question was, What mechanism accounted for these divergences from what would otherwise have been a uniform line of descent?

To answer this question, Lamarck introduced the notions of an interior sense and volition. Faced with a particular environment, which offers opportunities for an organism to thrive, an organism might choose to exercise those portions of its body that increase, for example, its ability to get food, while it might leave other parts of its body relatively unused. Those parts exercised most frequently would be strengthened and would develop a greater supply of fluids, while those exercised less often would atrophy. A fish living in a drying lake in which water-borne food is decreasing while there is a ready supply of worms on the damp land recently uncovered, for example, might choose to use its side fins to lever itself out of the water to get food, increasing the strength and size of its side fins, while its dorsal fin and tail fin decreased in size and strength through disuse. If, as Lamarck supposed, physical characteristics acquired in this way can be inherited by the organism's progeny, we can imagine over time the passage of our fish into an animal approaching something like a salamander, with legs, no dorsal fin, and a finless tail.

It was this notion of the inheritance of characteristics acquired through the willful exercise of certain organs that became both the most famous and one of the most troublesome aspects of Lamarck's theory, especially among socially and religiously conservative thinkers. If it were possible for an organism, through its own efforts, to transcend its given place in the natural order and to lift itself into a place higher in the chain of being, the implications for the stability of the human social order were staggering, as were the problems for any notion of preordained design. Thus, respectable persons came to associate Lamarck's version of transformism with working-class socialism, free-thinking materialism, and even atheism.[23] Conservative anatomists and natural historians, while they often admired Lamarck's descriptive work on invertebrates, rejected his notion of the inheritance of acquired characteristics, often explicitly because of its political implications. Joseph Henry Green (1791–1863), the conservative teacher of Richard Owen, for example, appealed to the ideas of Cuvier and von Baer to assure his audiences that they need not suppose, "that there is any power in the lower to become, or to assume the rank and privileges of, the higher, upon any such fanciful scheme as that proposed for the invertebrate animals by the laborious and otherwise meritorious naturalist, Lamarck,—a scheme in which the ground and cause is everywhere meaner and feebler than the effect, and in which blindness is made the source of light, and ignorance would be the parent of mind and thought."[24] Instead, Green assured his listeners, it was much more sound philosophy to assume "that the ascent is an indication of a law; and the manifestation of a higher power acting in and by nature."[25]

Given the combination of Lamarck's inability to establish the heritability of acquired characteristics, his eccentric chemistry, his presumption of multiple spontaneous generations of the same form in the face of mounting evidence against spontaneous generation, and the religious and social liabilities that his theory carried with it, it should hardly be surprising that Lamarck's transformism had few early supporters outside of the relatively small radical and religiously heterodox medical communities in Britain and France.

Evolution Becomes Popular: Robert Chambers's *Vestiges of the Natural History of Creation*

In 1844 an anonymously published work, *Vestiges of the Natural History of Creation,* appeared in London, incorporating a version of transformism into what was purportedly intended to be a respectable extension of the natural theological tradition represented by the Bridgewater Treatises. Drawing from the latest works in astronomy, geology, natural history, anatomy, and philosophical history, the author, later determined to be the Scottish publisher Robert Chambers, wove together a story in which all domains of nature reflected a common divinely ordained pattern of continuous development. The growth of a child becomes the model for understanding the origins of the universe and all that it contains. Solar systems and planets grow out of the condensation of nebular matter. Geological strata provide evidence of a progressively complex organic population, beginning with invertebrates and moving through fish, reptiles, mammals and ultimately, humans. And history provides a record of increasingly complex and morally advanced societies.

Ignoring the volitional element from Lamarck's transformist theory, *Vestiges* presents an account of the transformation of organisms that incorporates elements from both the Lamarck-based recapitulation theory and Karl von Baer's doctrine of the progressive differentiation from a general pattern to an increasingly specialized one. Chambers explicitly uses Charles Babbage's discussion from the *Ninth Bridgewater Treatise* to suggest that, while ordinary reproductive processes almost always produce progeny like the parents, God might have programmed into the reproductive process an occasional deviation from this principle, allowing a move from one form of being to the "next more complicated."[26] Fetal development is governed by a progressive and branching law that Chambers illustrates with the simple accompanying diagram:

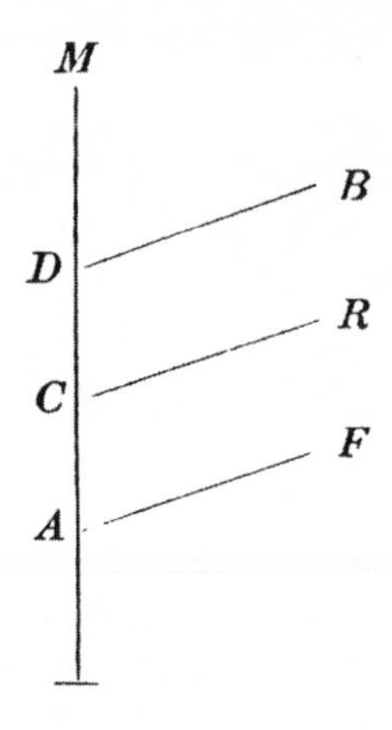

Fig. 7.2. Chambers's diagram indicating the divergence of foetal organisms from the main line of development. From Robert Chambers, *Vestiges of the Natural History of Creation* (London: John Churchill, 1844, p. 212).

Suppose that the line AM represents the normal development of a fetus over time. Fetuses of fish, reptiles, birds, and mammals all develop identically up to point A: "The fish there diverges and passes along a line apart, and peculiar to itself, to its mature state at F. The reptile, bird, and mammal go on together to C, where the reptile diverges in like manner and advances by itself to R. The bird diverges at D, and goes on to B. The mammal then goes forward in a straight line to the highest point of organization at M. . . . Limiting ourselves at present to the outline afforded by this diagram, it is apparent that the only thing required for an advance from one type to another in the generative process is that, for example, the fish embryo should not diverge at A, but go on to C before it diverges, in which case, the progeny will be not a fish, but a reptile. To protract the *straight-forward part of the gestation over a small space*—and from species to species the space would be small indeed, is all that is necessary."[27] Chambers was unable to specify just what conditions led to prolonging the path of fetal development of successive organisms along the line AM. Instead, he insisted that it was enough to show that there must be such conditions.

It is important to note that Chambers included humans within his developmental theory, seeing them as the pinnacle of the developmental process. To provide evidence for the continuity between the inferior animals and man, he drew from the works of the phrenological physiologist, G. J. Davey, who argued that the brains of human infants and lunatics are like those of healthy animals, "lower in the scale of organization."[28] Similarly, Chambers argued that Caucasians represented the peak of human development, concluding his discussion of human physiognomy with the italicized claim that "*the leading characters, in short, of the various races of mankind, are simply representations of particular stages in the development of the highest, or Caucasian type.*"[29] Finally, he insisted that the female of each species represents a condition in which fetal development is arrested before its completion. In this case Chambers draws his evidence from descriptive studies of bees and argues that what happens among insects should be taken as a model that extends even to humans, because, "there is a unity throughout nature which makes the one case an instructive reflection of the other."[30]

With respect to each of these central issues—the incorporation of humans into the natural order and the developmental accounts of racial hierarchy, sexual physiology, and even mental illness—Chambers brought to the attention of a broad public audience a series of issues to which later evolutionary theories would be applied, with major social impacts. Interestingly, among these issues, the only one on which Chambers received serious criticism was the basic assumption of continuity between animals and humans, which even religious moderates in England still found unacceptable.

Though it admittedly involved substantial speculation, especially in connection with its theory of the transmutation of species, *Vestiges* appealed widely to middle-class nonscientific audiences, selling over 23,300 copies in Britain between 1844 and 1860.[31] A middle-class liberal and a moderate in religion, Chambers took special care to insist that, even though the processes of development laid out in *Vestiges* were naturalistic, they were completely consistent with orthodox discussions of the action of God's ordinary providence through natural laws.[32] Moreover, he offered a doctrine of gradualist progress that was consistent with liberal reforming politics. Thus he sought to decouple transformist ideas from their atheistic and radical political associations, and for the most part he was successful except in the eyes of some spokespersons for religious orthodoxy.

Initial reviews in liberal religious and political periodicals were largely favorable, though a few extreme evangelical reviewers saw the naturalistic arguments of *Vestiges* as a support for atheistic materialism. To Chambers's surprise, harsher reviews came from the scientific community. The Cambridge group—including the geologist Adam Sedgwick, who wrote an eighty-five-page *Edinburgh Review* response; William Whewell, who produced a pamphlet response, *Indications of a Creator*, in 1845; and John Herschel, who attacked *Vestiges* in his presidential address at the June 1845 meeting of the British Association for the Advancement of Science—were particularly critical.

In general, scientists attacked the author for being too speculative and for failing to do the difficult observational work necessary for inductive science.[33] Sedgwick was particularly insistent that Chambers had overemphasized the continuity of the fossil record and he claimed that in reality there were substantial discontinuities, with whole new classes appearing with no obvious precursors. The consequence of Chambers's work was that evolutionary ideas became a subject for general polite discourse, although Sedgwick did think that it should be kept out of the hands of "our glorious maidens and matrons."[34] It became clear, however, that if any theory of biological evolution was to get a serious hearing within the scientific community in Britain, it would have to be built upon a collection of evidence vastly more compelling and comprehensive than the suggestive analogies that Chambers offered. More importantly, the gaps in the geological record would have to be more satisfactorily explained. These circumstances were not lost on the young naturalist Charles Darwin as he read the review of *Vestiges* by his mentor in geology while he grappled with formulating his own presentation of a theory of evolution.

Herbert Spencer and the Linkage between Biological Evolution, Energetics, and "Progressive" Social Theory

In Germany, the rise of the historical disciplines was associated with a turn away from analytic models of understanding toward models that likened historical processes to organic growth and development. In each case, the organism, whether biological or social, was understood to be undergoing a process guided by some internal force, principle, or *bildungstrieb.* In England, Herbert Spencer most effectively promoted a similar shift toward philosophies of history that appealed to organic images and presumed the existence of some kind of internal directing principle.

Born in 1820 to a scientifically inclined schoolmaster in Derby, Spencer drew his initial political orientation from an uncle who was a Chartist sympathizer and an intense believer in laissez-faire economic ideas. Almost entirely self-educated, Herbert began publishing on both the natural sciences and on social questions when he was just sixteen, a year before he became an engineer for the London and Birmingham Railway.[35] Though sympathetic to many radical goals, Spencer was committed to gradualist processes long before he began writing from an evolutionary perspective, before Chambers's *Vestiges* appeared, and before he learned of Comtean Positivism. He wrote in 1843: "I look upon despotism, aristocracies, priestcrafts, and all other *evils* that afflict humanity, as the necessary agents for the training of the human mind, and I believe that every people must pass through the various phases between absolutism and democracy before they are fitted to become *permanently* free."[36]

In 1840, Spencer, whose brother was a geologist, read Lyell's *Principles of Geology,* which had a doubly important impact on his social thought. First, it focused his attention upon the universality of change in the natural world. Second, it introduced him to Lamarck's version of biological evolution, to which he remained committed with respect to most issues for the remainder of his life. Thus, in his *Social Statics* of 1851, in spite of the title, Spencer emphasized what Mill had called the dynamic study of society. He stressed the ubiquity of change, drawing principally on geological images:

> Nature, in its infinite complexity is ever growing to a new development. Each successive result becomes the parent of an additional influence, destined in some degree to modify all future results. No fresh thread enters into the texture of that

> endless web, woven in "the roaring loom of Time," but what more or less alters the pattern. It has been so from the beginning. As we turn over the leaves of earth's primeval history, as we interpret the hieroglyphics in which are recorded the events of the unknown past, we find this same ever-beginning, never-ceasing change. We see it alike in the organic and the inorganic, in the decompositions of matter, and in the constantly varying forms of animal and vegetable life. Old formations are worn down; new ones are deposited. Forests and bogs become coal basins, and the now igneous rock was once sedimentary. With an altering atmosphere and a decreasing temperature, land and sea perpetually bring forth fresh races of insects, plants, and animals. . . . Strange indeed it would be if, in the midst of this universal mutation, man alone were constant, unchangeable. But it is not so. He also obeys the laws of indefinite variation. His circumstances are ever altering, and he is ever adapting himself to them.[37]

The ubiquity of change, especially of change in response to a changing environment was not the only lesson regarding humans and their social institutions to be learned from the changes of organisms over time, however. Lamarck had suggested that the transformation of organisms through time showed a pattern in which organisms became more complex and in which the parts of which they were composed became increasingly highly differentiated. Such a pattern was also observed in the growth of individual organisms by German developmental biologists, including Karl von Baer, whose views were promoted by Richard Owen in England and appropriated by Chambers and Spencer. Emphasizing the analogy between societies and living organisms, Spencer wrote:

> A . . . remarkable fulfillment of this analogy is to be found in the fact that the different kinds of organization which society takes on in progressing from its lowest to its highest phase of development are essentially similar to the different kinds of animal organization. . . . This union of many men into one community, this increasing mutual dependence of units which were originally independent, this gradual segregation of citizens into separate bodies with reciprocally subservient functions, this formation of a whole consisting of numerous essential parts, this growth of an organism of which one portion cannot be injured without the rest feeling it, may all be generalized under the law of individuation. The development of society, as well as the development of man and the development of life generally, may be described as a tendency to individuate—to become a *thing.* And rightly interpreted, the manifold forms of progress going on around us are uniformly significant of this tendency.[38]

All but one of the basic themes of Spencer's evolutionary social theories are already signaled in *Social Statics.* The notion that societies, like biological organisms, evolve from relatively homogeneous to increasingly heterogeneous forms; the notion that the ultimate goal of social evolution is a condition in which each person will be free to express most completely his or her own individuality; the notion, borrowed from both Lamarck and von Baer, that there is some kind of internal compulsion, or force, that drives social institutions to modify themselves in response to changing circumstances; and the related conviction, certainly borrowed from laissez-faire economic doctrines, that if evolution is to do its job in perfecting society, it must be left to a large degree to follow its own internal logic without undue interference—all were present in Spencer's thought in 1851.

Even the notion that competition among individuals or institutions might provide a mechanism for preserving those that are superior was hinted at in a discussion of improvements in the delivery of mail, which had been brought about by competition among private carriers. The positive claim that competition among similar entities for scarce resources (in this case, customers) played a minor role in the overall argument of *Social Statics.* It was used principally to extol the virtues of free trade vis-à-vis governmental monopoly.[39] The related negative claim, that is, that by insulating persons from competitive pressures we make it impossible for the evolutionary process to produce the progress that is its goal, plays an absolutely central role in Spencer's widely read and cited argument about the Poor Laws: "We must call those spurious philanthropists, who, to prevent present misery, would entail greater misery upon future generations. All defenders of the Poor Law must, however, be classed as such. That rigorous necessity which, when allowed to act on them, becomes so sharp a spur to the lazy and so strong a bridle to the random, these paupers' friends would repeal, because of the wailings it here and there produces. Blind to the fact that under the natural order of things, society is constantly excreting its unhealthy, imbecile, slow, vacillating, faithless members, these unthinking, though well-meaning, men advocate an interference which not only stops the purifying process but even increases the vitiation . . . thus in their eagerness to prevent the really salutary suffering that surrounds us, these sigh-wise and groan-foolish people bequeath to posterity a continually increasing curse."[40]

Spencer did suggest briefly that humans were gradually developing a greater capacity for sympathy, which Adam Smith had seen as the foundation for all morality. He then claimed that one of the negative consequences of the kind of public charity implicit in the Poor Laws was that it actually inhibited the growth of benevolence that was a product of individual acts of charity. It was Darwin, however, who turned the growth of sympathy into the driving force of human social evolution in the *Descent of Man,* and it was only after Darwin had done so that Spencer emphasized in his *Principles of Ethics* (1879), that altruism became a significant consideration for survival as societies advanced.[41]

For the next fifty years Spencer continued to explore and amplify these themes in a series of widely read essays and books, including *First Principles* (1862), *The Study of Sociology* (1873), and *The Principles of Sociology* (4 vols., 1876–85), all of which constituted parts of what he called his Synthetic Philosophy. He added only one major element to his Synthetic Philosophy after 1851. This addition gave Spencer's theories an extra claim to comprehensiveness and an extra appeal to those who were looking for the latest scientific fad. More importantly, it set evolution in a context drawn from the most important new development in the physical sciences during the mid nineteenth century, and it linked evolution to an inverse process that Spencer called dissolution, but which under the name of "degeneration," was to play a huge role in later nineteenth-century cultural developments.

Drawing from William Groves's *The Correlation of Forces,* a popular treatment of what we now call Energetics, Spencer linked his notion of progressive evolution to a source in the transformation of energy associated with the random motion of disconnected particles into an energy associated with ordered material structures. During the 1840s, several scientists, including Julius Robert Mayer, James Prescott Joule, and Hermann von Helmholtz, working in very different contexts, had demonstrated that there was a physical quantity called "*Kraft,*" later, "energy," that could be manifested in a variety of ways and that was conserved through transformations from one manifestation to another. By 1847, Helmholtz had expressed the new insight in a particularly useful way. For any system of particles, the total energy could be understood as consisting of the sum of two terms, one representing all of the masses of the particles times one-half of the squares of their respective velocities (this came to be called the kinetic energy of the system), and the other representing the "tensions" that existed by virtue of the fact that each particle might attract or repel some or all of the other particles with a force related to the distance between the two particles and to the specific kind of particles they were (electrically charged particles, for example, attracted particles of opposite charge and repelled particles with the same charge with a force that varied as the inverse square of the distance). Moreover, in any physical process in which the system of particles being considered was isolated from the environment, the total energy of the system remained unchanged. This relationship could be expressed by the following equation:

$$E = \Sigma \tfrac{1}{2} m_i V_i^2 + \Sigma\Sigma\, Y(r_i, r_i)$$

where m is the mass of a particle, v is its velocity, and Y (ri, rj) is some function of the inter-particulate distance. The energy described in the function Y came to be called potential energy.

Spencer took the conservation of energy in any natural process and defined evolution as a process in which energy is transferred from the random kinetic energy of the particles of a material system into energy stored as "tensions" in complex structures within the system: "Evolution is an integration of matter and concomitant dissipation of motion; during which the matter passes from an indefinite, incoherent homogeneity to a definite, coherent heterogeneity; and during which the retained motion undergoes a parallel transformation."[42]

Such a definition did little to suggest how particular evolutionary paths might be analyzed. But viewing evolution as an energy transformation implied that the processes associated with evolution would also be subject to the second law of thermodynamics that Rudolf Clausius and William Thomson, Lord Kelvin, had articulated in the early 1850s. This second law can be expressed in a wide variety of logically equivalent ways, one of which is to say that in natural processes there is a tendency for all energy in a system of particles to be transformed into kinetic energy, such that the velocities of all particles of the system are uniform, that is, such that there are no temperature differentials in the system and thus no possibility for converting the motions of the particles into useable work. Given this perspective, although Spencer, like Lamarck, felt that systems naturally evolved more highly differentiated and complex structures, he was forced to argue that there must be a natural process that is the inverse of evolution that leads to the creation of motion at the expense of structural organization. This process, he termed "dissolution": "evolution, under its simplest and most general aspect is the integration of matter and concomitant dissipation of motion; while dissolution is the absorption of motion and concomitant disintegration of matter."[43]

Spencer was vastly more interested in "progressive" evolutionary processes, and at least early in his career he was confident that the forces of dissolution could in many cases be overcome. After 1862, however, he was fully convinced that evolution represented only a temporary triumph over the competing natural processes of dissolution.

At the level of social transformation, Spencer linked the idea of the inevitability of dissolution and its possible extended avoidance to an element appropriated from Comte and Saint-Simon, probably from his reading of Martineau's translation of *The Positive Philosophy* in 1855 (Spencer had a long-term, close, personal relationship with Martineau). Both Comte and Saint-Simon had made a distinction between industrial and militant, or militaristic societies. Using that distinction, Spencer admitted that not all social transformations were necessarily positive and that under appropriate circumstances, progress toward industrial society might be diverted into a retrogressive militaristic path that constituted a form of dissolution. In such a case, individuals knowledgeable about social processes might act self-consciously, within the constraints implicit in the particular stage of the society, to reverse the retrogression and to promote progress once again. Thus, Spencer, like Comte and Marx, did argue that self-conscious reforming activity was possible and potentially valuable, but he insisted that it had to be both gradualist and guided by a sensitivity to the constraints imposed by particular social contexts.

CHAPTER 8

Darwinian Evolution: Natural Selection, Group Selection, and Sexual Selection

Both Robert Chambers and Herbert Spencer had presented sweeping theories of evolution that incorporated the idea that evolution extended fully to human physical and mental developments as well as to societal change well before Charles Darwin published anything on evolution. These theories, which were based largely on Lamarck's progressive development ideas, were being presented to extended audiences. And Spencer's evolutionary social theories were already being used to bolster the legitimacy of laissez faire economics and to justify political policies regarding such issues as the Poor Laws by the early 1850s, almost a decade before the publication of Darwin's first major work on evolution, *On the Origin of Species.* But Darwin's works gradually eclipsed all prior investigations of evolution in terms of their impact upon scientistic social theorizing. Given Darwin's key importance, we spend this chapter focusing on his writings and on those of a few of his most immediate friends and disciples, including Alfred Russell Wallace, the co-discoverer of natural selection; Thomas Henry Huxley (1825–95), "Darwin's Bulldog"; the psychiatrist Henry Maudsley (1836–1918); and the botanist, town planner, and sexual theorist, Patrick Geddes (1854–1932).

Darwin had already been working on the problem of how new species come into existence during the 1830s, well before either Chambers or Spencer began to write. He sent a lengthy summary of his ideas to Joseph Hooker (1817–1911) in 1844, and he had corresponded with a number of English and American naturalists about his ideas throughout the 1840s. Why Darwin chose not to publish until virtually forced into it when Alfred Russell Wallace sent a paper, "On the Tendency of Varieties to Depart Indefinitely from the Original Type," to Hooker for publication in the 1858 volume of the *Journal of the Proceedings of the Linnaean Society of London,* is a matter of contention. Wallace's paper contained the key features of Darwin's theory; so Hooker, who knew how long Darwin had been working on the problem of the transmutation of species, managed to get Darwin to permit him to publish extracts from his 1844 letter in the same issue. Within a year, Darwin had completed and published a nearly six-hundred-page "sketch" of his theory, *On the Origin of Species by Means of Natural Selection, or The Preservation of Favored Races in the Struggle for Life.*

At least early in life, Charles showed no hint of any great ability. Sent to Edinburgh at age sixteen with his older brother, who was studying medicine, Charles was appalled by the dissections he saw and was turned off by the scientific lectures he attended. He later wrote that as he became aware of how wealthy he was going to be, he gave up all scholarly ambition. After a family conference, Charles was sent to Cambridge to study for the ministry because it was socially acceptable, even though his father was an atheist and there is no indication that Darwin felt any special calling. At Cambridge, Charles showed increasing interest in natural history, guided by the botanist, John Stevens Henslow, and the geologist, Adam Sedgwick. Both of these mentors were deeply religious men, and they adopted the natural theological approach to nature that was associated with the writings of William Paley. Both were committed to the notion that species were specially created by God and to the catastrophist version of geology associated with the writings of Cuvier. There is every indication that Darwin shared their basic views in 1831 as he ended his Cambridge career.

As Darwin prepared to leave Cambridge, he was certain that he had no more ambition to become a clergyman than to be a physician; but he had become increasingly fascinated by natural history and natural philosophy, devouring John Herschel's *Preliminary Discourse on the Study of Natural Philosophy* (1831) and Alexander von Humboldt's *Personal Narrative of Travels to the Equinoctial Regions of the New Continent, During the Years 1799–1804* (English trans., 1814–29). Henslow recommended Darwin for a volunteer position as naturalist and companion to Captain Robert Fitzroy on the HMS *Beagle,* which was about to depart on a five-year, around the world voyage to chart the waters off the coasts of South America.

Darwin and the Voyage of the HMS *Beagle*

Darwin set sail from Plymouth on December 27, 1831. He returned to Falmouth on October 2, 1836, a man changed in almost every way. He had abandoned the natural theological perspective on natural history for a completely secular and materialist one. He had become both a convert to and a major contributor to the uniformitarian approach to geology associated with Charles Lyell. Furthermore, he had become convinced that present species are descended from earlier ones; though it was not until about a year after his return to England that he figured out to his own satisfaction how the transformations occurred and were propagated.

The first major step in Darwin's new career took place on the first leg of the *Beagle*'s journey. At Henslow's suggestion, Darwin began reading the first volume of Lyell's *Principles of Geology* (1830) as the ship sailed toward the Cape Verdi Islands. When the *Beagle* approached the coast of Saint Jago, one of the Cape Verdi Islands, Darwin was able to observe directly the dramatic effects of the ongoing processes of elevation and subsequent subsidence that he was reading about by looking at a cross section of the cliffs near a volcano on the island. Fossils located in a stratum of limestone well above sea level indicated that the stratum had once been below sea level and had elevated over time. Near the volcano, however, the weight of the mountain created by volcanic action had pushed the limestone back into the sea. Beginning from these observations, Darwin proposed a new explanation for the distribution and structure of coral atolls in the Pacific Ocean.

The two great puzzles about coral atolls were why such a huge number had their tops within a few feet of sea level and why so many appeared in the form of a shallow ring of land surrounding a lagoon. Darwin knew that corals grow only in shallow water, roughly within sixty feet of the surface, where water movement brings the stationary organisms adequate supplies of food from the open ocean. If, he argued, an island with its top initially well above sea level, slowly subsides, corals will build up a mass in the shallow sea surrounding the island and they will grow most rapidly on the outer boundaries where food is most plentiful. As long as the island sinks slowly enough, the corals can keep up with the subsidence, building up larger and larger masses until they form a ring around a shallow lagoon (see fig. 8.1). Within a matter of weeks after the beginning of his voyage, Darwin had thus begun to contribute to geological theory in a Lyellian vein. When Darwin wrote to Lyell about his new ideas, the great geologist invited him to present his theory to the Geological Society of London as soon as he returned to England.

When the *Beagle* made its first landfall in Brazil in late February of 1832, the second volume of Lyell's *Principles of Geology,* which considered the geographical distribution of organisms and fossils, was waiting for Darwin. Again, Lyell's reflections stimulated the young naturalist to focus on a topic that he had not considered seriously before. As Darwin observed and collected examples of South American fauna and fossils over the next three years, he discovered first that closely related

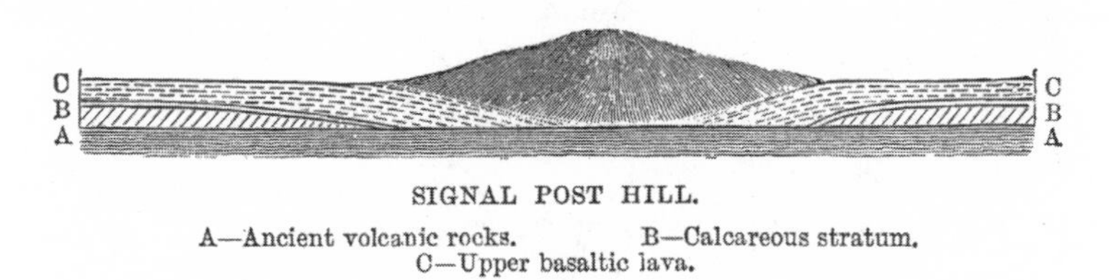

Fig. 8.1. Darwin's diagram of cliffs around a volcanic cone on the Island of St. Jago. From Charles Darwin, *Geological Observations on the Volcanic Islands and Parts of South America Visited during the Voyage of H.M.S. "Beagle"* (New York: D. Appleton and Company, 1896, p. 12).

species of animals succeeded one another over time (he was particularly impressed by the fossil evidence of giant armadillos related to present-day varieties). Second, he observed that in a number of cases in which two distinct but closely related present-day species occupied contiguous geographical regions, there was often a region of overlap, sometimes dominated by an intermediate form. Though he apparently continued to believe with Lyell in the special divine creation of species throughout the voyage, when he returned to England and began to reflect on his experiences during late 1836 and 1837, these two classes of observations convinced Darwin that, contrary to Lyell's opinion, what Lyell called "successive development" did occur and that there was some natural, law-bound, process by which one species was gradually transformed into another over a long period of time.

The case of overlapping geographical distributions provided for Darwin an insuperable problem for the traditional notion of design, with its presumption of perfect divine adaptation of organisms to their environments. If one of two closely related organisms was perfectly adapted to a particular set of circumstances, how could it be the case that one or more other closely related species were, as well? It made more sense to Darwin to think that an original parent species gave rise to varieties that were better adapted to conditions at the extreme ends of its range—where one extreme might be relatively dry and the other relatively wet, or one relatively low and one relatively high, for example—and that those varieties diverged until, for some reason, they ceased interbreeding and became separate species. In such a case, though the two divergent species might each come to dominate in some part of the region, they might continue to coexist with one another and possibly with the parent species in the area of overlap where moisture or elevation was more moderate.

Darwin became particularly fascinated by the puzzle of the relationship between organisms that had become physically isolated and those in nearby, but separated, regions. For a time he even became convinced that the only way that different varieties of a parent species could be blocked from interbreeding long enough to diverge to a point at which they could become mutually infertile was if they were at least temporarily isolated.[1] Once Darwin became convinced that the transmutation of species did occur and that it did so as varieties of some parent species gradually diverged from one another, he began to speculate about and to investigate possible mechanisms for transmutation, setting up special notebooks on the topic.

Darwin, Malthus, and Natural Selection

One key to Darwin's new theory came in the form of an insight gained from reading Malthus's *Essay on Population* late in 1837. Malthus's theory claimed that as humans compete with one another for scarce food, the weak die off, while those better able to compete survive. Lyell had pointed out that Augustin De Candolle had also claimed that in some sense, "all nature is at war, one organism with another."[2] Combining these two notions, Darwin argued that if slight random changes occur in the hereditary material passed on to an organism, that organism will either be advantaged or disadvantaged in its competition with other members of its species. If it is disadvantaged, it is likely to die before it has a chance to reproduce; so the changed hereditary material will not be passed on. If, however, it is advantaged, the individual has a greater likelihood of surviving and reproducing. The change is likely to be passed on to a number of progeny; each of which is more likely to survive and reproduce, and so on, until eventually a new variety of the parent species with the changed hereditary material comes to dominate in the population. By piling up many such small changes, eventually a completely new species may come into existence. This entire process, Darwin called "descent with modification"; and the mechanism for propagating favorable changes in hereditary material he called "natural selection."

Several features of this process deserve notice because of their implications for the subsequent acceptance of the theory. First, it is implicit in the theory that mental and behavioral characteristics as well as physical characteristics, to the extent that they are heritable, must be carried through the hereditary material transmitted from one generation to the next. As a materialist theory, Darwin's views challenged most traditional religious beliefs. Darwin was careful not to make the materialist implications of the theory explicit in *The Origin of Species,* in which his discussions of behavioral characteristics, instincts, intelligence, and so on, were carried out as if they might be inherited independent of bodily structures. But most readers understood Darwin as a materialist; and in *The Descent of Man and Selection in Relation to Sex,* published in 1871, he discussed at length the relationships between brain structure and mental functioning, especially in connection with the differences between ordinary humans and "macrocephalus idiots,"[3] reinforcing the notion that mental events have physical causes.

Second, insofar as the variations in hereditary material are assumed to be random, evolution by natural selection is a completely nonteleological process and its results are thus not designed. For many nineteenth- and early-twentieth-century readers, the cruelty and inefficiency associated with such a process seemed inconsistent with their understandings of a caring God or their hopes for secular progress. George Bernard Shaw expressed the revulsion of many to the apparently accidental character of the process in a particularly powerful way in the introduction to *Back to Methuselah:*

> The Darwinian process may be described as a chapter of accidents. As such, it seems simple, because at first you do not at first realize all that it involves. But when its significance dawns on you, your heart sinks into a heap of sand within you. There is a hideous fatalism about it, a ghastly and damnable reduction of beauty and intelligence, strength and purpose, of honor and aspiration, to such casually picturesque changes as an avalanche may make in a mountain landscape, or a railway accident in a human figure. To call this Natural Selection is blas-

phemy, possible to many for whom nature is nothing but a casual aggregation of inert and dead matter, but eternally impossible to the spirits and souls of the righteous. If it be no blasphemy, but a truth of science, then the stars of heaven, the showers and dew, the winter and summer, the fire and heat, the mountains and hills, may no longer be called to exhaust the Lord with us by praise: their work is to modify all things by blindly starving and murdering everything that is not lucky enough to survive in the universal struggle for hogwash.[4]

Alvar Ellegård has shown that Unitarians, some liberal Protestants, and the Catholic Church were able to come to some accommodation with Darwin's theories, at least insofar as they were not applied to humans. Moreover, a few Calvinists saw in the theory a support for the doctrine of special election; arguing that the apparent randomness of variations was actually a reflection of God's unknowable will. But for most who had started out believing in natural theology, the theory of evolution by natural selection was a bitter pill to swallow; and there is little doubt that it contributed both to the growth of agnosticism and to the notion that science and religion were in conflict with one another.[5]

Secular Criticisms of Evolution by Natural Selection

Through most of the nineteenth century, even though Darwin's work convinced most naturalists of the existence of descent with modification, only a minority completely accepted his account of the mechanism. In part this was no doubt because its implications were so inconsistent with traditional religious and political attitudes that they seemed unacceptable; but in part it was because natural selection faced substantial technical and methodological difficulties. The theory of natural selection was able to account for variations in such characteristics as coloration, which no Lamarckian theory of willful change could accommodate; and it could explain why some features of an organism that might have been useful in some earlier environment remain even when they are no longer needed. But Darwin could offer no account of how and why changes in hereditary material spontaneously arose. Moreover, it was very difficult for him to understand how varieties that stayed in contact with one another in the early stages of differentiation kept from interbreeding and blending with one another until sufficient changes had occurred to produce infertile crosses.

In addition, the theory of natural selection did not conform to the demands of either idealist philosophies of science, such as those promoted by Whewell, which insisted upon demonstrations of necessary causation, or to the demands of the most rigorous empiricist or "Baconian" philosophies of science, according to which any legitimately scientific claim must be induced directly from observational evidence.[6] Darwin was largely unmoved by such criticisms because he was quite self-conscious and self confident regarding his own methods. On the one hand, he agreed with the Positivists in opposition to idealists, that necessary causes were an unattainable goal in science and that the best that one could achieve was a knowledge of general facts. On the other hand, he was inclined to argue, contra the Baconians, that the very process of observation must be guided by some preexistent theory or hypothesis. Writing to the political economist, Henry Fawcett, in September of 1861, he stated: "About thirty years ago there was much talk that geologists ought only to observe and not to theorize; and I well remember some one saying that at this rate a man might as well go into a gravel-pit and count the pebbles and describe the colours. How odd it is that anyone should not see that all observation must be for or against some view if it is to be of any service."[7]

The theory of speciation by natural selection clearly began as a hypothesis suggested by the experiences of breeders and by Malthus's political economy. And Darwin chose what to look for based on the expectations that his theory established. While such methods were certainly consistent with a careful reading of Comte or Mill, they violated the methodological expectations of many scientists. A huge amount of evidence presented by Darwin was consistent with the hypothesis of natural selection, but many people remained (and remain) uncomfortable about its speculative origins. Until the rise of what biologists often call "the modern synthesis" between natural selection and Mendelian genetics in the 1920s, Darwinian natural selection remained a minority, though important, perspective even among biologists. Some variant of Lamarckian progressive evolution retained a dominant position, especially among American and Continental biologists and evolutionary social theorists alike.

When *On the Origin of Species* appeared in 1859, Darwin chose to temporarily duck the vexed issue of how humans are to be understood in relation to evolution except for a single cryptic remark that "much light will be thrown on the origin of man and his history," by the evolutionary approach to psychology taken by Herbert Spencer.[8] Since Darwin was fully committed to integrating humans into the evolutionary story in the long run, he probably hoped that by setting aside the human issue, he might allow his theory a chance to catch on before it faced its most severe challenge. There is some evidence that his strategy worked; for initial reactions to the *work* were much less hostile than those to the later *Descent of Man and Selection in Relation to Sex,* which integrated humans into evolutionary theory in 1871.[9]

The Contributions of Darwin's *Origin of Species:* Overwhelming Evidence for Evolution and a Focus on Natural Selection

Darwin articulated two distinct, though related, goals in writing *The Origin of Species:* "firstly, to show that species had not been separately created and secondly that natural selection had been the chief agent of change."[10] To accomplish the first goal, Darwin set out initially to establish the existence of variations in populations of organisms and to show that variations could be propagated to create increasingly divergent varieties under domestication.[11] Next, he documented the existence of variability in natural populations and showed that the frequency of observable variations was greater in populations that were relatively large, in populations that had greater geographical range, and in populations of species already known to have many varieties.[12]

Having securely established the existence of and conditions promoting variability in natural populations, Darwin sought to undermine the traditional rigid boundaries between varieties and species, so that evidence of the gradual creation of new, widely accepted, varieties could be interpreted as evidence for the creation of new species as well. "In determining whether a form should be ranked as a species or a variety," he argues, "the opinion of naturalists having sound judgement and wide experience seems the only guide to follow."[13] Unfortunately, in many cases, reputable naturalists disagree. Even the supposedly simple definition of Buffon, that when two organisms are of the same species they can interbreed, producing fertile offspring, Darwin shows to be problematic. He writes: "The sterility of various species when crossed is so different in degree and graduates away so insensibly, and on the other hand . . . the fertility of pure species is so easily affected by various circumstances, that for all practical purposes it is most difficult to say where perfect fertility ends and sterility begins."[14] Given the difficulty of distinguishing between varieties and species, Darwin argues that, in fact, there is no essential difference—that species can be considered nothing but "strongly marked and well defined varieties."[15] Thus the gradual creation of varieties becomes direct evidence for the gradual creation of species.

Against the alternative claim that each species is divinely created to be adapted to the environment into which it is set, Darwin offers several kinds of evidence and arguments. First are cases in which the inhabitants of widely separated but very similar environments turn out to be substantially different from one another but closely related to their nearest neighbor species. Evidence from the fauna of deep limestone caves in Europe and America exemplify this case. The two environments are virtually identical; but the inhabitants are substantially different and are clearly related to near neighbors on the surface rather than being specially designed for the underground environment.[16] Second are cases in which indigenous populations are shown to be imperfectly adapted when foreign organisms overwhelm them when they are introduced into the environment.[17] Most importantly, Darwin details the existence of "rudimentary organs," by which he means structures that have no use to the organism in its present form but that may have been valuable to an ancestral form. Such structures make no sense in terms of design; but they are completely understandable in terms of descent with modification.[18]

Having made his case for the gradual accumulation of changes producing diverging varieties of a species and then new species themselves, Darwin faces the central difficulty that had scuttled Lamarck's and Chambers's theories among scientists. Why don't we see innumerable transitional forms? Part of his answer to this question follows from the claim that natural selection is the mechanism for propagating variations; for, as Darwin shows, "the very process of natural selection constantly tends . . . to exterminate the parent forms and the intermediate links."[19] Since the new dominant form is more successful than its near relatives in the struggle for existence, it deprives them of food, and so on, and tends to drive them into smaller territories and to extinction. Thus, the parent species do not remain around long after their more successful progeny have emerged. This special feature of natural selection interacts with a second factor that is so important that we might understand why so few present species have left continuous fossil records of their descent even without the first. The imperfect geological record preserves evidence of only a minute fraction of living things.

Very few "soft" organisms are preserved at all in the fossilizing process. Among animals, only those with shells or bones, and among plants, only those that are relatively rigid and cellulose-rich ordinarily leave fossil imprints. Even organisms with rigid structures will not leave a trace unless sedimentation traps them and deprives them of oxygen soon after death; so the likelihood of *any* organism being present in the fossil record is very small and proportional to both the population at any time and the span of time during which it lived, all other things being equal. Since the near relatives of very successful organisms, the "intermediate forms," rarely live long, it is clear that they will seldom leave a trace.[20] This is especially true because the geological conditions that allow for the accumulation and persistence of sedimentary layers, the only structures in which fossils can be found, are rare. All exposed land is subject to the powerful forces of subaerial degradation. The chemical and mechanical action of air and water carrying corrosive materials constantly wears away the upper layers of the soil; and wave action produces much the same effect in shallow seas where most marine life is present. Sedimentation is likely to accumulate only during periods of local subsidence and only when the rate of sedimentation is greater than the rate at which the surface is scoured away, which is very rare.[21]

Given all of the conditions that have to be met in order for an organism to leave a fossil trace, it is easy to understand why the record does not contain a continuous sequence of minute transformations that could fill in the gaps between organisms that have left a trace. There is, however, positive evidence from embryological studies in favor of slow and gradual changes, since "the embryo . . . serves as a record of the past condition of the species," and the development of modifications in the embryo are always slow and continuous.[22]

In much the same way that Darwin addressed Sedgwick's powerful objections to evolution based on the gaps in the fossil record, he also addressed objections based on the inconceivability of random variations leading to such complex structures as the human eye.[23] At the same time, Darwin was able to demonstrate that the principles of evolution and natural selection could explain a number of previously discovered laws and eliminate the need for others. Thus, for example, von Baer's law of the Unity of Type, according to which all organisms of the same class agree in basic structure regardless of their varying habits of life, followed directly from the principle of evolution, according to which the Unity of Type is a consequence of an identity of ancestors at some earlier stage in evolution.[24] On the other hand, Lamarck's and Spencer's assumption that there is some "innate tendency toward progressive development" in all organisms was rendered unnecessary. Natural selection itself favored more highly differentiated parts, "because they could perform their functions more efficiently."[25]

Implications of Darwin's *Descent of Man*—1: Group Selection, the Evolution of Altruism, and the Character of Inherited Mental Illness

Darwin's emphasis in the *Origin of Species,* following the lead of Malthus and of Spencer, was on intra-specific competition and on the argument that variations come about primarily as a consequence of the selection of traits that enhance the survival of individuals acting in their own self-interest. Given its grounding in laissez-faire political economy, it should not be surprising that those evolutionary social theories that drew their Darwinian arguments primarily from the *Origin of Species,* tended to be ones that focused on individualism and that generally opposed governmental interference in the lives of citizens except for national defense and self-defense and to enforce contracts—that is, social theories that were "liberal" in the nineteenth-century meaning of the term.

When Darwin came to write about humans, some twelve years later, in *The Descent of Man and Selection in Relation to Sex,* he admitted that he had previously, "perhaps attributed too much to the action of natural selection or the survival of the fittest."[26] His emphasis now moved to two forms of selection that did not focus on the survival of the egoistically driven individual. Of these two forms, the first, labeled "the natural selection of groups" in 1875 by W. K. Clifford (1845–79),[27] functioned to preserve "variations which are beneficial to the community,"[28] rather than to the individual. Such variations still operated through the individual; but they functioned to preserve the group, while often placing the individual exhibiting them at substantial risk. Like the simpler individual form of natural selection, this new form of group selection also occurs as a consequence of competition; but now the competition is seen as between "tribes" or communities, rather than as among individuals of the same group: "When two tribes of primeval man, living in the same country, came into competition, if (other circumstances being equal) one tribe included a great number of courageous, sympathetic, and faithful members, who were always ready to warn each other of danger, to aid and defend each other, this tribe would succeed better and conquer the other."[29] In time, other tribes even more richly endowed with members who support one another would defeat the victor in competition, leading to an advance of cooperative qualities and extending them throughout the world.[30] It seemed clear to Darwin that, given the relatively weak physical constitution of humans, it must have been group selection, operating primarily to promote cooperation rather than competition within living groups, that had "sufficed to raise man to his present high position in the organic scale."[31] Moreover, he argued, the primary mechanism for this group selection involved the growth of a "moral sense" that is opposed to the instinct for self-preservation.

Just as Malthus's doctrines were a crucial key to Darwin's understanding of natural selection, Adam Smith's doctrine of "sympathy" from *The Theory of Moral Sentiments* is crucial to Darwin's understanding of the origins of morality and the role of morality in human evolution, though Darwin probably became aware of this doctrine through the Positivist writings of Harriet Martineau and George Eliot rather than through reading Smith himself. For Darwin, all social animals are endowed with certain social instincts, of which sympathy—the ability to share a feeling (especially of grief, fear, or joy) with another—is both the highest and most important.[32] It is sympathy for the plight of an endangered comrade, for example, that leads a social animal to risk its own life to go to the aid of another or to share its scarce food or other resources with an ill or disabled member of the community.

Initially, according to Darwin, instinctive sympathy extends only to members of a single community rather than to all members of a species or to more than one species; and it is this early selective character of sympathy that allows it to be the subject of the natural selection of groups. For, as Darwin argues, "Those communities which included the greatest number of the most sympathetic members, would flourish best, and rear the greatest number of offspring."[33] As time goes on, those who act out of sympathy come to be admired by other members of the community because of their altruistic acts; and a new motive, the desire to be approved and admired, begins to shape behavior in altruistic directions that may not be instinctual. At this point, when actions that involve either an indifference to or a violation of self-interest are consciously undertaken to serve the community, we say that those acts are "moral" and that the person engaging in such acts has a "moral sense." Finally, through the repetition of such acts, "Man . . . will through long habit, acquire such perfect self-command that his desires and passions will at last yield instantly and without struggle to his social sympathies and instincts, including his feeling for the judgement of his fellows."[34]

At some point, Darwin argues, as humans become increasingly capable of recognizing the extended consequences of their actions and as they consider not just the survival, but also the happiness of others, morality begins to extend beyond one's specific community to all of humanity and, eventually, even to other species: "His sympathies become more tender and widely diffused, extending to men of all races, to the imbecile, maimed, and other useless members of society, and finally to the lower animals."[35] Darwin was no less aware than Spencer that the extension of sympathy to the weaker members of society would have possibly negative consequences in allowing the reproduction of those who would probably not survive outside of advanced society. Unlike Spencer, however, Darwin saw these negative consequences as the necessary price to be paid for something more important: a "nobility" that was a condition for the emergence of civilized nations out of barbarism. "We must therefore bear the undoubtedly bad effects of the weak surviving and propagating their kind," he argued.[36] In spite of the apparent partial evils that morality engendered, Darwin was certain that, on balance, the levels of cooperation and coordination that morality made possible ensured that "at some future period, not very distant as measured by centuries, the civilized races of man will almost certainly exterminate, and replace, the savage races throughout the world."[37] He even argued that in a civilized individual, a revival of egoistic motives relative to moral ones was "one of the early symptoms of mental derangement."[38]

This suggestion that there might be a reversion from more altruistic to more egocentric bases for behavior among some individuals and that such a reversion might be linked to mental derangement became linked with both Spencer's notion of dissolution and with the Lamarckian assumption that some acquired traits or tendencies might be inherited. Together, these elements were used to develop a powerful and pervasive new understanding of heritable mental disease that raised the specter of increasing mental and moral "degeneration" and that specifically linked pressures toward degeneration to the complexity and conditions of modern industrial life.

Henry Maudsley and Darwinian Psychiatry

Though many mid-Victorian psychological theorists contributed and drew from this new understanding of mental illness, perhaps the most directly influenced by *The Descent of Man* and the most influential in Britain was Henry Maudsley. Trained at University College Hospital in London, in 1859, at age twenty-four Maudsley became medical superintendent of the Cheadle Royal Hospital for the Insane at Manchester. Three years later, he became editor of the *Journal of Mental Science,* the premier psychiatric journal of the time. Soon after, he began a series of book-length treatments of psychology and psychiatry that allowed him to dominate English discussions of mental illness for nearly half a century.

Maudsley put all discussions of mental defect within an evolutionary context. Following Chambers and others, he was inclined to identify those with subnormal intelligence as individuals whose development had been arrested at the stage of an earlier evolutionary ancestor: "When we reflect that every human brain does, in the course of its development, pass through the same stages as the brains of other vertebrate animals . . . and when we reflect further, that the stages of its development in the womb may be considered the abstract and brief chronicle of a series of developments which have gone on through countless ages in Nature, it does not seem so wonderful . . . that it should, when in a condition of arrested development, sometimes display animal instincts. . . . There is truly a brute brain within the man's; and when the latter stops short of its characteristic development as human . . . it may be presumed that it will manifest its most primitive functions, and no higher functions."[39]

More important for our present purposes is Maudsley's interpretation of other forms of mental illness, including those associated by the Victorians with epilepsy, insanity, and criminality. He did acknowledge that some forms of mental illness might be produced largely by environmental factors and others by injuries to the brain. But fifty percent or more, he argued, were caused by some inherited defect.[40]

Before 1873, Maudsley followed Spencer's basic line of argument. Humankind was evolving into a more complex and highly differentiated species suited to ever more highly specialized roles in industrial society. This process used energy; and it was basically inevitable that occasionally, either through "inherited weakness," or some other "debilitating cause," some persons would be "rendered unequal to the struggle for life." Though sociability and altruism were significant, the emphasis was on individual strength and competitiveness. Those who failed "are the waste thrown up by the silent but strong current of progress; they are the weak crushed out by the strong in the mortal struggle for development; they are examples of decaying reason thrown off by vigorous mental growth, the energy of which, they testify."[41]

It was also the case for Maudsley, as for Spencer, that as society became increasingly complex, there were increasing opportunities for a reversal of the direction of change in its members; so one had to expect to see an increased variety and quantity of mental illness with the progress of society.[42] But for Maudsley, this reversal for individuals could be read as the price to be paid for social progress in much the same way that for Darwin, physical weakness might be the necessary price to be paid for "civilization." Thus, wrote Maudsley, "Madness may be a waste of the individual to the profit of the race . . . a seeming evil which is truly a phase in the working out of a higher good."[43]

After the publication of *The Descent of Man*, sympathy and selflessness became the principle defining characteristics of advanced humanity and the prerequisites for all human progress for Maudsley as they were for Darwin. The loss of sympathy then became the first stage in all important forms of inherited mental illness: "Sympathy with his kind and well-doing for its welfare, direct or indirect, are the essential conditions of the existence and development of the more complex social organism. . . . If an individual fails to bring himself into sympathetic relations, either conscious or unconscious, with surrounding human nature, he becomes a sort of discord, and is on the road . . . which leads to madness or to crime."[44] The focus on self, rather than others, became the one thing that all cases of madness had in common. "One thing fails not to be brought forcibly home to those who live among the insane," Maudsley wrote, "namely how completely they are wrapped up in self, and what little hold the cares and calamities of those who have been living intimately with them ever take of them. . . . Living together for years, they . . . show no interest in, and no sympathy for one another."[45]

The complexity of industrial society increased the likelihood of mental illness of any kind, but it was the emphasis on competition for wealth associated with commercial society that tended to undermine altruism in particular. "The reaction upon character of a life spent solely in the business of getting rich is hurtful," Maudsley wrote, and the effects were particularly harsh on middle-class businessmen.[46] Given the lingering belief that acquired characteristics could be inherited and that processes of dissolution were natural, Maudsley followed the French prophet of degeneration, Benedict Morel (1809–73), in arguing that the relatively mild moral degeneracy of one generation was likely to be amplified in the next. In addition to and overriding Morel's emphasis on alcoholism and drug addiction as triggers to degeneracy, however, Maudsley emphasized the reversion to selfishness of the tradesman. The process was, for Maudsley, as for Spencer, the exact inverse of the progressive process of evolution, but it was usually much more rapid: "All the moral and intellectual acquisitions of culture which the race has been slowly putting on by organized inheritance of the accumulated experience of countless generations of men are now rapidly put off in a few generations, until the lowest human and animal elements only are left."[47]

Of all the implications of an evolutionary view of mental illness, that with the most immediate impact was more related to the natural selectionism of Darwin's earlier work than to the new arguments of *The Descent of Man.* Though the Lamarckian perspective associated with notions of progressive degeneration formed the foundation of a profound cultural pessimism toward the end of the nineteenth century, Lamarckism also offered the possibility of a cure for mental disease through effort. Natural selectionism, on the other hand, left no apparent escape from an inherited defect. "Multitudes of human beings come into the world weighted with a destiny against which they have neither the will nor the power to contend," wrote Maudsley in his early *Body and Mind,* "they are the step-children of Nature, and groan under the worst of all tyrannies—the tyranny of bad organization."[48] In the *Pathology of Mind,* in spite of the fact that he had adopted Larmarckist ideas for other arguments, he expressed the same Darwinian view with even greater eloquence: "There is a destiny made for each of us by his inheritance; he is the necessary organic consequent of certain organic antecedents; and it is impossible to escape the tyranny of his organization. . . . The dread, inexorable destiny which played so grand and terrible a part in Grecian tragedy, and which Grecian heroes are represented as struggling manfully against, knowing all the while that their struggles were foredoomed to be futile, embodied an instinctive perception of the law by which the sins of the fathers are visited upon the children unto the third and fourth generations."[49]

Against such a destiny Maudsley and his contemporaries could envision no protection. As a consequence, the hopeful environmentalism of psychiatry developed within the context of associationist psychology gave way to a much more pessimistic view linked to what Elaine Showalter has called "psychiatric Darwinism." At a time when the numbers of poor patients entering public asylums was dramatically increasing, Maudsley's views offered little hope for those suffering from inherited mental disease; but they offered great relief to those revolting against the increasing costs of treatment. Since inherited mental defects were incurable, there was no point in institutionalizing those whose disease did not make them a danger to themselves or to society. Responsibility for their care could be placed on their families rather than on the public. Those whose anti-social behaviors were more severe could basically be warehoused in mental institutions without concern for expensive attempts at treatment.[50]

Implications of *The Descent of Man*—2: Issues of Sex and Race

Darwin had argued that some social and mental dimensions of racial differences were a consequence of group selection processes and that the highly developed altruism of advanced civilized humans would certainly lead to the eventual extermination of less-advanced peoples. But he argued that the external physical differences among races (features such as skin color and the presence or absence of facial hair) could not be accounted for by any of the normal processes associated with either natural or group selection. As a consequence, he proposed that such features were a consequence of "sexual selection," which, he insisted, had acted powerfully upon humans as well as on many other animals.[51]

Approximately three fourths of *The Descent of Man and Selection in Relation to Sex,* an amount equivalent to the entire *Origin of Species,* was then devoted to discussing the processes of sexual selection. For our present purposes, only a few of Darwin's general claims are of great importance; but we should note that in the case of sexual selection as in the case of natural selection, it was at least in part the massive documentation underlying his arguments that made Darwin's claims so compelling and authoritative to many of his contemporaries.

Darwin begins by asserting that among most higher animals—he focuses on mammals and birds—while males frequently compete for females by engaging in contests of strength and stamina, females choose mates based on a variety of criteria. This fact, coupled with empirical evidence that the most vigorous females are almost always the first to choose and to breed, serves as the foundation of a claim that there will be a differential breeding and survival rate that favors the increase of male features that are attractive to females: "The females are most excited by, or prefer pairing with, the more ornamented males, or those which are the best songsters, or play the best antics; but it is obviously probable that they would at the same time prefer the more vigorous and lively males. . . . The more vigorous females, which are the first to breed, will have the choice of many males; and though they may not always select the strongest or best armed, they will select those which are vigorous and well armed, and in other respects most attractive. Both sexes, therefore, of such early pairs would . . . have an advantage over others in rearing offspring; and this has apparently sufficed during a long course of generations to add not only to the strength and fighting powers of the males, but likewise to their various ornaments or other attractions."[52]

Two aspects of this argument have important implications. First, in most animals, because the females are the selectors, sexual selection operates almost exclusively on males to enhance their appeal to females. Thus, for example, among birds many males develop extravagant plumage, spectacular songs, and elaborate displays of movement, while females remain relatively bland in their appearance and behavior, probably to fade into the background in the presence of predators. Second, because strength, vigor, and aggressiveness are almost always selected for by females, males are generally larger, stronger, faster, and more aggressive than females. To those who suggested that in man, especially, male strength and size might be a consequence of males doing more strenuous work, rather than a consequence of sexual selection, Darwin pointed out that in most hunter-gatherer societies, females work at least as hard as men. Only in recent and highly civilized societies, he argued, do men generally do more physically demanding work.[53]

Finally, within his general argument about the greater impact of sexual selection on males, Darwin claims that his own work on domesticated animals as well as anthropometric studies done by many others demonstrated that selective pressures operate more effectively on males because males have a greater intrinsic variability.[54] This so-called variability hypothesis, applied to human features such as intelligence, would come to have extremely important social policy implications. For example, it suggested that over time, men were becoming increasingly more intelligent than women, because intelligence is positively selected for by both natural selection and by sexual selection. As a consequence, according to Darwin, in whatever requires "deep thought, reason, or merely the use of the senses and hands," men are superior to women.[55] It thus made little sense to provide higher education to women, who were generally incapable of meeting the mental demands that advanced education provided, nor did it make sense to allow women to be involved in public political life, which was presumably governed by reason.

One of the many ironies of Darwin's life is that in spite of his theoretically based claims about female lack of intellect, two of his most frequent and admired guests were women intellectuals: Harriet Martineau and George Eliot. Moreover,

Fig. 8.2. Picture illustrating an extreme of male display among birds—the Bird of Paradise. From Charles Darwin, *The Descent of Man and Selection in Relation to Sex* (New York: D. Appleton and Company, 1896, p. 386).

it was to his daughter Henrietta, "Etty," that he turned for criticism of the ideas in *The Descent of Man.*

None of Darwin's conclusions regarding male superiority were original, of course. They drew heavily from earlier nineteenth-century craniometry and physical anthropology and from Spencer's discussions. But Darwin's suggestion that it was largely as a consequence of female preferences that the male was bigger, stronger, and smarter was certainly new; and the fact that his views appeared and promoted the traditional female roles of subservient mother, wife, and homemaker at a time when "the woman question" was a major social concern drew new support for Darwin's ideas from conservative quarters that had offered little support before.[56]

The theoretical foundation for late-nineteenth-century feminism in Britain and America had been laid in John Stuart Mill and Harriet Taylor's (1807–58) 1861 *The Subjugation of Women,* which opened by claiming that the subordination of one sex to the other was wrong, that it had become a barrier to human progress, and that it should be replaced "by a principle of perfect equality, admitting no power or privilege on the one side, nor disability on the other."[57] Following the earlier lead of Catherine Macaulay (1731–91) and Mary Wollstonecraft (1759–97), Mill and Taylor went on to insist that the obvious present differences between the sexes, aside from those directly relating to childbearing, were the consequence of education and environmental circumstances. Female emancipation and education had also been promoted by radicals influenced by Saint-Simonian and Owenite ideas and they seemed a natural consequence of the growing numbers of women, especially lower-class women, entering the labor force. Thus, between about 1860 and about 1890, there was a growing woman's movement demanding new educational opportunities, property rights, professional opportunities, and political rights.

Maudsley's Darwinian response to Mill's claims that education could bring about female and male intellectual equality was, as usual, particularly forcefully presented but typical in tone: "To my mind it would not be one whit more absurd to affirm that the antlers of the stag, the human beard, and the Cock's comb [all products of sexual selection] are the effects of education, or that, by putting a girl to the same education as a boy, the female generative organs might be transformed into male organs. . . . Because there is sex in mind, there should be sex in education."[58]

Darwin had admitted that the cause of greater male variability was uncertain; but he offered a suggestion that was to become amplified into the foundation of most subsequent Victorian theories of sexual differences. It is almost certainly the case that men and women expend roughly the same "matter and force," he argued. But they express their force in very different ways. While the female invests matter and force "in the formation of her ova," the male, "expends much force in fierce contests with his rivals, in wandering about in search of the female, in exerting his voice, pouring out odiferous secretions, &c."[59] Among the "&c." to which male energy might be differentially applied, Darwin hinted, was the mechanism of variation. At the hands of Herbert Spencer and Patrick Geddes, this speculation was turned into the claim that because females had to conserve their energy, "to meet the cost of reproduction," males had additional energy resources available for a whole range of tasks, ranging from additional psychic and intellectual development and activity to holding down strenuous jobs outside the home.[60]

One of the biggest fears of those who fused evolution with energy conservation was that women who chose to develop their intellect would have too little energy left for reproduction. Since those who already had the greatest mental abilities were most likely to seek education, there was a chance that the best breeding stock would produce fewer progeny and that the quality of the race would decline. "It would be an ill thing," Maudsley complained, "if it should so happen that we got the advantages of a quantity of female intellectual work at the price of a puny, enfeebled, and sickly race."[61] A statistical study of women who had attended Vassar, Smith, and Wellesley colleges in the United States between 1867 and 1879 was interpreted so as to promote such fears. While approximately 85 percent of women in the general population married by the time they were in their early thirties, fewer than 55 percent of the college women did; and of those who did marry, the college graduates had fewer than two children each, less than half of the average for the population at large. G. Stanley Hall (1844–1924), author of the study, concluded in a way that was logically suspect but psychologically comforting to traditionalists: "Absolute or relative infertility is generally produced in women by mental labor carried to excess."[62]

Just as Maudsley was the dominant British writer on psychiatric issues during the second half of the nineteenth century, Patrick Geddes was almost certainly the most famous writer on sexuality. After studying with Darwin's close follower, Thomas Henry Huxley, Geddes served as demonstrator in practical anatomy at University College in London. Moreover, he developed both a special interest in the organic physics movement associated with Helmholtz and Du Bois-Reymond and he became fascinated with Positivist sociology through meetings of the London Positivist Society and through a period of study in Paris. In 1889, he, with his student, J. Arthur Thomson, published *The Evolution of Sex,* an immensely popular text, that sought to synthesize evolutionary, energetic, and sociological considerations in accounting for sex differences.

Beginning from the notion that even at the level of the individual cell, masculinity involved the tendency to dissipate energy in activity (katabolism) and femininity involved a tendency to store energy (anabolism), Geddes argued that the most basic sexual characteristics—activity in males and passivity in females—were already established in single-celled organisms. Subsequent evolutionary developments had left this fundamental difference intact, and no attempt to deny or mitigate this dichotomy could ever succeed. In particular, feminists, who called for female equality with men, in the workplace, in education, or in the political domain, argued pointlessly, for, "what was decided among the prehistoric *Protozoa* can not be annulled by the act of parliament."[63]

Once again, Geddes's evolutionary and energetic arguments justified virtually all of the middle-class sexual stereotypes prevalent in late-Victorian England. Women were naturally passive, less courageous, less independent, and physically and intellectually inferior to men; but they had evolved in a way that made them more nurturing, more patient, more constant in their affections, and more intuitive and sympathetic than men. Thus, women were situated to provide the moral leadership of society and to become the agents of moral change, as Comte had suggested.

According to Darwin, many features initially acquired by sexual selection remained tied to the male sex. Thus, antlers, or horns, in many of the four-footed mammals, as well as beards, both in some species of monkeys and in humans, appeared only in males. Other features could be partially transferred. Such, for example was intelligence; for as Darwin pointed out, if some element of male intelligence increases was not transferred to women, "man would have become as superior in mental endowment to women, as the peacock is in ornamental plumage to the peahen."[64] Neither Darwin nor Spencer and Geddes were inclined to see human sex differences in intelligence or morality as this divergent, even though they were deemed significant. On the other hand, some features apparently acquired by sexual selection seemed to be easily transferred across sexes.[65] Such features were skin color, facial form, and hair texture in humans, that is, the defining characteristics of traditional racial taxonomies. As a consequence, argued Darwin, those racial attributes not accounted for by group selection were attributable to sexual selection and subsequent transference.

In humans, Darwin argued, women probably chose mates very early in the evolutionary history of the species—thus, the prominence of beards in some races. But over time, men, through their greater strength and intellect, came to keep women, "in a far more abject state of bondage than does the male of any other animal."[66] Males rather than females are thus the selectors in most, though not all, advanced human societies; and it is the female, rather than the male, whose form defines standards of beauty and who is more often adorned. In this case, however, almost all those features selected for are transferred to both sexes; "so that the continued preference by the men of each race for the more attractive women, according to their standard of taste, will have tended to modify in the same manner all the individuals of both sexes belonging to the race."[67] The relatively recent process of male selection accompanied by transference accounts for one of the unique and puzzling features of humans—their relative lack of body hair—which, Darwin argues, is almost certainly opposed by natural selection, because hair provides protection. In all cultures, women have less body hair than men and lack of body hair is viewed as an aspect of female beauty. If, then, women were systematically selected to decrease body hair and they at least partially transferred this characteristic to their progeny of both sexes, that process would have led to the present situation in which only a few races still exhibit some body hair on males while there is little left on females.[68]

The absence of body hair brings us back to issues of racial differences. Although some German and American race theorists in the 1880s and 1890s did try to use Darwinian arguments to support discriminatory and paternalistic policies toward non-Europeans, neither Darwin nor the other two major Darwinian spokesmen for natural selection in biology, Alfred Russell Wallace and Thomas Henry Huxley, were racists by Victorian standards. Darwin and Huxley were convinced that the physical attributes of the recognized human races were produced almost exclusively by sexual selection for aesthetic reasons. Moreover, Wallace, who rejected sexual selection, was inclined to view "savage" peoples as physically and morally superior to Europeans. Describing the Aru islanders of the Maylay Archipelago, with whom he lived for several months, he wrote: "Here, as among the Dyaks of Borneo and the Indians of the upper Amazon, I am delighted with the beauty of the human form, a beauty of which stay at home civilized people can never have any conception. What are the finest Grecian statues to the living moving breathing forms which everywhere surround me. . . . A young savage handling his bow is the perfection of physical beauty."[69]

Yet Wallace, like Darwin and Huxley, was convinced that civilized men were likely to be invincible in clashes of culture because of their superior organizational skills and technologies. Those savages who were likely to be willing to accept European rule would probably be able to adapt to a modern agricultural existence, he felt; but those who would resist would be crushed. Speaking specifically of the Papuans, he wrote: "A warlike and energetic people, who will not submit to slavery or to domestic servitude, must disappear before the white man as surely as do the wolf and the tiger."[70] Darwin was not as admiring of savage peoples as Wallace, nor did he join Wallace in actively promoting abolitionist and anti-imperialist movements, but neither did he publicly promote either slavery or colonization.

In this chapter, we have seen that Darwin and a handful of closely associated biologists and medical men initiated scientistic extensions of Darwin's perspectives on nonhuman biological evolution to major issues of public concern, including questions of how societies evolve, how mental illnesses should be conceptualized and treated, what differences there are between sexes and what policies those differences should entail regarding such topics as education and suffrage, and what differences exist among races and how those differences should guide policies regarding slavery and imperialism. In the next chapter we turn to a collection of post-Darwinian scientistic social theorists who promoted the extension of evolutionary perspectives as the foundation for comprehensive social theories that have played a major role in directing a wide range of European public policy discussions and decisions into the present.

CHAPTER 9

"Social Darwinisms" and Other Evolutionary Social Theories

The term "Social Darwinism," or initially, "*Darwinisme sociale,*" came into use in the early 1880s almost exclusively as a term to condemn theories that claimed to find in the evolutionary notion of the survival of the fittest a justification for extreme individualism and laissez faire economic policies. To some degree at least, the initial users of the term were creating a straw man; for it is unlikely that any serious scholars or political figures held the rigid views imputed to them by their critics, who often appealed to evolutionary arguments drawn from *The Descent of Man* in condemning "Social Darwinism."[1] But over time, the term has been used in a wide variety of ways to cover virtually any appeal to evolutionary arguments, whether Lamarckian or Darwinian in tone and whether focused on individual or group dynamics, to argue on behalf of any social policy. It is in this broader sense that I will use the term; though I will try to assess the extent to which various forms of evolutionary social theory drew from specifically Darwinian, as distinguished from other evolutionary perspectives.

It is particularly important to remember that at least until the early 1890s, when August Weisman's (1834–1914) studies of cell division focused attention on the transmission of "germ plasm" from one generation to the next, even the most ardent and exacting followers of Darwin continued to acknowledge the possibility of the inheritance of acquired characteristics. These characteristics were then viewed as propagated by natural selection. In general, I will emphasize European rather than American developments, though in a few cases I will discuss American theorists who had a particularly important impact on European social movements or American social policies whose foundations were laid by particular European authors. Before turning attention to evolutionary social theorists, however, it may be worthwhile to briefly characterize some of the important events, trends, and issues in late-nineteenth-century European society that those social theorists were concerned to understand and to address.

The Social and Political Context for the Rise of Social Darwinisms

The first, and by far the most important trend in late-nineteenth-century Europe was economic. It was the intensification and spread of industrial capitalism, which had begun in Britain during the late eighteenth century. According to Walter W. Rostow, the takeoff toward industrialism began in France by the 1820s, in the United States and Germany by the 1830s, in Sweden during the 1860s, and even in Russia by the 1880s.[2] On the one hand, industrialization allowed for huge gains in productivity and capital accumulation, fueling ever more industrialization and urbanization. On the other hand, it led to cycles of overproduction and to income concentration in the hands of the capitalists. One consequence was a severe economic depression during the 1870s. The depression fed ongoing Europe-wide worker unrest that produced many local strikes and a rising working-class interest in socialist, anarchist, and communist movements.

Another consequence of industrialization was that it led to a scramble to find new markets for the output of European and American factories and new sources of raw materials for the production of finished goods. This eventuated in an intensification of imperialist policies and competition—often violent—for colonial empire. Jules Ferry, Prime Minister in the French Third Republic, spoke for the leadership of almost every industrialized state when he spoke to the French Chamber of Deputies on March 28, 1884: "In the area of economics I am placing before you, with the support of some statistics, the considerations which justify the policy of colonial expansion as seen from the perspective of a need, felt more and more urgently by the industrialized population of Europe, and particularly those of our rich and hard-working country of France: the need for export markets. Is this a fantasy? Is this a concern [that can wait] for the future? Or is this not a pressing need, one might say a crying need of our industrial population? . . . Yes, what our major industries, irrevocably steered by the treaties of 1860 into exports, lack more and more, are outlets."[3]

Technological innovations associated with capitalism produced dramatic changes in the life of most of those Europeans not engaged in primary agriculture, and virtually all of those changes increased the pace of events. At the beginning of the century, communications from one end of almost any country to the other was likely to take weeks. Letters sent by stage from London to Edinburgh, for example, often arrived two weeks after they were sent. By the end of the century, the telegraph had created the possibility of almost instant communication, and the telephone was making instant voice communication possible. Transportation saw an almost equally dramatic acceleration. Roads were greatly improved with the use of iron bridges and "Macadamized" surfaces, but the biggest change involved the creation of railway networks that linked virtually every major city and town in Britain and on the Continent, and that allowed the cheap transport of goods to almost any market. By the end of the century, private motor vehicles were even beginning to become more than a rich man's novelty. At the beginning of the century, water and steam powered almost all machinery and when public areas were lighted at all at night, they were lit by gas lamps. By the end of the century, electric motors and electric lights were spreading rapidly.

Capitalism also changed European social and political dynamics. As wealth moved from a traditional landowning aristocracy into the hands of entrepreneurs from a wide range of social backgrounds, social mobility—both upward and downward—increased. Status gradually came to be associated less with birth than with success in economic competition. Power in Britain gradually shifted away from the landed aristocracy and toward the House of Commons in Parliament; and by the early 1870s the conservative government of France finally became republican for good. Though strong "absolutist" monarchies continued to exist in eastern Europe, political power moved into the hands of liberal republicans in the French Third Republic. Germany was finally unified under Prussian leadership, and though dominated by conservative forces under Otto von Bismarck's (1815–98) reign as chancellor, the German Empire began introducing democratic elements through an elected *Reichstag,* or Parliament, in 1871. Both liberal and socialist parties began to grow in strength in Germany through the remainder of the century.

Finally, on a less immediately dramatic, but equally profound level, nineteenth-century Europe was going through a period of dramatic secularization that may have begun during the previous century, but that accelerated throughout Europe during the nineteenth century. Church attendance declined everywhere, and religious arguments were decreasingly important in public life except to the extent that they provided a complement to economic arguments to justify European imperialism—that is, our obligation to bring Christianity to the heathen was often appealed to just after economic arguments for colonial expansion were articulated.

Social Darwinisms

As social theorists and spokespersons for various groups tried to understand and shape European events and trends in the latter part of the nineteenth century, they thus had to come to grips with competitive capitalism and imperialism and their consequences, whether they applauded or opposed them. Similarly, they faced the problem of accounting for how pervasive and powerful religious institutions have been throughout history, while explaining why they were on the decline during the nineteenth century.

Darwin's emphasis in *The Descent of Man* on how the increase of altruism and cooperation as consequences of natural selection applied to group competition rather than to individual competition fit beautifully with socialist ideas that had been promoted on the Continent by the Saint Simonians and in Britain by the inheritors of Owenite cooperative attitudes. As a consequence, a highly cooperative form of evolutionary socialism was advocated by numerous left-leaning theorists, including Leslie Stephen (1832–1904) in Britain and Edward Bellamy (1850–98) in the United States. Linked to Darwin's explicit complaint that warfare served to expose the finest young men to death, leaving the "shorter and feebler men, with poor constitutions" at home to procreate,[4] this cooperationist strand within Darwin's thought, led to a long-standing tradition of what Paul Crook has recently called "peace biology."[5]

On the other hand, Darwin's emphasis on inter-group competition, especially on tribal competition, provided the central arguments of militaristic, imperialistic, and overtly racist ideologies in the writings of men such as Walter Bagehot (1826–77) in Britain, and Ernst Haeckel and Friedrich von Bernhardi (1849–1930) in Germany. Finally, evolutionary arguments were used very effectively by opponents of socialism and extreme individualism alike to reject all "radical" suggestions regarding social policy and to explore the conditions under which moderate and evolutionary, rather than revolutionary, reforms might mitigate the negative consequences of class warfare and cutthroat capitalist competition. In their most extreme form, as in Peter Kropotkin's (1842–1921) *Mutual Aid,* such theories promoted anarchism and the rejection of all collective coercion.

In what follows, we will be concerned with those who used evolutionary rhetoric, positive or negative, to promote particular social and political agendas. Though some of these figures, such as Ernst Haeckel, were academic biologists or geologists, many had no academic affiliation, and few had very sophisticated understandings of the biological doctrines from which they borrowed. Nonetheless, it was through their efforts that evolutionary ideas helped to shape the social and political mindscapes and landscapes of the late-nineteenth-century world, both within and beyond the boundaries of Europe. Two subjects will be avoided at this point, to be dealt with later. First, there will be almost no consideration of fictional works that explored the consequences of evolutionary theories. One exception, Edward Bellamy's *Looking Backward* (1888), will be discussed here because its literary form was completely subservient to its blatantly didactic aims. Second, consideration of the organized eugenics movement, which may have been the most enduring form of Social Darwinism, will be left largely for a future work because it was vastly more important in the opening decades of the twentieth century than in the closing decades of the nineteenth. A brief discussion of the very early years of eugenics and race-hygiene will appear in the next chapter, however; because they emerged primarily in response to doctrines of degeneration that will be treated there.

At least in Britain and Germany, evolutionary social thinkers in the late nineteenth century did nearly all share one critically important perspective regardless of their political and social agendas. They agreed that, as a matter of fact, economic life in particular, was increasingly governed by the law of the survival of the fittest. Unregulated economic competition was leading to the accumulation of vast wealth in the hands of a few large-scale capitalists and to the ever deeper impoverishment of the masses of working poor. This situation in turn was leading to open conflict between the wealthy capitalists and the workers, evidenced by ever more violent strikes, and to collusion among capitalists to break up strikes with open violence.

What differentiated the various forms of evolutionary social theory were the prognoses that this stereotyped situation called forth and the prescriptions, if any, for mitigating the evils that it entailed. On the far right and far left there was a sense that the situation should simply be allowed to play itself out until the unfit individuals or nations were eliminated (one right prognosis) or until capital became so concentrated that every industry became a monopoly that obviously had to be nationalized, leading to a peaceful, classless world (one left prognosis). Between these extremes, an array of proposals to mitigate the evils through either voluntary or governmental action appeared, all of which saw at least some significant level of competition as essential to progress and all of which argued for the importance of guiding reform efforts through a scientific, that is, evolutionary, understanding of circumstances.

Walter Bagehot's Emphasis on War and "Animated Moderation"

Beginning in January 1867, well before the composition of Darwin's *The Descent of Man,* the *Fortnightly Review* in London began publication of a series of five essays that offered one of the first comprehensive interpretations of the role of evolution by natural selection in human social development. The author of these essays was Walter Bagehot, a banker, editor of the *Economist,* highly respected political journalist, and close friend to many liberal politicians. Lightly edited, and with an additional summary essay, the series was published in book form in 1872 as *Physics and Politics, or: Thoughts on the Application of the Principles of "Natural Selection" and "Inheritance" to Political Society.* Bagehot's goal was quite clearly to promote what some commentators call "conservative liberalism," what he called "animated moderation," or what the followers of Comte, whose works Bagehot knew well, called "progress with order." But in the process of explaining why only slow and deliberate progress, as opposed to complete stasis or radical change, was viable in contemporary England, Bagehot offered a view of human evolution that seemed to many to authorize militaristic nationalism.

Bagehot says in his summary essay, which ties together the whole series, that he has sought both to offer a definition of human progress that is "verifiable"—by which he means, one that the overwhelming majority of humankind will agree to—and to explain how such progress has occurred from human prehistory to the present.[6] To answer the first question, he offers a contemporary example that reflects a central theme of his entire work: "Let us consider, in what a village of English colonists is superior to a tribe of Australian natives who roam about them. Indisputably in one, and that a main sense, they are superior. They can beat the Australians in war when they like; they can take from them anything they like, and kill any of them they chose. As a rule, in all the outlying and uncontested districts of the world, the aboriginal native is at the mercy of the intruding European. Nor is this all. Indisputably in the English village there are more means of happiness, a greater accumulation of the instruments of enjoyment, than in the Australian tribe. The English have all manner of books, tinsels, and machines, which the others do not use, value, or understand. And in addition, and beyond particular inventions, there is a general strength which is capable of being used in conquering a thousand difficulties."[7] Whatever our various views of progress may be today, Michael Adas has argued quite compellingly that Bagehot was correct in claiming that the vast majority of mid-Victorians did view their technology, especially their military technology and the natural scientific knowledge that they understood to produce technological innovation, as the sources of a superiority of civilized European peoples over all others.[8]

It was clear to Bagehot that technological innovation could not have developed and propagated outside of societies; nor could individuals or small kin groups have survived conflicts with larger groups during human prehistory. So he argued that the formation of cooperative groups, or relatively stable polities, extending beyond blood ties was the first precondition for human progress. Moreover, he argued that once polities formed, it was natural selection through warfare that determined which groups survived and thrived, thus providing a key insight to be appropriated by Darwin in his *Descent of Man:* "Whatever may be said about 'natural selection' in other departments, there is no doubt of its predominance in early human history. The strongest killed out the weakest, as they could. And I need not pause to prove that any form of polity is more efficient than none; that an aggregate of families owning even a slippery allegiance to a single head would be sure to have the better of a set of families acknowledging no obedience to anyone, but scattering loose about the world and fighting where they stood."[9] The great questions for the early period of human development then became: How did human groups come to cohere and what particular characteristics were selected for?

Bagehot believed that only what we call religion, which he believed was grounded initially in omenology and superstition, was capable of commanding the kind of allegiance necessary to hold early groups together. Speaking of primitive religions, Bagehot wrote, "They fixed the yoke of custom thoroughly on mankind. They were the prime agents of the era. They put upon a fixed law a sanction so fearful that no one could dream of not conforming to it."[10] Thus, in the earliest stages of civilization, groups in which the institutions of state and religion are identical are most coherent and militarily successful. Even in later times, strong religious beliefs tied to the state provide a powerful military advantage. "People laughed at Cromwell's saying, 'Trust in God and keep your powder dry,'" Bagehot wrote, "but we now know that the trust was of as much use as the powder, if not more. That high concentration of steady feeling makes men dare everything and do anything."[11] Cromwell's comment suggests two final considerations concerning religion. Those religions that give greater "confidence in the universe" and those that are monotheistic rather than polytheistic are best suited to survival because fewer actions are prohibited when the universe is not viewed as filled with dangers that must be avoided and because the emotional power of religion can best be focused when its single object "is not confused by competing rites or distracted by miscellaneous deities."[12]

Just as Bagehot explains the growth of positively oriented monotheistic religions out of ritually based polytheisms as a consequence of natural selection through inter-group warfare, he also explains the present dominance of patrilineal family structures out of an early stage in which matrilineal social organization was common: "The first great victory of civilization was the conquest of nations with ill-defined families having legal descent through the mother only, by nations of definite families tracing descent through the father as well as the mother, or through the father only. Such compact families are a much better basis for military discipline. . . . [So] the nations with a thoroughly compacted family system have 'possessed the earth'; that is, they have taken all the finest districts in the most competed for parts; and the nations with loose systems have been left to mountain ranges and lonely islands."[13]

Even intelligence and morality, which Bagehot divorces from religion, are, he argues, selected for by warfare: "There are cases in which some step in intellectual progress gives an early society some gain in war; more obvious cases are when some kind of *moral* quality gives some such gain. War both needs and generates certain virtues; not the highest, but what may be called the preliminary virtues, as valor, veracity, the spirit of obedience, the habit of discipline . . . a race with reason and high moral feeling beats a race with reason but without high moral feeling."[14]

Having established that many of the key features of civilization, including the dominant family structure, the presence and dominance of certain forms of religion, and the emergence of morality, are the consequence of a largely unconscious process of natural selection effected by early warfare between human tribes, Bagehot goes on to argue that the "cake of custom," which provides a military advantage to early societies, could easily stifle all possibility of change and bring progress to a halt unless some mechanism or mechanisms for continuing to produce variations emerge.

He thus argues that, in a second stage of human evolution, variability in societies becomes increasingly important.[15] Such variability is produced in part by what Bagehot calls "race mixture" or what later anthropologists call the cultural contacts that occur as conquered and conquering tribes intermingle. When this happens, features of the culture of the conquered become available for adoption by the conquerors and vice versa.[16] More importantly, variability is promoted in groups that have "government by discussion," rather than by the kind of restrictive theocracies or kingships that are most valuable in the early history of humankind. This is so because when issues are brought before assemblies for discussion, the customs that were unquestioned in traditional societies are broken down. Tolerance is encouraged. And progress is promoted both because fewer options are automatically excluded from consideration and because the habit of reflection and forethought that government by discussion engenders produces science and technical innovation.[17] Under these circumstances, Bagehot argues, "War has ceased to be the moving force in the world."[18] The stage of government by discussion, is, however, much more complex than the early stages of human social evolution because the virtues of discussion must augment rather than replace the more "savage" virtues. Those nations that become too intellectual are sure to be destroyed. "History is strewn with the wrecks of nations that have gained a little progressiveness at the cost of a great deal of hard manliness and have thus prepared themselves for destruction," Bagehot insists, and, "those savage virtues that tend to war" remain the essential foundation of even modern human nature.[19]

Organizational progress, moral progress, and intellectual progress that are not applied to military exploits may have come to supplement and to some extent supplant war-based natural selection. Nonetheless, for Bagehot, when all is said and done, military conflict remains a major engine of progress. He may have denied that war was any longer the chief moving force in history; but he also continued to claim that "since the long-headed men first drove the short-headed men out of the best land in Europe, all European history has been the history of the superposition of the more military races over the less military."[20]

Darwin drew in his *Descent of Man* from Bagehot's argument that "civilization begins, because the beginning of civilization is a military advantage."[21] But Darwin insisted that though sympathy was initially selected for because it provided military benefit in early tribal society, it gradually expanded to incorporate all humankind and negated the need for war. Bagehot, on the other hand, insisted that cooperation remained an in-group characteristic that was valuable, even in the modern world, largely because it offered a competitive advantage in war. Both interpretations drew numerous followers.

Wallace and Huxley: The Uniqueness of Human Beings and the Triumph of Ethics over Nature

There is a peculiar irony in the fact that Alfred Russel Wallace and Thomas Henry Huxley became the intellectual leaders among those late-Victorians who argued that though natural selection may have played some role in the early development of human society, civilized humans now had it within their power to take control of their own evolution as no other species could. The irony lies in the fact that as naturalists, both Wallace and Huxley were more deeply committed to natural selection than Darwin himself and they did more to publicly promote Darwinian notions of evolution than the more reticent Darwin. Huxley became the symbol of the purported Darwinian "reduction" of humans to mere animals through his famous confrontation with Bishop Samuel Wilberforce at the 1860 Cambridge meeting of the British Association for the Advancement of Science. Wallace was so adamant about natural selection that he denied any significant role to sexual selection among animals or humans throughout his long career, which reached into the twentieth century. Yet both men rebelled against the deterministic application of natural selection to human society.

As early as 1864, Wallace began to argue that there is substantial evidence that human evolution was subject to changed rules because, unlike other animals, humans were able to use their intelligence to modify their environment to suit their needs, rather than adapting themselves to their environments. In man's earliest history, tribal struggle had provided a selective pressure. But through the invention of clothing, artificial shelters, sophisticated weapons for hunting, and so on, humans replaced the natural physical evolution that all other animals were subject to with a social evolution open, at least to some degree, to their own conscious direction.[22] Just a year before, in *Man's Place in Nature,* Huxley too had insisted that though humankind was a product of evolution, civilized man was radically different from all predecessors: "No one is more strongly convinced of the gulf between civilized man and the brutes; or is more certain that whether *from* them or not, he is surely not *of* them. No one is less disposed to think lightly of the present dignity, or despairingly of the future hopes, of the only consciously intelligent denizen of this world."[23]

Wallace, who had grown up in a radical, working-class environment and who had received his education in Mechanics' Institutes and Owenite Halls of Science, remained a social reformer who participated in a wide range of anti-slavery and anti-imperialist organizations. He continued, throughout his life, to believe that humans were capable of self-directed improvement; and he had a life-long commitment to spiritualism, which reinforced his belief in the Christian doctrine of life after death for human souls. Thus, while he was inclined to accept a completely naturalistic, materialistic, and nonteleological account of evolution up to the point of human evolution, he claimed that human development was guided by a higher power and that natural selection could not account for it.

He argued strongly that the great gap in brain size and brain power between humans and their nearest primate relatives provided compelling evidence for his views, because natural selection alone could only account for an incremental difference.[24] Drawing on his Owenite background, as well as on evidence of the inverse correlation between intellectual development and fertility, he argued that universal advanced education for all, once implemented, would eliminate the Malthusian problem of overpopulation and promote universal cooperation, leading to a Utopia in which distinctions of class and race would eventually disappear along with the need for coercive governments.[25] Wallace did not believe that humanity was far advanced along this path. He railed against the exploitation of workers at home and the exploitation of colonized peoples abroad, insisting that future historians would record "that we of the nineteenth century were morally and socially unfit to possess and use the enormous powers for good or evil which the rapid advance of scientific discovery had given us; that our boasted civilization was in many respects a surface veneer."[26] Nonetheless, he saw humanity as on its way to a divinely ordained bright socialist future.

Huxley's views were considerably less optimistic, especially after his daughter's death in 1887. Prior to that time, he had viewed evolution by natural selection in much the way that the early Spencer had seen his more Lamarckian version—as a fundamentally progressive, though not consciously directed, process. But in 1888 Huxley's reflections changed direction. In all pre-human and pre-social evolution, he began to argue, as Spencer did around the same time, "retrogression" was as likely as "progression." Appealing to Kelvin's thermodynamic arguments, he acknowledged that natural, "cosmic" processes were actually moving toward an ultimate collapse of the universe as we know it.[27] As a consequence, Huxley insisted in *Evolution and Ethics* (1893), that no moral and ethical lessons could be read out of the evolutionary process and that, in an important sense, human ethical advance depended on fighting against the cosmic process of evolution by natural selection, rather than acceding to it: "Social progress means a checking of the cosmic [read natural] process at every step, and the substitution for it of another, which may be called the ethical process; the end of which is not the survival of those who may happen to be the fittest, in respect of the whole of conditions which obtain, but of those who are ethically the best . . . the ethical progress of society depends, not on imitating the cosmic process, still less in running away from it, but in combating it."[28]

Huxley admitted with Bagehot and Darwin that in the prehistory of man, the early stages of morality emerged as a consequence of selection through tribal warfare; but he argued that in civilized man the search for a naturalistic ethics had to be abandoned in favor of a self-directed morality that rejected violence and promoted a universal love among humans. Its goal would be, "to help fellows rather than tread them down, to fit as many as possible to survive rather than to aim at survival of the fittest."[29] This goal would not be easy to achieve, however, because the long evolutionary history of humankind had produced a fierce creature, akin to the ape and the tiger in cunning and ferocity. To tame such a creature and hold its natural instincts at bay would demand constant effort.

For Huxley, that effort would have to be directed by some kind of strong central governmental authority. Frustrated by Spencerian laissez-faire arguments because they blocked British governmental attempts to support education and scientific investigation, Huxley called for more government, rather than less.

Kropotkin and Novicow: Mutual Aid, Naturalistic Ethics, and Pacifism

On almost every conceivable issue, Peter Kropotkin opposed Huxley in a series of eight articles published between November 1890 and June 1896 in the English periodical, *Nineteenth Century* and subsequently published as *Mutual Aid* in 1902. Born into an aristocratic Russian family in 1842, Kropotkin became associated with Mikhail Bakunin's anarchism and eventually emerged as one of the great anarchist theorists. During the 1860s and 1870s he served in diplomatic and military posts in Siberia and Manchuria, where he led several expeditions to study animal behavior. He was struck in 1880 by a lecture delivered by Karl Kessler of St. Petersburg University, in which Kessler argued that there was in nature a law of mutual aid that was every bit as important as the law of mutual struggle; and he turned the theme of that lecture into the foundation of his own biologically based social theory.

Kropotkin denied the ferocity of prehistoric man, the role of intra-specific tribal conflict in creating the beginnings of morality, and the need to posit some form of instinct-transcending love or sympathy as the foundation for civilization. The latter he considered some kind of Rousseauean sentimentalism. He insisted that, in general, the struggle for existence was one against a hostile environment and that nowhere in nature did one find "that bitter struggle for the means of existence, *among animals belonging to the same species,* which was considered by most Darwinists as the . . . main factor of evolution."[30] Instead, Kropotkin argued that an instinct for mutual aid and support evolved among nearly all animals and that this instinct was not related to love or personal sympathy: "It is not love, and not even sympathy (understood in its proper sense) which induces a herd of ruminants or of horses to form a ring in order to resist an attack of wolves; not love which induces the wolves to form a pack for hunting; . . . and it is neither love nor personal sympathy which induces many thousand fallow-deer scattered over a territory as large as France to form into a score of separate herds, all marching toward a given spot, in order to cross a river. It is . . . an instinct that has been slowly developed among animals and men in the course of an extremely long evolution, and which has taught animals and men alike the force they can borrow from the practice of mutual aid and support, and the joys they can find in social life."[31] Though he referred to descriptions of animal behavior provided by Darwin and others in supporting this claim, Kropotkin's arguments gained in plausibility from his ability to draw from his own extensive and detailed observations of animals living in an extremely stressful environment.

It was true, Kropotkin agreed with Darwin, that higher ethical and moral principles did develop out of love and sympathy; but the emergence of society depended on an instinct of "human solidarity" that was not dependent on tribal conflict for its propagation. In fact, warfare produced a kind of counter-morality that allowed for the most revolting cruelties to be perpetrated.[32] Furthermore, because mutual support is a natural tendency, Kropotkin argued that Huxley and his ilk were completely misguided in arguing for a separation of nature and ethics. Naturalistic ethics promoted mutual aid, it did not have to be imposed against human nature, and it could be completely voluntarist, rather than supported by a coercive governmental structure. Capitalism was a temporary aberration rationalized by advocates of struggle and it would disappear or be overwhelmed as groups who promoted communitarian ideals—peasants, urban poor, and trade unions, for example—prevailed.[33]

Like many other anarchists at the end of the nineteenth century, though Kropotkin did not view struggle as the key to human advancement, he was perfectly willing to argue that capitalism might have to be overthrown through revolutionary, rather than evolutionary processes. One of his self-styled followers, Jacques Novicow (1849–1912), however, developed Kropotkin's emphasis on mutual aid into a pacifist doctrine. Novicow was, himself, a victim of Tsarist coercion, emigrating to Paris as a young man. Equally appalled by the coercive elements in reactionary and in socialist politics, Novicow published a series of works: *Les Luttes entre sociétés humaines* (1893), *La guerre et ses preténdus bienfaits* (1894), and *La critique du darwinisme social* (1910). In each of these, he argued that the development of human reason coupled with the instinct for mutual support had created an alternative to the "collective homicide" that was a consequence of the aggressive nationalism sweeping Europe. This nationalism had produced the Franco-Prussian war of 1870, and seemed poised to lead to further conflicts throughout Europe. War had an inherently negative selective impact because it was the physically and morally superior males who went off to die, leaving the weak and cowardly to propagate. The peaceful process of forming ever wider federations to deter aggression saved the best breeding stock, prevented the waste of material as well as human resources, and generally represented a vastly more reasonable and efficient way to provide protection than war itself. That is, rational ethics and naturalistic ethics were identical. Virtually all of the French and German peace movements leading up to the First World War, with their promotion of a European Federation, seem to have been grounded in Novicow's "scientific pacifism."[34]

Darwinism and Anglo-American Imperialism: Benjamin Kidd's *Control of the Tropics*

In 1894, a junior clerk in the British Civil Service, Benjamin Kidd (1858–1916), published *Social Darwinism,* in part through the patronage of Alfred Milner (1854–1925), soon to be Lord Milner, proconsul in South Africa, and one of the chief architects of British policy leading to the Boer War. The book was an overnight sensation, with first-year sales in Britain of more than 7,500 copies, and first-year sales in the United States of 18,958 copies.[35]

Kidd's work was important in part for the way in which it highlighted the role of "ultra-rational," religious, concerns in social evolution. Both Darwin and Bagehot had argued that religion was crucial in providing for social cohesion and the growth of altruistic relative to egoistic values in society; but Kidd expanded their analyses, demonstrating to his own satisfaction and that of a substantial readership, both that though "rational" policies could be articulated to serve the interests of particular individuals or classes of individuals, such policies could never promote progressive social evolution. That is, as long as the interests of elements of the current population were taken to supersede the interests of future populations, social progress would be subordinated to factional interests. For Kidd, religion in general and Christianity in particular inculcated those altruistic concerns, especially for future generations, that made positive social evolution possible.

Kidd was also one of the first social theorists to completely embrace Weisman's germ theory of inheritance and to reject the Lamarckian notion that acquired somatic characteristics could be inherited. Indeed, Kidd was so impressed by Weisman that he visited the German biologist at Freiburg in 1890.[36] There he discussed what he saw to be critical consequences of Weisman's views for his understanding of human evolution. First, since the effects of training and education in one's parents could have no impact on one's inherited characteristics, if individuals from the lower classes were given a fair start at birth, they would be "the equals in natural inheritance with the classes above them."[37] With respect to individual human evolution Kidd was a hyper-competitionist; so he argued that evolution would proceed more rapidly with more competitors and that minimizing the starting advantages of the wealthy would improve competition.[38] By asserting human equality of opportunity, Christianity thus promoted fair competition. Socialism, on the other hand, by promoting equality of outcomes, rather than opportunities, thwarted competition and retarded evolutionary progress.

For our present purposes, the second implication that Kidd drew from Weisman's theory was that virtually all of human evolution was accounted for by social and cultural development and not by somatic divergence. The latter was far too slow in the absence of the inheritance of acquired characteristics to account for the huge behavioral differences among humans. As a consequence, Kidd agreed with Darwin that the major differences between "civilized" men and "savages" were to be attributed to "social efficiency," rather than to individual heritable features.[39] The necessity of organizing for confrontation with a particularly hostile natural environment had led inhabitants of climates north of the 50th parallel, especially English-speaking and Russian-speaking populations, to develop a particularly high state of social efficiency.[40] On the other hand, inhabitants of the tropics, where nature provided subsistence with little effort, had little need for advanced culture and had less highly evolved social organizations.

Given their low level of social efficiency, the indigenous populations of the tropics were incapable of effectively exploiting the rich resources of the regions in which they lived. Yet, argued Kidd in an 1894 interview for the New York *Outlook,* "With the advance which science is making . . . it is in the tropics . . . that we have the greatest food-producing and material-producing regions upon the earth; . . . the natural highways of commerce in the world are those which run North and South, and we have the highest possible interest in the proper development and efficient administration of the tropical regime, and in an exchange of products therewith on a far larger scale than has been attempted or imagined."[41] It followed for Kidd that Europeans and Americans had an obligation to colonize and to manage the resources of the tropics. But Kidd's vision of European and American imperialism differed from that of most Europeans; and he laid out that vision in detail in a series of articles for the London *Times,* which were published in 1898 as *The Control of the Tropics.*

While it was true that the indigenous populations of the tropics were socially less advanced and thus presently incapable of efficiently managing their own lands, it was also true that they were full-fledged human beings deserving of humane treatment and that they were completely capable of relatively rapid social development under the guidance of Europeans. Consequently, Kidd condemned the exploitative "plantation" system of colonization, which de facto conscripted native labor to work for the northern "owners" of agricultural plantations and natural resources that was a typical pattern of French and Spanish imperialism.[42] Similarly, he opposed the permanent settlement and acclimatization of large numbers of European colonizers, as happened with the Boers in southern Africa, both because it completely subordinated native interests and because he was convinced that Europeans, outside of their natural environment, would have trouble maintaining their culture. He was convinced that settlement-based colonization would rapidly revert to the plantation system.

In opposition to the plantation and settlement models of colonization, Kidd proposed a "trusteeship" model in which the advanced northern nations, led by England and America, would manage tropical resources as a trust for civilization. Imperialist nations would respect the interests of indigenous populations at the same time that they brought civilized ideals into intimate contact with existing lower cultures, offering them models to emulate. Existing native cultures, which represented in cultural development the stage corresponding to childhood in individual growth, would then progress, "by bringing them under a programme of administrative efficiency . . . divorced from profiteering and exploitation, masterminded from the temperate European [or American] homeland," and carried out by a white administrative elite, rather than by a settler group.[43] The ultimate goal would be to gradually turn over management and government functions to indigenous peoples as their cultures became more like those of the Europeans.

While Kidd's ideas stimulated some debate in Britain and shaped the views of politicians including Joseph Chamberlain (1836–1914), they played a much greater role in American debates about how to treat Cuba and the Philippines. Franklin Giddings (1855–1913), whose *Democracy and Empire* (1900) provided a blueprint for American governmental policy, cited Kidd in laying out its central argument: "[Cuba and the Philippines] must be held as territorial possessions, to be governed firmly, in the interest both of the world at large and of their own native inhabitants, by administrative agents appointed and directed by the home governments of the northern nations . . . if the civilized world is not to abandon all hope of continuing its economic conquest of the natural resources of the globe."[44]

It was probably true that Kidd thought himself to be promoting the interests of indigenous populations. It is even possible that in promoting the notion of trusteeship in connection with imperialism, Kidd and his acolytes mitigated some of the potentially most egregiously exploitative consequences of U.S. imperialism in Cuba and the Philippines. Nonetheless, one must have some sympathy for a socialist critic, writing in the Rochester *Post Express,* who argued that Kidd simply put a veneer "on the old predatory instinct for land grabbing," to make it more palatable to the general population, and who concluded that, "he who proposes such a scheme gravely is simply a hypocrite, if he be not a visionary."[45]

Edward Bellamy's "Nationalism"

In 1935, when a number of distinguished American intellectuals were asked to name the most important books written in the past half century, both the philosopher, John Dewey, and the historian, Charles Beard, listed a "fanciful romance" written by Edward Bellamy, a man of otherwise extremely modest accomplishments, as number two, immediately behind Karl Marx's *Capital.* Little known now, Bellamy's *Looking Backward* was one of the three best-selling American works of the nineteenth century, trailing only *Uncle Tom's Cabin* and *Ben Hur* in American sales. It was translated into twenty languages, stimulated the establishment of more than 165 American "Bellamy Clubs" devoted to promoting its aims in the four years following its publication, and provided most of the platform of the short-lived Populist Party.[46] More surprisingly, in its German translation as *Im Jahre 2000,* Bellamy's novel seems to have had a greater role in forming German working-class attitudes toward socialism than those of Marx and Engels combined.[47]

Looking Backward was, in the author's own words, "a forecast, in accordance with the principles of evolution, of the next stage in the industrial and social development of humanity."[48] Pulling together themes associated with Saint-Simonian socialism, Comte's Religion of Humanity, Marx's *Capital,* and Darwin's *Descent of Man,* Bellamy crafted a vision of a future society that appealed to a very substantial following. It was one in which immense material wealth, complete equality, and extensive human freedom emerged through the peaceful evolution of the nineteenth-century economic system of competitive free-market capitalism into the kind of completely nationalized, cooperative, and centrally managed "industrial system" envisioned by Saint-Simon.

For Bellamy, it was the confluence of two evolutionary trends that allowed for a peaceful path to a completely cooperative, centrally managed, yet nonoppressive society. One was the trend toward increasing selflessness and love of all humankind, which he viewed as the consequence of Darwinian sexual selection. The other was the trend toward the establishment of ever larger economic enterprises, under the pressures of capitalism, which had been predicted by Marx.

Julian West, the upper-class protagonist of Bellamy's novel, goes to sleep in Boston in 1887 to wake up in the Boston of 2000. He is immediately impressed by the material splendor of the city and his host reflects on the central difference between cities in the late nineteenth century and those of his own: "The cities of that period were rather shabby affairs. If you had the taste to make them splendid . . . the general poverty resulting from your extraordinary industrial system would not have given you the means. Moreover, the excessive individualism which then prevailed was inconsistent with much public spirit. What little wealth you had seems almost wholly to have been lavished in private luxury. Nowadays, on the contrary, there is no destination of the surplus wealth so popular as the adornment of the city, which all enjoy in equal degree."[49]

When it becomes obvious that the new Boston is subject to none of the intense social troubles associated with the great inequalities of wealth and concomitant poverty that had preoccupied citizens of the nineteenth century, West asks how the central "labor question" was solved. He is told that, "The solution came as a result of a process of industrial evolution which could not have terminated otherwise. All that society had to do was to recognize and cooperate with that evolution, when its tendency had become unmistakable."[50]

While agreeing with the reform Darwinists that policy should take cognizance of natural laws, Bellamy rejected their notion that humanity needed to overcome natural law. Instead, our salvation lay in welcoming rather than resisting the consequences of economic competition. There follows a brief sketch of the process of capital concentration that Marx had presented in *Capital.* By the late nineteenth century: "Small businesses, as far as they still remained, were reduced to the condition of rats and mice, living in holes and corners, and counting on evading notice for the enjoyment of existence. The railroads had gone on combining till a few great syndicates controlled every rail in the land. In the manufacturies, every important staple was controlled by a syndicate. These syndicates, pools, trusts, or whatever their name, fixed prices and crushed all competition, except when combinations as vast as themselves arose."[51]

Under these conditions, individual laborers were powerless; so they were forced to unionize in self defense; and severe strikes, often violent, followed. As much as people hated the tyrannical power of the great corporations, however, they had to acknowledge their economic efficiency. And as scholars tried to figure out how to gain the benefits of capital concentration without forcing workers to become the unwilling subjects of a ruling plutocracy, a new recognition emerged. Instead of resisting the trend toward monopolies, that trend should be driven to its logical conclusion. Early in the twentieth century "the evolution was completed by the final consolidation of the entire capital of the nation. . . . The nation, that is to say, organized as the one great business corporation in which all other corporations were absorbed; it became one capitalist in the place of all other capitalists, the sole employer, the final monopoly in which all previous and lesser monopolies were swallowed up, a monopoly in the profits and economies of which all citizens shared. . . . In a word, the people of the United States concluded to assume the conduct of their own business, just as one hundred years before, they had assumed the conduct of their own government."[52]

In this new state-managed economy, cooperation, rather than competition was the driving force. All persons received equal credit allocations annually to be spent in any way they pleased. All persons received the same basic education, served three years in the industrial army doing jobs assigned by the bureaucracy, and then chose their own career path, training for which was at state expense. In order to ensure that there were adequate volunteers for every essential occupation, hours of service required to warrant the annual credit for various jobs were adjusted to make all appropriately appealing to some segment of the population. Careers in literature and the arts, including music, depended on the artist's ability to get people to voluntarily shift credits to the artist as a form of royalty. Finally, motivation for all economic activity shifted from differential monetary rewards to a combination of satisfaction felt in public service, public recognition, honor, and (for males) the admiration of females.

This last factor was particularly important in maintaining evolutionary progress in morality and altruism, according to Bellamy, because since marriage could now be independent of economic considerations, sexual selection was given free reign; and the fact that women valued the altruistic over the egoistic male created a population in which cooperation prevailed over competition:

> For the first time in human history the principle of sexual selection, with its tendency to preserve and transmit the better types of the race, and let the inferior types drop out, has unhindered operation . . . not only is one of the great laws of nature now freely working out the salvation of the race, but a profound moral sentiment has come to its support. Individualism, which was in [the nineteenth century], the animating idea of society, not only was fatal to any vital sentiment of brotherhood and common interest among living men, but equally to any realization of the responsibility of the living to the generation to follow. Today this sense of responsibility . . . has become one of the great ethical ideas of the race, reinforcing, with an intense conviction of duty, the natural impulse to seek in marriage the best and noblest of the other sex. The result is, that not all the encouragements and incentives of every sort which we have provided to develop industry, talent, genius, excellence of every kind, are comparable to the effect on our young men with the fact that our women sit aloft as judges of the race and reserve themselves to reward the winners.[53]

Given the fact that sexual selection was a uniquely Darwinian, as opposed to Lamarckian or Spencerian, notion, Bellamy was one of the most distinctively Darwinian among American evolutionary social theorists.

On the other hand, Bellamy was distinctly un-Darwinian as well as opposed to the later Spencer in his theistic and teleological interpretation of evolution. Bellamy made it clear in *Looking Backward* that he was well aware of the long-term implications of the second law of thermodynamics when West's rescuer argues that had he not discovered West, the young man's body "might have remained in a state of suspended animation til, at the end of indefinite ages, the gradual refrigeration of the earth had destroyed the bodily tissues."[54] Similarly, he indicated his awareness of Spencer's notion that dissolution must eventually follow evolution. But his own view was distinctly teleological and providential. At the end of *Looking Backward,* Bellamy offers a millenarian interpretation of evolution, claiming, "Twofold is the return of man to God, 'who is our home,' the return of the individual by the way of death, and the return of the race by the fulfillment of the evolution, when the divine secret hidden in the germ shall be perfectly unfolded."[55] Bellamy's religious views were certainly heterodox; but by casting them in a dominant Christian idiom, he helped to expand the appeal of his version of Social Darwinism immensely.

Social Darwinisms in Germany

Nowhere else did Darwinism, both as a biological theory and as the foundation for social thought, become as deeply rooted and broadly accepted as in Germany. The reasons are many. First, throughout the nineteenth century, in almost every domain of intellectual life, German thinkers emphasized change, transformation, and historical processes. Whether they were concerned with Goethe's metamorphoses, Hegel's gradual unfolding of reason, or von Baer and Müller's studies of embryological development, Germans were preoccupied with change and open to evolutionary ideas. Lamarck was widely admired, and Chambers's *Vestiges of Creation* appeared in two German translations during the 1840s.

Second, in Ernst Haeckel, Darwin found a disciple who was an outstanding professional biologist and even more effective than T. H. Huxley in promoting Darwinian ideas. A rabid German cultural nationalist, Haeckel was equally an admirer of such materialists as David Strauss, Lewis Feuerbach, and Ludwig Büchner on the one hand and such Romantic scientists and *Naturphilosophen* as Goethe and Schelling on the other. So he was able to appeal equally to materialists and to naturalists and ordinary citizens with tendencies toward *Naturphilosophie.* Moreover, unlike Huxley, Haeckel built an entire philosophy of life around evolution, explicitly using it to ground ethics, educational practices, a nontheistic religion, attitudes toward imperialism, war, German racial superiority, and so on. With respect to all of these topics, Haeckel managed to link Darwinism to powerful preexistent tendencies in liberal German culture in a way that seemed to many to make Darwinian theory their central uniting thread.

According to the historian of biology, Erik Nordenskjold, more persons learned about evolution from Haeckel's *Natürliche Schopfungsgeschichte . . .* , or *History of Creation: or the Development of the Earth and Its Inhabitants by the Action of Natural Causes. A Popular Exposition of the Doctrine of Evolution in General, and That of Darwin, Goethe, and Lamarck in Particular* (first German edition, 1868, American edition, 1876), which went through nine German editions and translation into twelve languages before 1899,[56] than from any other source, including *The Origin of Species.*[57] Furthermore, *Die Welträtsel* (1899) or *The Riddle of the Universe at the Close of the Nineteenth Century,* with German language sales of over three hundred thousand before World War I and translation into more than twenty other languages, brought Haeckel's evolutionary philosophy to more persons than had read any other German nonfiction work to that date.[58]

Third, Darwinism, especially as promoted by Haeckel, offered scientific materialist popularizers such as Ludwig Büchner a way to understand the origin and development of the entire universe without appeal to a divine designer-creator. This was tremendously valuable given their opposition to established religion, and it led scientific popularizers and journalists to make Darwinism the subject of more popular scientific texts and articles than any other topic from 1870 to World War I.[59]

Fourth, Darwinism offered German socialist leaders a language and set of analogies that were vastly more accessible and appealing to working-class audiences than the esoteric language of Hegel and the complex political economy of *Capital*, with which to promote their own political and economic agendas. But attempts to use Darwinism to promote socialism had a transformative effect on German working-class culture, making it less militant and revolutionary and more patient and evolutionary.[60]

Haeckel's Evolutionary Monism

Though Germany saw a powerful evolutionary socialist movement, which drew heavily from both Bellamy and Kropotkin, the version of social Darwinism that ultimately had the greatest impact on German culture (though it almost certainly did not have the largest immediate audience) was probably that due to Haeckel;[61] and Haeckel's Darwinism was as anti-socialist as it was anti-conservative. In fact, wrote Haeckel, "Socialism demands equal rights, equal duties, equal possessions, equal enjoyments for every citizen alike; the theory of descent proves, in exact opposition to this, that the realization of this demand is a pure impossibility.... The theory of selection teaches that in human life as in all animal and plant life everywhere, and at all times, only a small and chosen minority can exist and flourish, while the enormous majority starve and perish miserably and more or less prematurely.... [Evolution] is aristocratic in the strictest sense of the word."[62]

Haeckel had begun to read Darwin's *Origin of Species* in 1860 while he was still a student of Carl Gegenbauer (1825–1903), the Jena comparative anatomist. Like Gegenbauer, he was deeply impressed with Darwin's empirical support for a theory that he saw as prefigured by another great Jena scientist, Goethe. Working primarily with the development of marine invertebrates, Haeckel became convinced of the validity of the "biogenetic law," that is, that the development of the individual organism and the evolutionary development of the species followed essentially the same pattern. This notion had already been discredited by von Baer, among others; but Haeckel modified its claims slightly so as to relax the requirement that every step in phylogenetic development be paralleled in individual growth. He nonetheless maintained the claim of overall close similarity: "Ontogenesis is a brief and rapid recapitulation of phylogenesis," he wrote.[63] And in a series of works, including *General Morphology* (1866), *Biology of the Calcispongiae* (1872), and *Studies on the Gastraea Theory* (1873–84), he provided extensive evidence to support the biogenetic law, which became a key to his entire evolutionary philosophy.

Like many English and American Social Darwinists, Haeckel argued that most of the social and political blunders made by states and their representatives were a consequence of the failure to embrace a scientific view of humankind: "We can only arrive at a correct knowledge of the structure and life of the social body, the state, through the scientific knowledge of the individuals who compose it, and the cells of which they are in turn composed. If our political rulers and our 'representatives of the people' possessed this invaluable biological and anthropological knowledge, we should not find our journals so full of the sociological blunders and political nonsense which at present are far from adorning our parliamentary reports and even many of our official documents."[64]

Unlike most of his English and American colleagues, however, Haeckel joined the majority of German liberals in two critical ways. First, from its inception, German liberalism was much more open to a strong and interventionist state than that of England and America. The chief architect of German liberal economics, Friedrich List, carried cameralist interest in the efficient management of economic and natural resources into his *National System of Political Economy* (1841), and Haeckel joined in this support of a strong state, supporting Bismarck's efforts at creating a centralized German state. This meant that Haeckel promoted state-managed eugenic policies long before the eugenics movement developed. As early as 1868, for example, he proposed that the state consider implementing the ancient Spartan practice of infanticide for children who were "weak, sickly, or affected with any bodily infirmity."[65]

Second, German liberalism was largely anti-Christian; and Haeckel joined the majority of liberals in directly confronting organized Christianity as a major source of error and calling for a totally secular state: "Never will our government improve until it casts off the fetters of the church and raises the views of the citizens on man and the world to a higher level by a general scientific education. Whether a monarchy or a republic be preferable, whether the constitution should be aristocratic or democratic, are subordinate questions in comparison with the supreme question: Shall the modern state be spiritual or secular? . . . The first task is to kindle a rational interest in our youth and to free them from superstition."[66]

At this point, Darwinian evolution by natural selection became centrally important for Haeckel, for in his eyes Darwin produced, for the first time, compelling evidence that humans were part of nature rather that set aside from it, that they had evolved most recently from old-world anthropoid apes,[67] and ultimately from some single-celled organism, and that the evolutionary process that had produced man was not part of any divine plan. In fact, it was a process that involved no design whatsoever—either by some transcendent being or by any of the creatures involved in the process. "Nowhere . . . in the evolution of animals and plants do we find any trace of design, but merely the inevitable outcome of the struggle for existence, the blind controller, instead of the provident God, that effects the changes of organic forms by a mutual action of the laws of heredity and adaptation."[68]

For Haeckel, then, Darwinian evolution provided the key argument against basing social and political decisions on the traditional religious notions of morality. Indeed, Haeckel argues that, contra the Darwin of *The Descent of Man,* there is no moral order in nature, or that if there is, it is one in which the strongest is morally the best.[69] Just as Darwin had produced a theory that could account for human physical development and apparent purposiveness in the universe without any need for a transcendent providential designer, he had also produced a theory that demonstrated the continuity between humans and other natural organisms with respect to a variety of features that traditional religions had posited as uniquely human, features including the will, reason, and consciousness.

With respect to the will, Haeckel argues that we recognize will in humans by virtue of their preference for certain experiences and aversion to others, which leads them to approach what they find pleasurable and flee from what they find painful. But precisely those features are to be found in entities as simple as chemical elements, which show differential affinities for one another,[70] and they can be shown to grow increasingly complex by gradual stages through the simplest plants and animals, which show "tropisms" or tendencies to approach sources of light or heat, all the way to the mammals, including man. For this reason, Haeckel argued that all natural entities have a "psychic life," or "soul," and that matter is not inert but filled with active forces, down to its most primitive elements. Following Schelling, he even argues that all of these individual souls are part of an overall World Soul that is completely inseparable from the matter of the universe. Against the traditional Cartesian, Kantian, and Christian "dualism" that viewed matter and soul as two quite separate kinds of substance, one of which (matter) was subject to deterministic laws while the other (spirit) could exercise free choice, Haeckel offered "Monism," which identified matter, spirit, and the divine: "We adhere firmly to the pure, unequivocal monism of Spinoza: Matter, or infinitely extended substance, and spirit (or energy), or thinking and sensitive substance are the two fundamental attributes or principle properties of the all-embracing divine essence of the world, the universal substance."[71] As a consequence, Monism insisted that God was not only not outside of nature, but that God and nature were, in fact, one single entity that was subject to deterministic laws. Human free will was thus shown to be another Christian illusion: "We now know that each act of the will is as fatally determined by the organization of the individual and as dependent on the momentary condition of its environment as every other psychic activity. The character of the inclination was determined long ago by *heredity* from parents and ancestors; the determination to each particular act is an instance of *adaptation* to the circumstances of the moment . . . according to the laws which govern the statics of emotion."[72]

As individual material entities become increasingly complex, their psychic life also complexifies in a way that depends upon their organization; so when an organism dies and decomposes, its soul dies with it, proving the scientific impossibility of the immortality of the individual soul.[73] The Christian belief in the immortality of the soul was particularly pernicious, according to Haeckel, because it was part of a worldview that overemphasized the uniqueness and value of the individual soul and led to a kind of artificial selection that protected the weak. This false view, Haeckel, argued, "explains the sad fact that in reality, weakness of the body and character are on the perpetual increase among civilized nations, and that, together with strong, healthy bodies, free and independent spirits are becoming more and more scarce."[74]

Christian doctrines also came into conflict with a healthy naturalistically based morality in connection with their overemphasis on altruism and the renunciation of self interest. Like Darwin and Walter Bagehot, Haeckel acknowledged the importance of social instincts in creating the feelings of duty on which the defense of society and the creation of civilization rests.[75] But he argued that continued progress depended on the proper balancing of egoism and altruism rather than the subordination of one to the other; and he valued egoism much more highly than Darwin and Huxley, for example. The mistake of Christian ethics was to exaggerate the love of one's neighbor at the expense of self-love, going beyond the "golden rule," which was supported by Monist philosophy to the "turn the other cheek" doctrine, which was pernicious. "Translated into the terms of modern life," Haeckel wrote, "that means: 'when some unscrupulous scoundrel has defrauded thee of half thy goods, let him have the other half also.' Or, again in the language of modern politics: 'When the pious English take from you simple Germans one after another of your new and valuable colonies in Africa, let them have all the rest of your colonies also—or, best of all, give them Germany itself.'"[76]

The fact that one of the most important and widely read evolutionary theorists and popularizers explicitly used evolution to justify overt and aggressive opposition to Christianity in what he described as a *"cultur-kampf,"*[77] almost certainly played some role in intensifying conservative Christian opposition to Darwinian evolution during the early twentieth century.

In connection with the evolution of psychic life of the will, Haeckel introduces a second theme that he repeats over and over. On the one hand, "We can attain a clear knowledge of the long scale of psychic development which runs unbroken from the lowest, unicellular forms of life up to the mammals and man at their head." On the other hand, Haeckel insists that within mankind there is a continuing development that creates a huge difference between primitive and "civilized" man. "The psychic distance between the crudest savage and the most perfect specimen of the highest civilization is colossal—much greater than is commonly supposed," he wrote.[78] Returning to the theme of racial differences among humans in connection with reason, he argues: "The difference between the reason of a Goethe, a Kant, a Lamarck, or a Darwin, and that of the lowest savage, a Veddah, an Akka, a native Australian or Patagonian, is much greater than the graduated difference between the reason of the latter and that of most of the rational animals, the anthropoid apes, or even the papiomorpha, the dog, or the elephant."[79]

Similarly, when he comes to discuss the levels of consciousness in organisms, he first demonstrates a continuous development of consciousness from simple organisms through the higher vertebrates, then asserts the distance between the "lowest" and "highest" races of men: "The consciousness of the highest apes, dogs, elephants, etc., differs from that of man in degree only, not in kind, and the graduated interval between the consciousness of these rational placentals and the lowest races of men (the Veddahs, etc.) is less than the corresponding interval between these uncivilized races and the highest specimens of thoughtful humanity (Spinoza, Goethe, Lamarck, Darwin, etc.)."[80]

Through all of these passages, Haeckel joined his version of Darwinism to his long-standing beliefs in German racial superiority and to the Aryan superiority theories of count Gobineau.[81] He clearly believed that members of the "lower" races were less highly evolved along many dimensions. And he argued that since they were "psychologically nearer to the mammals (apes and dogs) than to civilized Europeans, we must, therefore, assign a totally different value to their lives."[82]

Haeckel was probably not aggressively anti-Semitic early in his career. Indeed, Spinoza belonged to his small pantheon of great thinkers. He was, however, a believer in the racial superiority of Northern Europeans, and he justified a colonial policy, especially in Africa, that discounted the interests of indigenous populations on the grounds that their "value" was vastly less than that of the colonizers. Haeckel's racism became increasingly virulent and extensive as the First World War approached, but it was already present in 1868, when he wrote in his *History of Creation,* that "no woolly-haired nation has ever had an important history,"[83] and that "all attempts to introduce civilization among [the African and Australian tribes] and many other tribes of the lowest human species, have hitherto been of no avail: it is impossible to implant human culture where the requisite soil, namely the perfecting of the brain, is wanting."[84]

Though Haeckel's extension of Darwinian biology did not create German racism, it certainly encouraged it. Moreover, it provided updated scientific support for both a ruthless racist colonial policy that had long-term consequences in South Africa and a domestic focus on racial purity that ultimately fed the later views of the National Socialists.

Overall Assessment of Social Darwinisms

As a consequence of the extreme flexibility that evolutionary theories seemed to have had in bending to so many uses, Bert Lowenberg was led to argue that: "Charles Darwin was all things to all men. . . . Phrases such as 'struggle for existence' and 'survival of the fittest' were gayly ripped from the mosaic of evolution and made to serve the wishful thinking of aristocratic theorists. While evolution gave comfort to thinkers of this stripe, it also inspired collectivist dreamers with scientific hope. Advocates of laissez-faire, mindful of vested interests under attack by proletarian philosophies, pointed to the immutable sway of natural law, but reformers preached social sermons based on identical texts. Pleaders for universal peace and the brotherhood of man were stirred by instructions found in evolutionary monographs, but militarists studied the same manuals with opposite results. Certain philosophers claimed Darwin as their authority for materialism and hedonism; others found metaphysical and ethical idealism confirmed by Darwinian biology."[85]

At some level, the discussions of social Darwinisms in this chapter confirm Lowenberg's insights; but they do not justify the further assumption that such opposed arguments simply cancelled one another out so that the world would be much the same as if evolutionary scientism was never applied to social developments. Darwinian arguments lent scientific authority and credibility to social doctrines. They inflamed passions and provided a popular and emotional appeal that helped to push people into actions that they might otherwise not have countenanced. Moreover, evolutionary theories were not infinitely flexible; and the rhetorical impact that appeal to or condemnation of evolutionary arguments had did come at the cost of the shaping of understandings and goals to fit evolutionary or anti-evolutionary rhetoric. It is very difficult to imagine, for example, the development of "trusteeship" colonial policies without the evolutionary justifications given to them by Kidd.

PART IV

The End of the Century

CHAPTER 10

The Problem of Degeneration in the Late Nineteenth Century

In each of the preceding three chapters, progressive notions of evolution—in biology, psychology, and society—played a central role, with nonteleological evolutionary ideas playing a distinctly secondary role, except in the writings of Darwin and his immediate supporters, Huxley and Wallace. In each case I also briefly mentioned a third option: that of physical, moral, or social degradation, often linked to popular understandings of the second law of thermodynamics. Huxley identified this process with the dominant "cosmic process" and argued that humans might stave off decay indefinitely by intentional collective action. Spencer called the process "dissolution" and saw it as inevitable, though avoidable in the relatively short timescale of human existence. Maudsley called it "degeneration," and alone among the figures we have discussed so far, he saw it as an immediate problem exacerbated by the growth of industrial and commercial capitalism.

Toward the end of the nineteenth century, the notion of degeneration became increasingly powerful, because it seemed that poverty, criminality, alcoholism, immorality, and insanity were becoming ever more pervasive features of life in the most civilized portions of the world. Degeneration became a dominant literary theme, and fear of degeneration gave rise to a series of social movements, from the temperance movement through the nudist movement to the race-hygiene movement, aimed at reversing degenerative trends. The central feature of all such movements was that they tended to see degeneration, not as a primarily economic or ethical or religious or philosophical problem, but as a problem that had its roots in medical and biological facts and that thus had to be addressed through medical and biological interventions.

The reality of degeneracy was widely accepted in Britain by the 1880s. In fact, the evolutionary zoologist Edwin Ray Lankester published a work entitled, *Degeneration, A Chapter in Darwinism* in 1880; Maudsley began emphasizing degeneration in his psychiatric writings of the mid 1880s; and Thomas Hardy began to explore degenerationist themes in his novels of the late 1880s. But degeneration was not to have a major impact on English and American life until the early twentieth century, when it played a central role in the eugenics and the temperance movements. The situation was quite different on the Continent. There the medical writings of the Frenchman, Bénédict Augustin Morel (1809–73) and the criminal anthropology of the Italian, Cesare Lombroso (1835–1909), focused attention on issues of degeneration much earlier. The young socialist, Alfred Ploetz (1860–1940) of Breslau, concerned with growing social and economic problems in Germany, sought to solve those problems by biological means. Hippolyte Taine's (1828–93) immensely popular historical analyses of the French Revolution and its aftermath located one of the great driving forces of history in human degeneration. And the literary works of the single most widely read Continental novelist of the century, Émile Zola (1840–1902), demonstrated a near obsession with degeneracy.

By 1892, the Hungarian physician and social theorist Max Nordau (1849–1923) could argue in *Entartung* (*Degeneration*) that, "We stand now in the midst of a severe mental epidemic; of a sort of black death of degeneration and hysteria."[1] Moreover, he could claim that fin-de-siècle Continental literature and art was virtually defined by its fascination with the degenerate and decadent. Certainly there were many who disagreed with Nordau and saw his own reaction to circumstances as hysterical; but it also seems to be the case that while concern with degeneration has been endemic in Western culture, it did reach something like epidemic proportions between 1860 and the beginning of the twentieth century almost everywhere in Continental Europe. Nordau differed from most of those who were concerned with degeneration at the end of the nineteenth century in thinking that it might be a temporary phenomenon and that since degeneracy leads to sterility, only those who could withstand the mental exhaustion felt in response to the frantic pace of late-nineteenth-century urban industrial life would procreate. Looking just a hundred years ahead, he writes: "The end of the twentieth century, therefore, will probably see a generation to whom it will not be injurious to read a dozen square yards of newspapers daily, to be constantly called to the telephone, to be thinking simultaneously of the five continents of the world, to live half their time in a railway carriage or in a flying machine, and to satisfy the demands of a circle of ten thousand acquaintances, associates, and friends. It will know how to find its ease in the midst of a city inhabited by millions, and will be able, with nerves of gigantic vigor, to respond without haste or agitation to the almost innumerable demands of existence."[2]

Morel's Psychology and Taine's Histories

The views of the first major theorist of degeneracy, Augustin Morel, were quite different from those of Nordau. Morel, whose *Traite des Dégénérescences Physiques, Intellectuelles et Morales de l'Espece Humaine* had appeared in 1857, was, like Henry Maudsley, a physician who spent his career in treating mental patients with frustratingly little success. Moreover he was working at a time when statistical evidence seemed to show that there was a rapid increase in those afflicted by "all the abnormal states which have a special relation with the existence of the physical and moral in humanity."[3] These states included not only insanity, but also alcoholism, poverty, unemployment, drug use, illicit sexual activity, sterility, communism, and crime, all of which Morel viewed as symptoms of an overarching *dégénérescence* infecting society. All such symptoms he identified particularly with the "dangerous" urban under-classes, though individuals from any class could be affected. His goal was to explore both the characteristics and the etiology of the various forms of degeneration so that efforts at treatment or prevention might be directed appropriately. From his perspective, the great problem was that "ignorance . . . of the distinctive character of these morbid varieties introduces a deplorable confusion in treatment. Where moral therapeutics should exert its influence, reigns the repressive force of the law; and from another angle, hopes, which must be cruelly deceived in practice, direct all the activity of medicine towards the cure of unmodifiable beings."[4]

Taking a point of view drawn from Buffon's degenerative model of evolution, Morel argued that social pathologies such as alcoholism in one generation of a family left the next generation susceptible to minor mental disorders, which in turn led to more severe disorders, such as criminality and idiocy in the next, eventuating in sterility.[5] What was perhaps more important for Morel, this process was essentially irreversible and its human products, incurable. Unlike the generation of psychiatrists, including Pinel, whose training in *Idéologie* or associationist psychology led them to believe that education and exposure to a healthy environment could cure mental disease by replacing unhealthy associations with healthy ones, Morel could suggest no cures. Thus, in writing of cretinism, a form of insanity, Morel insisted that once the diagnosis has been confirmed, "all the pedagogic procedures, and the best hygienic influences are in vain . . . [The Cretin] will remain what he is: a monstrous anomaly, a typical representation of the state of dégénérescence."[6] Morel did recommend isolation for the insane, for while isolation might not be able to cure their condition, it could at least keep them from infecting the next generation.

Among the first major supporters of Morel's views was Philippe Buchez (1796–1866), physician, ex-Saint-Simonian, author of *Introduction à la science de l'histoire* (1842), and inept president of the short-lived Legislative Assembly following the overthrow of the July Monarchy in 1848. Prior to 1848, Buchez had acknowledged the existence of human degeneration; but like Buffon, he had seen human degeneration as a process associated with races of humans far from the European cradle of civilization and living under conditions of savagery. Describing these distant peoples, he wrote: "They are as ugly as weak; as unintelligent as wretched. They are short-lived, apparently conserved only to present us with an example of the state of degradation into which the human species can fall, when it abandons the light of duty and succumbs to the unique temptations of its animal nature."[7]

The events of 1848–49 dramatically changed Buchez's views, however, and none more than the events of May 5, 1849, when a group of radical demonstrators led by Louis Blanc (1811–82) overran a session of the Assembly that Buchez was chairing. According to his close friend and literary executor, as Buchez sought vainly to bring order to the session, he cried out that the actions of the demonstrators belonged in the *Bicêtre,* a Parisian insane asylum. Though he remained hopeful about the possibilities of social progress, he increasingly saw the failures of all of the revolutionary movements, beginning with that of 1789, as a consequence of their being taken over by groups of men "of semi-savage natures, uneducated outcasts, who understood nothing but brutality and coarseness."[8]

For Buchez, the problem now became one of understanding how to account for such persons in France, the most civilized of all nations. Morel's degeneration theory offered him an answer. It was from Morel's degenerates that the "incendiaries," the "false spirits," and "people of evil instincts who are a scourge and danger to society," came. Perhaps even more important, in the aftermath of the disastrous (for France) Franco-Prussian war of 1870–71, Morel suggested that social disorders and insurrections could produce mental alienation and trigger pathological behaviors. Thus, degeneration became the consequence as well as the cause of historical events.

The notion that modern revolutionary movements as well as their outcomes can be understood in terms of social-psychology and that degeneracy might be both a cause and effect of social disruptions moved out of the professional psychological literature, where Buchez's views were confined, to become late-nineteenth-century commonplaces through the works of the self-styled scientific historian, Hippolyte Taine. Many younger intellectuals, including Caesar Lombroso, Émile Zola, Émile Durkheim (1858–1917), and Sigmund Freud (1856–1939), acknowledged that Taine's work had a major influence on their own.

Taine was educated at the *École Normale Supérieure,* where he became particularly interested in both literature and philosophy. Because he insisted on lecturing on Spinoza's moral system to a group of conservative examiners for his *agrégation* in 1851, he failed and was forced to take a job as a substitute teacher in a provincial College in Nevers. There he refused to sign a loyalty oath in favor of Louis Napoleon, so he was demoted to teaching in an elementary school, an assignment that he refused. For a time, beginning in 1852, Taine studied medicine and psychiatry, attending the lectures of, among others, the radical Lamarckian physiologist Geoffroy Saint-Hilaire. Taine hoped to get his doctorate in philosophy, writing a dissertation on topics from physiological psychology. Unfortunately, his views, which drew heavily from those of the intellectual Positivists, were anathema to the Eclectic philosophers who held academic positions in Paris, so his hopes were thwarted. Taine then turned to literature, writing a dissertation on La Fontaine.

For a time he wrote as a freelance author and literary critic, promoting such realist authors as Stendhal (1783–1842), and thumbing his nose at Victor Cousin and his Eclectic philosophical school in *The Nineteenth Century French Philosophers* (1857). In 1864, his *A History of English Literature* appeared, and he was appointed professor of Aesthetics and Art History at the *École des Beaux Arts* in Paris. For the next twenty years he remained in this position. Through the 1860s, Taine continued to write primarily literary and art criticism and worked on finishing his psychological treatise, *On Intelligence* (1870). In the aftermath of 1870–71, however, his interests shifted to an analysis of French history treated in the form of a medical diagnosis. France was a patient, and the events of 1793, 1848, and 1870–71 were symptoms of a severe disease, which Taine likened to the progressive degenerative disease, syphilis.

The Origins of Contemporary France consisted of three sections. The first volume, *The Ancien Régime,* came out in 1875. Three volumes on *The French Revolution* appeared in 1878, 1881, and 1884, and two volumes on *The Modern Regime* appeared in 1891 and, posthumously, in 1894. It was in this historical writing on France, especially on the French Revolution, that Taine's concern with the role of psychology and degeneration in history became most overt. But before turning to that history we need to look briefly at the basic psychological theory that is applied in his historical writings. At bottom, Taine argues with the evolutionists that, "Man is an animal by nature and by structure, and neither nature nor structure ever loses a single hold. He has the canine teeth like the dog and the fox, he once fastened them in the flesh of his own kind; his descendants cut each other's throats with stone hatchets for a morsel of raw fish. Even now he is not changed; he is only softened; . . . each one still fights for his morsel of raw fish, only it is under the eye of the policeman and not with stone hatchets. The foundation of the natural man are irresistible impulses, anger, appetites, greed; all blind."[9]

Taine accepts from the Lamarckian tradition that the instincts and aptitudes associated with this animal component of humans can be modified by environmental forces over extremely long periods of time. Different races may thus have slightly different core characteristics. In the short run, however, this part of man is impervious to change. The elements of the animal component of humanity constitute "a class of instincts and of aptitudes over which revolutions, decadence, civilizations have passed without having affected them. These aptitudes and these instincts are in the blood and are transmitted with it."[10]

Overlaid on this bundle of animal drives is a personality and a consciousness constructed out of associations that are often inconsistent with one another. For example, our visual sensations associated with an oar dipped in water suggest that the oar is bent, while our tactile sensations suggest that it is straight. At any time, our consciousness consists of a dynamic and only temporarily stable equilibrium among competing elements, delicately balanced and subject to disruption: "Madness is always hovering near the mind, as illness is always hovering near the body; for the normal condition is a temporary victory only; it results from and is renewed by the continual defeat of the contrary forces. Now these last are always present; an accident may give them preponderance; there is little required to enable them to assume it; a slight alteration in the proportion of the elementary affinities and the direction of the constructing process would bring on a *degeneracy*."[11]

In some respects Taine's history of the French Revolution and its aftermath was unexceptional. In *The Ancien Régime,* for example, Taine drew from a long-standing liberal tradition in assessing the material causes that established conditions favorable to revolution. First were the abuses of privileges by a nobility and clergy who no longer served the state effectively. Second was the wastefulness of the French Crown. Third was the tax system imposed on ordinary citizens to support the profligacy of the state and its noble dependants, a system that reduced the peasantry and working classes to a state of poverty and susceptibility to famine and fears of famine. Coupled with these material causes were a set of intellectual developments that Taine viewed through the eyes of conservative interpreters, such as Edmund Burke. There was a general decline of respect for authority that Taine associated with a combination of what he called the classical spirit of the eighteenth century and the premature attempt to apply scientific methods to human affairs. The classical spirit, which Taine identified with "rhetorical reasoning," emphasized logical coherence and orderly structure and totally failed to acknowledge the complexity and messiness of actual life. The obsession of intellectuals with scientific methods is something that Taine sympathized with, but following Comte, he saw the analytic methods of eighteenth-century science as fundamentally inadequate to understanding human and social phenomena. Thus he saw late-eighteenth-century attempts at imposing scientific order on society as fundamentally misguided and unable to deal with local particularities. Finally, Taine saw the Rousseauist confidence in the fundamental goodness of the people that infused supporters of revolution as completely unjustified

In his volumes on the French Revolution itself, Taine broke with tradition, focusing on the psychology of both the leaders and the followers in the revolution. He saw most of the revolutionary leaders as egomaniacs who had internalized the worst features of classical and scientistic attitudes. Describing Robespierre, for example, he wrote: "Under the pressure of his faith and his egotism, he had developed two deformities, one of the head and the other of the heart; his common sense is gone, and his moral sense is utterly perverted. In fixing his mind on abstract formulas, he is no longer able to see men as they are; through self-admiration he finally comes to viewing his adversaries, and even his rivals, as miscreants deserving of death. On this downhill road nothing stops him, for, in qualifying things inversely to their true meaning, he has violated within himself the precious conceptions which bring us back to truth and justice."[12] Taine went on to explain how ordinary people got caught up in the revolutionary fervor. In doing so, he initiated a powerful tradition of interest in crowd, or mob, psychology, as well as a tradition that saw in the revolution the beginnings of a degenerative decline in the French citizenry that led naturally to the tragedy of the Paris Commune in 1871. Exposed to the fear of famine and to the violence of mob actions, the unstable civilized veneer of the normal citizen is shaken and cracked: "His anger is exasperated by peril and resistance. He catches the fever from contact with those who are fevered. . . . Add to this the clamors, the drunkenness, the spectacle of destruction, the nervous tremor of the body strained beyond its powers of endurance, and we can comprehend how, from the peasant, the laborer, and the bourgeois, pacified and tamed by the old civilization, we see all of a sudden spring forth the barbarian, and still worse, the primitive animal, the grinning, sanguinary, wanton baboon, who chuckles while he slays, and gambols over the ruin he has accomplished."[13]

Degeneracy and the Naturalistic Novels of Émile Zola

Among those who were deeply influenced by both Morel's emphasis on the degenerative effects of drunkenness and immorality and by Taine's views on the psychology of degeneration was Émile Zola, France's most popular nineteenth-century literary figure. Zola was born at Paris in 1840, the son of a French mother and an Italian civil engineer who moved his family to Aix-en-Provence, where he was in charge of a major municipal water project when his son was about five years old. Two years later Zola's father died of pneumonia, leaving his wife and son with little more than their personal belongings and debts. Zola was able to start his secondary education at Aix, where he and Paul Cézanne became close friends. By 1857, Zola's mother had sold off virtually everything owned by the family and she was forced to leave Aix for Paris, where a family friend helped Émile get a scholarship to the Lycée Saint-Louis. For the next two years the family moved into increasingly inexpensive lodgings while Émile attended school and wrote poetry, short stories, and literary essays. Failing his bachelors examinations in both 1858 and 1859, Émile briefly took a low-paying clerk's job at the Paris docks, but he soon quit and tried to live on what he could get from his journalistic writing.

Between 1860 and 1862, Zola ended up living in a room in one of the cheapest districts of Paris, with lower-class prostitutes and indigent alcoholics as his neighbors. Soon he was involved in an affair with a poor prostitute, Berthe, that left him depressed and disillusioned about love. Writing to one of his friends back in Aix, he observes, "the prostitute is irremediably lost, the widow frightens me, and the virgin does not exist."[14] The experiences of this time in his life would play a major role in shaping the content of his art.

Zola's first volume of short stories, *Contes à Ninon,* appeared in 1864. Most of these stories had traditionally romantic themes, but Zola's correspondence indicates that as early as 1860, he was becoming converted to the idea that because advances in the sciences were so central to contemporary culture, they needed to become part of the substance of literature. Zola completed his first novel, *La Confession de Claude,* based on his affair with Berthe, in 1865. Like Zola, Claude hopes to save a prostitute through his love. Describing their lovemaking, he expresses a deep sadness at his failure to move his lover, Laurence: "I hugged her tight, felt her body abandon itself disdainfully to my embrace. But nothing on earth could make her open up her soul to me. Her heart and her thought shunned me. I was holding nothing but a lifeless rag, a body so weary, so worn out, so inert, it was incapable of answering my embrace with anything at all."[15] Ultimately, though Claude cannot help Laurence, he is strengthened by the experience.

La Confession de Claude was only a modest critical success and a financial failure (total royalties came to only 450 francs), but its publication did have a major impact on Zola's life. The novel's notoriety led his conservative employer to insist that he look elsewhere for work. In response, Zola became a full-time freelance art critic and author. Over the next two years Zola wrote a couple of popular mystery novels just for the money (two sous a line) and a series of essays in which he identified himself more and more with scientific and even Positivistic attitudes. Writing in the literary periodical, *L'Événement,* in May of 1866, he argued, "Whether we like it or not, the trend is toward the exact study of facts and things . . . the movement of our times is certainly realistic, or rather, positivistic."[16] In an address to the Congrés Scientifique de France, held in Aix that December, he offered the following new and scientistic definition of the novel: "The novel is a treatise of moral anatomy, a compilation of human facts, an experimental philosophy of the passions. Its object is, with the help of some verisimilar action, to portray mankind and nature as they really are."[17] Using Honoré de Balzac (1799–1850) as an exemplar, he describes his art in terms of scientific method: "His only concern is with truth, and he exhibits our heart to us on the operating table. Modern science has presented him with the instrument of analysis and the experimental method. He proceeds like our chemists and our mathematicians: he decomposes men's acts, determines the causes, explains the effects; he operates according to fixed equations, in a factual study of the influence of environment on individuals."[18]

Throughout early 1867 Zola worked on an ambitious attempt to put his new ideas about the novel into effect. The result was serialized in the August through November issues of *L'Artiste* and brought out in book form in November. The story of the rise and fall of a lower-class laundress who eventually commits joint suicide with her alcoholic murderer husband, *Thérères Raquin* again drew from Zola's experiences among the dregs of society, and it again called forth the outrage of respectable critics, who labeled it a "puddle of mud and blood."[19]

This time, however, the novel was a modest commercial success, and Zola used the preface to the second edition, published in May of 1868, to respond directly to his critics, expanding his notion of what the modern novel should be and accepting identification as a "naturalistic" writer, a term that had been coined by Taine in 1858.

Denying that his aims were pornographic or immoral, Zola insisted that he wrote out of "pure scientific curiosity" and claimed that, "I have simply done on living bodies the work of analysis which surgeons perform on corpses." In particular, he had chosen to study temperaments, rather than character: "I have chosen persons completely dominated by their nerves and blood, devoid of free will, pushed into each action of their lives by the fatality of their flesh."[20]

Throughout his career, Zola would continue to insist that his work was "scientific" and accurate in its portrayals of situations and events. In fact, in 1885, he published a kind of explication and justification of his method in the form of an essay on "The Experimental Novel," which appeared first in a Russian literary journal. The essay failed to explore aspects of Zola's artistry and his special emphasis, elsewhere attributed to Taine, on drawing characters dominated by some singular, "excessive," trait. But it served as a touchstone for the development of Russian and German Realism through its appropriation by the German Darwinian popularizer and novelist, Wilhelm Bölsche (1861–1939), who patterned his own *Die naturwissenschaftlichen Grundlagen der Poesie: Prolegomena einer realistischen Äesthetic* (1887) on Zola's essay. Furthermore, it did catch some of the flavor of the systematic and meticulous way in which Zola planned his novels and the painstakingly researched and documented features of the physical and historical environment into which he places his characters. One of the important points made in "The Experimental Novel" was that the ultimate goal of scientific knowledge and of the experimental novel was *not* merely to satisfy the curiosity, but rather to acquire the ability to control events. On this issue, Zola was in complete agreement with both Positivist and Saint-Simonian aesthetic doctrines.

In 1868 Zola proposed a tremendously ambitious project: to write ten novels unified by a single theme over a five-year period. *The Rougon-Maquarts: A Natural and Social History of a Family under the Second Empire* would follow several generations of the progeny of a single neurotic woman with her respectable gardener husband, Rougon, and with her smuggler-lover, Maquart, during the rule of Napoleon III. Descendants of the former would rise to the heights of Second Empire society, while descendants of the latter would descend to its depths. The series, moreover, would involve two parallel themes: "(1) the purely human, physiological element, the scientific study of a family with the inevitable consequence of the fatalities of its lineage; (2) the effect of the modern era on this family, its breakdown through the ravaging passions of the epoch, the social and physical action of the environment."[21] That is, the novels would deal with the hereditary and environmental causes of degeneration.

Zola's publisher, Albert Lacroix, accepted the proposal and guaranteed Zola a minimum of 500 francs per month while he worked on the project. So Zola began on the first of what eventually became a twenty-novel series written over a twenty-two-year period. Though some of the individual novels were relatively optimistic in their tone, the overall thrust of the series was to highlight the degenerative tendencies of the age. As he wrote for *La Tribune* at the time that he was developing his proposal, "When a society is rotting, when the social machine is getting out of order, the role of the observer and thinker is to note every new wound, every unexpected new blow . . . to study them boldly and candidly, without fear or prevarication, in order to draw from them the elements of the world of tomorrow."[22]

The first six novels of the series were modestly successful, but with the 1877 publication of *L'Assommoir,* another very dark novel about the urban working classes, Zola's life changed dramatically. Up to the early 1870s he had to keep writing journalistic pieces just to keep food on the table for his family. Things improved somewhat after the beginning of 1876. But *L'Assommoir* went through ninety-one printings in less than four years, made him the literary lion of Paris, and made him a wealthy man. Subsequently, at least four more novels in the series, *Nana* (1880), *Germinal* (1885), *La Bête Humaine* (1890), and *La Débâcle* (1892), were comparable or even greater commercial and critical successes. Each of these, with the exception of *Germinal,* which shows a hint of optimism at the end, was darker than the last, leading Max Nordau to see Zola as "a high class degenerate" himself, and as the leader of those engaged in "the premeditated worship of pessimism and obscenity."[23]

At one point in *Nana,* the story of a high-class courtesan, Nana asks what one of her wealthy aristocratic clients thinks of a story in *Figaro,* which is, in fact, her story in a nutshell, one that epitomized Morel's theories of degeneration: "It was the story of a girl descended from four or five generations of drunkards, her blood tainted by an accumulated inheritance of poverty and drink, which, in her case, had taken the form of a nervous derangement of the sexual instinct. She had grown up in the slums, in the gutters of Paris; and now tall and beautiful, as well made as a plant nurtured on a dung heap, she was avenging the paupers and outcasts of whom she was a product. With her, the rottenness that was allowed to ferment among the lower classes was rising to the surface and rotting the aristocracy. She had become a force of nature, a ferment of destruction, unwittingly corrupting and disorganizing Paris between her snow-white thighs."[24] Her aristocratic lover shivers because he recognizes his own place in the story, but he cannot stop himself from accepting the ever more extreme levels of submission and degradation that she forces upon him.

Degeneration has gone even further in Jacques Maquart, one of the central figures of *La Bête Humaine,* Zola's novel about the railroad's impact on French society. Jacques is a full-blown psychopath who is driven to kill women. Terrified by his own feelings, he reflects early in the novel about why these urges come on him in a way that reflects Taine's theory that men are violent animals with a tenuous overlay of civilized personality: "[His] family was really not quite normal, and many of them had some flaw. At certain times he could clearly feel this hereditary trait, not that his health was bad, for it was only nervousness and shame about his attacks that had made him lose weight in his early days. But there were sudden attacks of instability in his being, like cracks or holes through which his personality seemed to leak away, amid a sort of thick vapor that deformed everything. At such times, he lost all control of himself and just obeyed his muscles, the wild beast inside him. . . . He was coming to think that he was paying for others, fathers, grandfathers who had drunk, generations of drunkards, that he had their blood, tainted with a slow poison and a bestiality that dragged him back to the woman-devouring savages in the forests."[25]

Finally, in both *La Bête Humaine* and *La Débâcle,* the degeneration theme moves beyond the Maquarts to society as a whole. Zola addresses this theme most explicitly in *La Débâcle,* his novel of the Franco-Prussian War and the Paris Commune of 1871. One of the central characters reflects on the process that has led him to take up the cause of the Communards in 1871 after the French defeat at the hands of the Germans: "The degeneration of his race, which explained how France, virtuous with the grandfathers, could be beaten in the time of the grandsons, weighed down on his heart like a hereditary disease growing steadily worse and leading to inevitable destruction when the hour came."[26]

But Zola's most powerful symbol of France rushing toward its own collapse appears at the end of *La Bête Humaine.* Jacques is the engineer of a troop train filled with soldiers heading to Paris to fight the encircling German armies. His fireman, the drunken Pecqueux, who believes that Jacques has seduced his woman, continues to stoke the firebox after Jacques tells him to stop. Zola describes the train: "The huge mass, eighteen trucks chock full of human cattle, was careering across the black countryside with a ceaseless roar. These men being carted off to the slaughter were singing, singing their heads off, making such a din that it drowned the noise of the wheels." Jacques and Pecqueux get into a fight and fall off the train and under its wheels to die. Now, completely out of control, the train roles on: "The train, like a wild boar in the forest, held to its course, heedless of red lights and detonators. . . . What did the victims matter that the machine destroyed on its way? Wasn't it bound for the future, heedless of spilt blood? With no human hand to guide it through the night it roared on and on, a blind and deaf beast let loose amid death and destruction, laden with cannon-fodder, these soldiers already silly with fatigue, drunk and bawling."[27]

Germinal, Zola's novel about the conflict between capital and labor, explores two dimensions of degeneration that had particularly concerned Taine: the tendency for revolutions to be taken over by increasingly radical and violent factions and the tendency of crowds to break down the inhibitions of their members.

Though not as immediately successful as *Nana, Germinal* has had a tremendous appeal for working-class and leftist audiences throughout Europe over a long period of time. It offers a powerful portrayal of the plight of the worker; a plight that is not simply the consequence of the viciousness of capitalists, but rather the consequence of an economic system that traps and dehumanizes both workers and owners, producing its own form of social degeneration. His basic point of view was articulated quite clearly in the notes that he penned while getting prepared to write the novel: "To get a broad effect I must have my two sides as clearly contrasted as possible and carried to the very extreme of intensity. So that I must start with all the woes and fatalities that weigh down on the miners. Facts, not emotional pleas. The miner must be shown crushed, starving, a victim of ignorance, suffering with his children a hell on earth—but not persecuted, for the bosses are not deliberately vindictive—he is simply overwhelmed by the social situation as it exists. On the contrary, I must make the bosses humane as long as their direct interests are not threatened; no point in foolish tub-thumping. The worker is the victim of the facts of existence—capital, competition, industrial crises."[28]

Early in the novel, the leadership of the miners comes principally from Rasseneur, a retired miner turned restauranteur, who hopes to be able to fight a reduction in the miners' pay rate without a strike. At a meeting at his café, the miners decide to strike, and the central character of the novel, Étienne Lantier, who is a Marxist, begins to contend for leadership and convinces the miners to join the Workers International. When the International cannot provide more than a few days' support for the strike, Étienne calls for a mass meeting. There the crowd of three thousand miners, fired up by Lantier's rhetoric, finally fully renounces Rasseneur's evolutionary socialist vision and vows to shut down all of the mines in the area. The miners agree to assemble the next morning at one of the mines that is still working.

Zola follows the mob as it becomes ever more frenzied and violent. First, the miners merely destroy property at a mine that the owner refuses to shut down. At the next mine, they beat the workers who had gone down into the mine. Zola describes the increasing violence of the crowd: "In the growing ferocity of this age-old craving for revenge that was lashing them all into madness, the yelling went on ever more raucous, as they called for death to all traitors, screamed their hatred of ill-paid toil, and roared the desperate need of empty bellies for bread. They began cutting cables, but now the file did not bite quickly enough, for everyone was mad to get on, ever onwards. . . . Years and years of hunger were now torturing them with a lust for massacre and destruction."[29] Next, the crowd surges off to lay siege to a manager's home and Étienne reflects on what is happening: "Even while he was hoarsely urging them on against Montsou, he could hear another voice within him, the voice of reason, asking in amazement what was the meaning of all this? He had not meant any of this to happen . . .

and he was dismayed at the sight of these brutes whom he had unleashed, for if they were slow to anger, they were terrible when roused, and their ferocity was implacable."[30]

In a final paroxysm of violence, the mob, spurred on by Étienne, who is not immune to the crowd mentality, attacks the grocery store of a venal man who is not connected to the mines at all. Crying "we want bread," the crowd invades his shop. When he dies, trying to escape, the maddened crowd castrates his dead body and marches away with his testicles and penis raised on a stick: "Ma Brûlé stuck the whole thing on the end of her stick, raised it on high, and carried it like a standard down the street, followed by a rout of shrieking women. Drops of blood spattered down, and this miserable bit of flesh hung down like an odd piece of meat on a butcher's stall."[31]

Finally, because the strike has engendered repressive military measures, leadership of the miners' movement moves briefly on to the most degenerate figure of all, the Russian anarchist, Souvarine, for whom sabotage and violence are intentional tactics.

Nordau's Critique of Zola

Many critics were angered by Zola's claims to be a "scientific" novelist because he sought to appropriate the authority of science in support of views that they found offensive, while he failed to meet any reasonable standards of what one means by "science." None were more adamant nor more offended than Max Nordau, whose chapter on "Realism" in *Degeneration* contains an extended attack on Zola as the paradigmatic realist or naturalist and a degenerate as well. To some extent, the bitterness of Nordau's attack is that of a disillusioned disciple, for, as he admits, "I allowed myself for thirteen years to be led astray by his swagger, and credulously accepted his novels as sociological contributions to the knowledge of French life. Now I know better."[32]

One of Nordau's complaints is that while purporting to explore the lives of ordinary people, Zola has, in fact, created a family of characters lying far outside the range of ordinary experience: "He who ridicules the 'idealists' as being narrators of 'exceptional cases,' of that which 'never happened,' has chosen for the subject of his *magnum opus,* the most exceptional case he could possibly have found—a group of degenerates, lunatics, criminals, prostitutes, and 'mattoids,' whose morbid nature places them apart from the species."[33]

There is a significant kernel of truth in what Nordau says, but Zola's concern with criminals, drunkards, prostitutes, and lunatics unquestionably mirrored contemporary trends in the medical and social-scientific literature. Whether or not there was a dramatic rise in drunkenness, crime, and mental illness during the nineteenth century—and there almost certainly was, because of the rapid growth of urban population and poverty—there was certainly a dramatic rise in *reported* cases of all three that came about as a consequence of intensified and improved methods of gathering statistical information. And this new information regarding the extent of social problems seemed to justify the sense of degeneration underlying Zola's exposé.[34] Furthermore, there was a concomitant rise in interest within the scientific community, whose concerns did respond to perceived public needs. As Nordau himself acknowledges, by 1895 the French literature on criminality and medical therapeutics had been filled for almost fifty years with the history of a single family, the Kérangal family, which produced in two generations "seven murderers and murderesses, nine persons who have led an immoral life . . . and besides all these, a painter, a poet, an architect, an actress . . . and one musician."[35]

It was true that Zola self-consciously tended to seek out "excessive" characters to write about. In this sense his subjects were, by definition, atypical. But since the time of Bacon and before, scientists had often hoped to learn about the normal by studying the pathological. So Zola's concern with social pathologies hardly undermines his claim to have science as a methodological model in mind as he writes; indeed, it rather supports that claim.

The Beginnings of the Eugenics and Race-Hygiene Movements

The eugenics and race-hygiene movements were vastly better organized and involved far more people during the early decades of the twentieth century than during the last two decades of the nineteenth century. Indeed, the first major eugenics organizations, the Racial Hygiene Society in Berlin, the Eugenics Education Society in London, the French Eugenics Society, and the American Eugenics Society, were not established until 1905, 1907, 1912, and 1923 respectively. Nonetheless, the eugenics movement, without the name, did begin in the 1860s with Clemence Royer's prefatory comments to her 1862 French translation of the *Origin of Species.* This introduction included a call to eliminate "the weak, the infirm, the incurable, the wicked . . . and all the disgraces of nature . . . which they perpetuate and multiply indefinitely."[36]

The term "eugenics" was coined by Francis Galton, Charles Darwin's cousin, in 1883, in order to characterize the attempt to improve humanity by allowing "the more suitable races or strains of blood a better chance of prevailing speedily over the less suitable."[37] Galton's primary concern was with encouraging particularly able persons to have more children; but Anglo-American eugenics rapidly switched emphasis to controlling the reproduction of the "unfit." This tendency was to a very significant extent the consequence of a study of inheritable social pathology reported by a New York merchant and amateur sociologist, Richard Dugdale. Dugdale first reported his results of a study of the Jukes family, which produced 709 persons in the course of six generations, of which no fewer than 128 were prostitutes; 18, brothel keepers; 26, convicted criminals; and 200 persons who had received poor relief. Dugdale's 1875 report to the Prison Association of New York[38] was soon commercially published as *The Jukes: A Study in Crime, Pauperism, and Heredity,* and it was widely read. Dugdale was a Lamarckist, who insisted that in succeeding generations the environment shaping the Jukes progeny was so abominable that it accelerated the degeneration. On the other hand, because environment played a significant role even in heritable mental and moral degeneration, he argued that it might be possible to intervene in the downward spiral by removing children from the family environment. As a consequence, his work was important in stimulating the movement to establish reformatories in which children of the unfit might avoid their negative influences.

Soon after August Weisman demonstrated that inheritable traits could not be influenced by environment in the early 1890s, however, any optimism for the possibility of environmental intervention died, leaving no strategy for dealing with the unfit other than isolation or sterilization. Moreover, it seemed clear in both England and America that the unfit were increasing rapidly. J. Arthur Thomson reported that the number of "defectives" in Britain more than doubled as a percentage of the total population between 1874 and 1896, and a similar doubling was reported for America between 1890 and 1904.[39] Thus, as early as 1891, Victoria Woodhull Martin wrote in her *The Rapid Multiplication of the Unfit* that "if imbeciles, criminals, paupers, and otherwise unfit are undesirable citizens, they must not be bred."[40]

Though Dugdale had not emphasized alcoholism as a feature of the Jukes family history, the late nineteenth century also saw American and English physicians amplify Morel's insistence that alcoholism was a prime heritable cause of degeneracy. Thomas Crothers, editor of the American Association for the Study and Cure of Inebriety's *Quarterly Journal of Inebriety,* between 1876 and 1918, for example, reported that alcoholism in one generation manifested itself in subsequent generations as a variety of degenerate diseases. It was passed on as "insanity in one, epilepsy in another, intemperance in a third, idiocy in a fourth, hypochondria in a fifth, hysteria in a sixth, and so on. . . . each generation increasing in numbers, and contributing in a direct ratio to the filling of our jails, penitentiaries, inebriate asylums, insane retreats and poorhouses."[41] For a time during the mid 1880s, the Women's Christian Temperance Union even published a *Journal of Heredity,* seeking medical support for their moral crusade. It is highly improbable that, without the presumed association between alcoholism and degeneracy, reformers could have pushed through the American experiment with Prohibition that had its own noxious consequences.

One final important element connected to late-nineteenth-century fears of degeneracy in America built upon a long-standing tradition of scientific racism in the United States. Up until the 1869 publication of the U.S. Sanitary Commission report, *Investigations in the Military and Anthropological Statistics of American Soldiers,* nominally authored by Benjamin Apthorp Gould Jr., a clear racial hierarchy with Caucasians at the top was generally maintained; but it was widely expected that racial crossing, especially among the lower races, would improve the human stock in much the way that hybridizing plant species produced a common phenomenon called "hybrid vigor." The results of Sanitary Commission anthropometry, as well as those of a number of subsequent Civil War studies, on the other hand, seemed to clearly demonstrate that mixed-race subjects were inferior to those of either parent race with respect to such important issues as "vitality" (measured by the play of the chest, that is, the difference between the circumference of the chest taken during a deep breath and during a full exhalation) and intelligence (measured by brain weight at time of autopsy).[42] Racial mixture subsequently came to be seen as a critical cause of degeneracy, and anti-miscegenation legislation was passed in a number of states. Even those few authors who thought that there might be some evidence for increasing mental ability by mixing breeds that were close to one another in the racial hierarchy, for example, Italians and the Irish, were convinced that mixing whites and blacks led to physical and moral degeneration.[43]

Many of the features of very early Anglo-American eugenics were repeated in the nearly simultaneous German race-hygiene movement initiated in the writings of Alfred Ploetz and Wilhelm Schallmayer (1857–1919). Ploetz was fascinated with the cultural nationalism spawned by the pan-Germanic followers of Herder, and drawing heavily from Ernst Haeckel's racial theories, he initiated the League to Reinvigorate the Race in 1879. During the late 1880s, Ploetz traveled for six months in the United States, then he returned to Europe, where he turned from economics, which had drawn his initial interest because of its applicability to social problems, and retrained himself as a physician specializing in mental diseases and alcoholism. For a brief period he returned to open a medical practice in the United States, where he became familiar with American arguments about degeneration and about race. Then, once again back in Europe in 1895, he wrote *Die Tüchtigkeit unserer Rasse und der Shutz der Swachen* (The Fitness of Our Race and the Protection of the Weak), which introduced the term *Rassenhygiene.* Ploetz used the term "race" in this work in multiple ways. On the one hand, he intended it to refer to the entire human race. On the other, he identified the well-being of all of humanity with the fortunes of the Aryan race, of which Germans represented the superior segment: "The hygiene of the entire human race converges with that of the Aryan race, which apart from a few small races, like the Jewish race—itself quite possibly overwhelmingly Aryan in composition—is the cultural race *par excellence.* To advance it is tantamount to the advancement of all humanity."[44] Like the American evolutionary race theorist, Joseph LeConte, who argued that the mixing of closely related races would improve the human stock, Ploetz actually advocated the intermarriage of Aryans and Jews on the grounds that it would create a biologically superior stock. Thus, he was far from supporting the growing anti-Semitic sentiment in Germany; but Ploetz's ideas certainly promoted the idea of medical management of racial health.

Wilhelm Schallmayer, Ploetz's near contemporary, was also a socialist who studied economics before switching to medicine in 1881. Like Maudsley and Ploetz, he practiced as a psychiatrist and became convinced that most mental illnesses were inherited and medically incurable, leaving the control of mental health largely in the domain of reproduction. So Schallmayer switched the focus of his practice from psychiatry to obstetrics and gynecology. In 1891 he published the first German book on eugenics, *Über die drohende korperliche Entartung der Kulturmenscheit* (Concerning the Threatening Physical Degeneration of Civilized Humanity). Schallmayer's first work does not seem to have had a major impact on policy, though he did propose a number of measures that would subsequently become important in Germany, including the regulation of marriage, the creation of medical genealogies, and the development of educational strategies focused on race health and of propaganda to promote eugenics.

In 1900, at the beginning of a new century, the eugenics and race-hygiene movements entered a new stage when Friedrich Krupp offered a prize of thirty-thousand marks to the winner of a contest aimed at discovering "what can we learn from the theory of evolution about internal political development and state legislation?"[45] First prize went to Schallmayer's *Vererbung und Auslese im Lebenslauf der Volker* (Heredity and Selection in the Life-processes of Nations), which argued strenuously that for its own survival, the state must play a major role in promoting the biological efficiency of its citizens, moving European concerns with degeneration out of the realm of the voluntary and into the realm of official governmental policy.

Summary

Through the first two thirds of the nineteenth century, scientistic thought, though ostensibly "objective" and value neutral, clearly had an optimistic tone. Saint-Simonian socialism, Owenite socialism, Positivism, Marxism, and even many forms of Social Darwinism were fundamentally progressive. That is, they all looked forward to using scientific knowledge of society to create a more perfect world. By just after mid century, however, a countervailing trend, stimulated by a strong perception that crime, drunkenness, mental illness, and immorality were increasing in society, began to emerge. This pessimistic scientism was linked to the second law of thermodynamics and Lamarckian evolutionary notions. Most closely associated at first with medical practitioners, especially those who treated the insane, it posited the existence of a natural tendency toward degeneration—both physiological and societal—that countered evolutionary progress. Through the literary works of figures such as Émile Zola, degenerationist views began to pervade the arts during the last decades of the nineteenth century. Furthermore, fears of degeneration began to stimulate movements to stave off degeneration as long as possible. Generally speaking, in the late nineteenth century these nascent movements, including temperance movements and race-hygiene movements were aimed at slowing down the deterioration of the human species rather than at improving humans and society. The beginnings of a more positive science, eugenics, which sought to improve humans as well, and which came into prominence during the early decades of the twentieth century, can, however, be seen in the writings of a small number of scientists including Francis Galton.

CONCLUSION

We have seen that throughout the early years of the nineteenth century every major tradition of natural science—and there were major differences in approach across both subject matters and national boundaries—spawned efforts to extend scientific ideas, methods, practices, and attitudes to matters of human social and political concern. That is, they spawned scientisms. In Britain, prior to mid century, as we saw toward the end of the introduction, the analytic strategies associated with mechanics generally led to a presumption that society could be treated simply as an aggregation of individuals, whose collective well-being could be understood as the summation of the happiness of those individuals, taken independently. Moreover, it was generally assumed that the happiness of an individual could be understood in terms of some kind of pleasure-pain calculus. This presumption promoted the classical and neo-classical economic theory associated with laissez-faire liberalism in politics and it promoted Utilitarian moral and political theories that crossed the political spectrum.

Early-nineteenth-century scientistic thinkers in Britain continued to build on eighteenth-century psychological and economic ideas, as we saw in chapter 6, but they pushed the social and political implications in several different directions. Those who asserted that human reason was capable of discovering what actions would produce desirable outcomes in terms of self-interest were most likely to be associated with radical and egalitarian political movements. The socialist, Robert Owen, and the economist and Radical member of Parliament, David Ricardo, exemplify this point of view. Those who were inclined to see happiness associated more with altruism than with uncomplicated self interest tended toward the middle of the political spectrum. Such were the Whig politicians associated with Henry Brougham and his *Edinburgh Review* friends. On the other hand, those who saw happiness as associated with the satisfaction of multiple, complex, and fundamentally incalculable emotional needs followed David Hume and Edmund Burke into sympathy with Tory Politics. Regardless of where one stood politically, however, as we saw in chapter 7, it had become clear by mid century that neither classical economics and Utilitarian morality nor traditional, established Christianity could provide coherent guidance for a social policy that took distributive justice into account. On the one hand, the initial expectation that some kind of invisible hand would produce a just society out of an aggregation of self-interested individuals or that one could design a set of laws that could bring each individual's well-being into alignment with the general good appeared to most to be vain hopes. On the other hand, it was clear to many intellectuals and working-class people, that the established church was complicit in defense of a status quo that left most citizens disenfranchised and economically disadvantaged.

Most of the radical and moderate republican political thinkers in France leading up to the French Revolution, including Claude Adrienne Helvetius and Anne-Marie-Nicolas Caritat de Condorcet, had also drawn upon assumptions from sensationalist psychology, and as a consequence, had developed utilitarian social theories. They presumed a basic inherent equality of all human beings, that nearly everything we are comes from our experiences, and that the general good could be discovered as a consequence of the linear summation of the well-beings of all individual citizens, each of which might, in principle, be calculated independent of the well-beings of others. But, as we saw in chapter 1, the experience of many moderate republicans during the Revolutionary Terror, along with the infusion of new models of scientific thinking drawn from changes in medical education, led the early generations of post-revolutionary social theorists to challenge the assumptions of their immediate predecessors.

Scientistic theorists turned against the previous generation's egalitarian assumptions, drawing from medical studies to justify their rejection of inherent equality. It now appeared transparently obvious that different people had different levels of physical and mental abilities. Physiologists, led by Marie-François Xavier Bichat, had demonstrated that different tissues in the body had different functions and suggested that different individuals might be dominated by different categories of tissues, some connected with physical strength, some with intellectual ability, and some with artistic ability. Thus, different individuals were naturally suited to play different and complementary roles in society. Indeed, the division of labor, which had been so important for economic progress, now seemed grounded in the innate differences among people rather than solely in the ways in which they had been socialized.

The new approach to medical education, which developed during the mid eighteenth century at Montpellier, and which informed social theorizing, also rejected attempts to provide causal accounts of physiological phenomena and diseases in terms of mechanical, alchemical, or other theories because those theories had proven of almost no value for therapeutics. Turning back for guidance to the writings of the Hippocratic school, Montpellier physicians emphasized the need for careful descriptions of diseases rather than for causal analyses. They supplemented individual case studies with epidemiological studies that suggested that a huge number of interacting, and not independent, factors, especially factors associated with age, sex, temperament, and climate, might be implicated in diseases, or even normal bodily functioning. They even suggested that the complexity of living organisms might give rise to emergent "vital" forces that were not understandable in terms of underlying physical laws at all. All of these notions were transferred to considerations of human societies by Pierre Cabanis and promoted by his *Idéologue* followers and those who learned from them.

J. B. Say, for example, in introducing his notion of utility into political economy, insisted that there were far too many interacting factors involved in making economic choices to allow one to provide a satisfactory causal account of such choices. The only way to determine the utility of any good to a person was to place that person in the context in which a choice must be made and describe what the person was willing to give up to acquire the good. In the same way, Say and his colleagues rejected the old natural law claim that persons have certain rights. They only have desires, or needs, the intensity of which can be measured by what they are willing to give up to meet them. Among those needs emphasized by the *Idéologues,* and appropriated by Continental socialists, including Henri Saint-Simon and Karl Marx, was a need for self-expression, which, like the "vital" physiological forces, was an emergent need that could not be accounted for as a calculable consequence of physical laws.

For some *Idéologues,* such as Destutt de Tracy and later advocates of extreme *individualisme,* the abandonment of a doctrine of natural rights authorized a completely individualistic social and economic theory, releasing each individual to ignore any needs and desires but their own. Others had no right to expect any consideration. On the other hand, for Saint-Simon and his socialist followers, who formed the focus of chapter 2, the meeting of many human needs and desires was deeply dependent on the existence of a society organized to maximize human well-being, especially economic well-being. Since the competitive character of capitalism necessarily implied the failure and consequent suffering of unsuccessful competitors, it was inevitably inefficient. Socialists, both on the Continent and in Britain, focused on the advantages of a cooperative rather than a competitive society, returning to an understanding of society as an organism in which the different parts benefited by their cooperation.

Given their assumption that different persons had different needs and abilities, most scientistic socialists—as distinguished from working-class radicals—were willing to argue that the vast majority of working-class people would be quite happy to give up any desire for a direct say in governmental affairs in favor of what seemed to be universal desires for adequate food, sex, housing, clothing, and self-expression through work. Thus, early-nineteenth-century theoretical socialisms showed relatively little interest in democracy, emphasizing participation in economic progress over participation in civic life. They urged that society be managed in the interest of all by an elite capable of understanding the desire for economic progress and how to spread the benefits of that progress across all classes of society.

Auguste Comte, the central figure of chapter 3, offered an alternative scientistic view of society for those liberal republicans who saw capitalism, with its offer of substantial rewards for those entrepreneurs willing to take risks, as the chief and probably necessary engine of economic progress. Not admittedly less concerned for the well-being of all classes than the socialists out of whose ranks he came, Comte was convinced by the neo-classical arguments of Malthus and Ricardo that the economic woes of the working class were more a consequence of their own over-breeding and the consequent over-supply of labor than of the rapacity of the owners of capital. His emphasis on the need to develop a full-fledged science of society—a sociology not built upon psychological assumptions about individual needs and behaviors—stimulated the development of an important academic tradition of sociology by the end of the century, especially in France and America. At the same time, it offered a plan for progress that was far less threatening to those in political and economic power than that of the socialists. Thus, while it proved appealing to middle-class intellectuals as well as to middle-class, reform-minded politicians such as Edwin Chadwick in Britain and Jules Ferry in France, it found little favor among the working classes.

One thing that differentiated the scientistic theories of both the Saint-Simonians and the Positivists from those of even their *Idéologue* predecessors was their dynamic, or historical, character. Saint-Simon had emphasized the oscillation between critical and synthetic periods of social development. Similarly, building on his Law of Three Stages, Comte had emphasized that the new science of sociology would have to understand social developments historically, even before they could be understood statically, or structurally. This historical perspective also informed Comte's proposal for a Religion of Humanity, which would effect the culmination of the historical tendencies of religions to expand the domain of love and caring, which had initially focused on family, then tribe, then nation, to all of humankind. Though Comte's Religion of Humanity had little impact on the Continent, its offer of a self-consciously subjective replacement for traditional religions drew small but exceedingly influential numbers of adherents in Britain, where such public intellectuals as George Eliot, Henry Lewes, and the "Iron Chancellor," Edwin Chadwick, promoted its values. Different varieties of Positivism were also important in both India and South America, where its appropriation by intellectual leaders tended to preempt more radical reform attitudes.[1]

In chapter 4, we explored how the political and intellectual traditions of the roughly three hundred early modern German princedoms and the political events associated with Napoleon's successful invasion of Germany produced a much different set of sciences and a different appropriation of natural scientific ideas and methods for ideological purposes than was the case in Britain or France. Throughout the eighteenth century, Germany had remained more fully controlled by a traditional nobility than its western neighbors. German social and economic theory, manifested in what came to be called the cameral sciences, reflected this situation by focusing directly on the wealth, security, and well-being of the prince and his state, rather than on that of his individual subjects. Of course, cameralist thinkers admitted that if the prince's subjects were more prosperous, the prince was likely to be more prosperous as well; but they did not share the common western expectation that a free market was likely to optimize the wealth of either the subjects or the state. Centralized management of the economy and of virtually all social life was the widely accepted ideal. In several states, including Prussia and Baden, state bureaucracies dominated by camerally trained, middle-class functionaries, rather than by the traditional nobility, had emerged, and these men generally promoted both scientific education and scientifically based improvements of industry. Thus, there was a substantial audience for and enthusiasm for science in Germany by the beginning of the nineteenth century, even though the economy was still largely agricultural and the population largely rural.

Germany had not been a scientific leader through most of the eighteenth century, so even the Berlin Academy of Science was dominated by foreigners, and the common language of discussion was French. Some German intellectuals, among whom Johann Herder was most important, resented the growing French influence and promoted the idea that every group with a common linguistic heritage has its own unique and important culture, which is most fully conveyed in the folk songs and folktales preserved where foreign influences have been minimal. With the French invasion by Napoleon, antagonism to French hegemony and fascination with a presumed unique German culture grew. With these sentiments, preparations to liberate Germany from France and to unify Germany became important among many young intellectuals. Within the natural sciences, though "international" science no doubt continued to be dominant in most places, a uniquely German science, often associated with the ideas of Goethe and Schelling and with the term *Naturphilosophie,* developed and spread to become a powerful force at such places as Jena and Berlin.

In considering its long-term ideological significance, four features of this science are most important. First, it was highly speculative and grounded in idealist rather than materialist philosophical assumptions. Second, it focused on transformation and change over time rather than on unchanging elements in the world. Third, in connection with the second feature, it emphasized organic phenomena as models rather than mechanical ones, because mechanical systems tended to be cyclical and unchanging, while organisms grow and decay over time. Fourth, practitioners of *Naturphilosophie* trended to be Christians and often understood the ideal sources of phenomena in terms of thoughts of God. Thus, for example, when Louis Agassiz, a very important biologist with idealist leanings, was challenged to account for the great similarities among groups of organisms, he argued that they were simply variations on a single theme in the mind of God.

As we saw in chapter 5, when Napoleon was finally driven from Germany, the students and young intellectuals who had joined together in patriotic singing societies, gymnastics societies, and university fraternities hoped to move rapidly to unify Germany, and a few *Naturphilosophen,* such as Lorenz Oken, promoted that ideal. The traditional nobility, however, managed to regain political control, and aided by most German clergy and by a number of *Naturphilosophen,* including Johannes Müller, they sought to suppress the nationalist movement through strongly repressive measures, outlawing fraternities and German cultural festivals, and expelling or imprisoning both students and faculty who persisted in promoting German unification based on the claim of a common culture grounded in the historical uniqueness of the German V*olk.* Such moves tended to radicalize many young intellectuals, who rallied around the materialist philosophy of Ludwig Feuerbach and those who extended his ideas. That philosophy was grounded in empirical scientific practices and in an inversion of idealist philosophy. For most idealists, the material world was a kind of epiphenomenal reflection of the world of ideas; whereas for Feuerbach, the Ideal world was a reflection of the more fundamental material reality.

In the hands of such self-styled "scientific materialists" as Ludwig Büchner and in those of the more radical dialectical materialists, including Karl Marx and Friedrich Engels, the most politically progressive elements of German culture tended toward an atheism that was seen as the natural consequence of scientific activity. It was probably the case that most nineteenth-century German scientists continued to be both conventionally Christian and politically conservative or uninterested. Thus, the tactical materialists associated with von Helmholtz and Du Bois-Reymond were at least nominally orthodox Christians and attached themselves to whatever political group was in the ascendant. However, the rhetorical abilities of the more radical thinkers, together with the German appetite for popular scientific writings, combined to make it seem to many sympathizers and opponents alike as if the scientific enterprise promoted both atheism and liberal to radical politics.

Marx and Engels were particularly insistent that communism, their brand of working-class socialism, was grounded in a scientific analysis of social development. Moreover, they used that claim to great effect in undermining the claims of competing socialist groups, who were derided as "utopian" in their plans and programs. To distinguish themselves from the more liberal theorists who looked to the scientific materialism of authors such as Büchner for their inspiration after the failed liberal revolutions of 1848, Marx and Engels adapted dialectical elements from *Naturphilosophie,* but insisted on the priority of material life over ideas, as Feuerbach had.

Political theories and ideologies during the second half of the nineteenth century, especially in Britain and Germany, were dominated by dynamic historical understandings drawn largely from arguments associated with Lamarckist interpretations of biological evolution, which were explored in chapter 7. Starting primarily with Spencer in Britain, evolutionary understandings of social development initially offered three great advantages over the analytically and individualistically based utilitarian theories that had dominated the first half of the century. First, for liberal and radical sympathizers especially, they could offer a naturalistically based morality suited to the growing commercial and industrial society to replace the unworkable pleasure-pain calculus of the Utilitarians and the backward-looking and ineffective morality associated with traditional established Christianity. Second, and perhaps more important, they offered a promise of progress that situated and justified the often unacceptable present condition of society as a necessary stage in the path toward a vastly better future. Third, and ironically, evolutionary notions could be adapted convincingly to virtually any political ends imaginable; so they became an almost universal model for both formal and informal social thought.

Following Darwin's publication of *The Origin of Species* and *The Descent of Man*, which formed the focus of chapter 8, there appeared a range of evolutionary social theories, explored in chapter 9, often labeled as "Social Darwinisms" in spite of the fact that many supplemented the mechanism of natural selection with Lamarckist notions of the inheritance of acquired characteristics. At the hands of those liberal thinkers whose attachment to evolutionary ideas continued to follow the pattern set by Spencer, including such American ideologists as Andrew Carnegie, present inequalities of wealth and privilege could be justified as a stage in progress toward a better society. At the hands of socialists such as Edward Bellamy, evolutionary arguments could be used to buttress and transform communism into a nonrevolutionary doctrine that made it possible for the communistic Social Democratic party to become the ruling political party in Germany without violence. At the hands of anarchists such as Kropotkin, evolutionary ideas could justify the claim that cooperation, or "Mutual Aid," was so natural that government was completely unnecessary. In the hands of aristocratic militarists and imperialists such as Bagehot and Haeckel, evolutionary theory could justify traditional hierarchies, wars of conquest, and the subjugation of "less-fit" races, although Darwin had insisted that war was dysgenic because it led to the death of the most physically fit and patriotic men. In the hands of those such as Kidd, for whom virtually all human evolution was social rather than biological, evolutionary theory could be used to justify a "trusteeship" model of imperialism, in which advanced nations supposedly managed the resources of less advanced peoples until they became capable of managing their own. This idea was adopted by both Britain and the United States to guide their imperialist policies.

For each of the persons named in the previous paragraph, evolutionary theory offered an optimistic vision of society in which the future was guaranteed to be better than the past if the lessons of evolution were learned and attended to. Such progressive evolutionary theories continued into the twentieth-first century, but several factors shifted the impact of evolutionary thought during the last two decades of the nineteenth century. First, the growing collection of social statistics seemed to suggest that insanity, alcoholism, prostitution, and criminality were all increasing dramatically. Such increases were certainly in some part artifacts of increasing sophistication in diagnosing and recording conditions that were simply invisible previously; but they were also partly the real consequences of rapidly growing urban poverty. Darwinian psychiatry, as practiced by cultural leaders such as Henry Maudsley in Britain, Bénédict Morel in France, and Cesare Lombroso in Italy, began to see these trends as the consequence of the pressures of industrial society working on less-fit segments of the population. In fact, as we saw in chapter 10, these men all developed theories of "degeneration," according to which atavistic tendencies accelerated over time. Such views seemed consistent with the second law of thermodynamics, which suggested that in the long run chaos must triumph over order.

At the hands of literary figures including Émile Zola and Hippolyte Taine in France and Thomas Hardy in Britain, degeneration became a major cultural theme, creating deep anxiety and pessimism. This pessimism was exacerbated as the mechanism of inheritance was linked to the "germ plasm" by Weisman, undermining the Lamarckian idea that one could will inheritable positive biological change. Additional fears were produced as evidence seemed to accumulate that non-Caucasian races, and even more importantly, mixed races were subject to more rapid degeneration.

Looking forward into the twentieth century, the apparently deterministic cast of nonteleological evolutionary theory, along with fears of degeneration, created or enhanced two major trends. On the one hand it led to the beginnings of the race-hygiene, temperance, and eugenics movements intended to counteract degenerationist trends. These movements became increasingly important, especially in Germany and the United States, during the first three decades of the twentieth century. Indeed, the United States, and to a lesser extent Germany and Russia, became the new loci of movements that continued and even at times amplified many of the scientistic movements of the nineteenth century. On the other hand, fears of determinism and degeneration added to a generally anti-scientific sentiment that was growing in Western Europe among the religious opponents of scientific naturalism, the Romantic opponents of accelerating technological change, and the philosophical movements culminating in Existentialism, which sought to carve out greater opportunities for human agency in a world that scientific trends depicted as increasingly deterministic. The century that began in Western Europe with such high hopes that scientific knowledge of society might help to identify features of the good life and to procure them for all humans thus ended with a growing sense that science could do neither.

In spite of the fact that nineteenth-century scientisms did not generally survive in their original forms into the twentieth century, it is nonetheless the case that they have left extremely important traces in the ideologies that direct almost all of our social and political lives today. Under the new name of conservativism in the United States and Britain, that version of nineteenth-century liberalism associated with Positivist visions of the importance of capitalism and Social Darwinist emphases on competition still forms the core of one of the dominant political ideologies of the early twenty-first century. Similarly, the focus on social justice and cooperation associated with Saint-Simonian and Owenite socialism and with cooperationist versions of Social Darwinism forms the core of early twenty-first-century social democratic ideologies on the Continent, labor ideology in Britain, and liberal ideologies in the United States. Finally, though Marxist communism is currently in retreat in the industrialized West, it seems alive and well in China and much of the developing world today.

NOTES

Introduction

1. Richard Olson, *Science Deified and Science Defied: The Historical Significance of Science in Western Culture,* 2 vols. (Berkeley: University of California Press, 1982 and 1990) and *The Emergence of the Social Sciences, 1642–1792* (New York: Twayne Publishers, 1993). Hereafter cited as *Science Deified and Science Defied* 1, *Science Deified and Science Defied* 2, and *Emergence of the Social Sciences,* respectively.

2. Olson, *Science Deified and Science Defied* 1:7–8.

3. F. A. Hayek, *The Counter-Revolution of Science: Studies on the Abuse of Reason* (New York: Free Press, 1952), 24.

4. Jürgen Habermas, *Toward a Rational Society: Student Protest, Science, and Politics* (Boston: Beacon Press, 1970), 50–51.

5. Ibid., 52.

6. Sandra Harding, *The Science Question in Feminism* (Ithaca: Cornell University Press, 1986), 16.

7. H. Stuart Hughes, *Consciousness and Society: The Reorientation of European Social Thought, 1890–1930* (New York: Alfred Knopf, 1958).

8. Friedrich Nietzsche, *The Birth of Tragedy,* ed. Oscar Levy (New York: Russell and Russell, 1964), 112.

9. See Stephen Kern, *The Culture of Time and Space, 1880–1918* (Cambridge, Mass.: Harvard University Press, 2003), for a valuable analysis of the transformation of notions of space and time following from the works of Husserl, Mach, and, eventually, Einstein.

10. Dorothy Ross, ed., *Modernist Impulses in the Human Sciences* (Baltimore: Johns Hopkins University Press, 1994), 1.

11. See Robert C. Bannister, *Sociology and Scientism: The American Quest for Objectivity* (Chapel Hill: University of North Carolina Press, 1987).

12. Quoted from Garland Allen, "Science and Society in the Eugenic Thought of H. J. Muller," *Bio Science* 20 (1970): 349.

13. Condorcet, *Report on Public Instruction,* translated in François de la Fontainerie, ed., *French Liberalism and Education in the Eighteenth Century* (New York: McGraw Hill Book Co., 1932), 337.

14. Isaac Newton, *The Principia,* trans. Andrew Motte (Amherst: Prometheus Books, 1995), 442–43.

15. David Hume, *An Inquiry Concerning Human Understanding with a Supplement: An Abstract of a Treatise of Human Nature* (Indianapolis: Bobbs-Merrill Co., 1955), 198.

16. Newton, *Principia,* 320.

17. Hume, *Inquiry Concerning Human Understanding,* 183–84.

18. Etienne Bonnot, Abbé de Condillac, *Philosophical Writings,* trans. Franklin Phillip and Harlan Lane (Hillsdale, N.J.: Lawrence Earlbaum Associates, 1982), 464.

Chapter 1: Ideology

1. Olson, *Science Deified and Science Defied* 2.

2. See Charles Coulston Gillispie, "The *Encyclopedie* and the Jacobin Philosophy of Science: A Study in Ideas and Consequences," in Marshall Clagett, ed., *Critical Problems in the History of Science* (Madison: University of Wisconsin Press, 1969), 255–90.

3. See Henry Guerlac's "Commentary," 317–20, in Clagett, *Critical Problems in the History of Science.*

4. See David Noble, *The Religion of Technology* (New York: Alfred A. Knopf, 1997), 181–83.

5. See Margaret Jacob, *Scientific Culture and the Making of the Industrial West* (New York: Oxford University Press, 1997), 178–86.

6. Cited in *The Doctrine of Saint-Simon: An Exposition,* ed. George G. Iggers (New York: Schocken Books, 1972), 168, n. 2, from "Memoire sur la Science de'lHomme," in *Ouvres de Saint-Simon et d'Enfantin* (Paris: E. Dentu, 1865–78), 40:16.

7. Olson, *Science Deified and Science Defied* 2:274–79.

8. See Robert Fox and George Weisz, *The Organization of Science and Technology in France: 1808–1914* (Cambridge: Cambridge University Press, 1980), 1–3.

9. See especially L. Pearce Williams, "The Politics of Science in the French Revolution," 291–308, in Clagett, *Critical Problems in the History of Science.*

10. As Elizabeth Williams has argued in her *The Physical and the Moral: Anthropology, Physiology, and Philosophical Medicine in France, 1750–1850* (New York: Cambridge University Press, 1994), some of the key early figures in establishing a "science of man" had been educated at the medical school at Montpellier. See especially chap. 1, "Montpellier Vitalism and the Science of Man."

11. See Williams, *Physical and the Moral,* chap. 2, for the central role of Cabanis in the growth of the medico-moralist tradition in Paris.

12. On the moral dimensions of associationist psychology, see either Olson, *Science Deified and Science Defied* 2:236–83, or Olson, *Emergence of the Social Sciences,* chap. 8, 96–117.

13. Martin S. Staum, *Cabanis: Enlightenment and Medical Philosophy in the French Revolution* (Princeton: Princeton University Press, 1980), 161.

14. On the methodological views of Helvetius and Condorcet, as well as their political implications, see *Science Deified and Science Defied* 2:256–83.

15. Pierre-Jean-George Cabanis, *On the Relations between the Physical and Moral Aspects of Man* (Baltimore: Johns Hopkins University Press, 1981), 2 vols., trans. Margaret Duggan Saidi from the 1805 2nd edition of *Rapports du physique et du moral de l'homme,* 50.

16. Ibid. 1:44.

17. In "Bureaucracy, Liberalism, and the Body in Post-Revolutionary France: Bichat's Physiology and the Paris School of Medicine," *History of Science* 19 (1981): 115–42, John Pickstone has made the argument that, whether consciously or not, Bichat formulated his physiological work to reflect particular political interests, and Bichat was an admirer of Cabanis.

18. Destutt de Tracy, *A Treatise on Political Economy, to Which Is Prefixed a Supplement*

to a Preceding Work on the Understanding, or Elements of Ideology (Georgetown, D.C.: Joseph Milligan, 1817), 254.

19. On the sources of Cabanis's medical epistemology and physiological ideas see especially Sergio Moravia, "Medicine e epistemologia nel giovane Cabanis," published as the introduction to the Italian translation of *Du dergre' de certitude de la medicine* (Bari, 1974) and chaps. 1–4 of Martin S. Staum, *Cabanis: Enlightenment and Medical Philosophy in the French Revolution* (Princeton: Princeton University Press, 1980).

20. Cited in Staum, *Cabanis*, 108.

21. On this issue, see John Yolton, *Thinking Matter: Materialism in Eighteenth-Century Britain* (Minneapolis: University of Minnesota Press, 1983).

22. Staum, *Cabanis*, 65.

23. Cabanis, *Relations* 1:51.

24. Ibid., 20.

25. Ibid., 43. Emphases mine.

26. Claude Adrienne Helvetius, *Treatise on Man*, trans. W. Hooper (New York: Burt Franklin, 1969; from 1774 original), 2:279.

27. Cabanis, *Relations* 1:72.

28. Ibid., 66–67.

29. Ibid., 52.

30. Cabanis's views on the difference between natural and artificial inequalities are nicely discussed in Staum, *Cabanis*, 118–20. The citation is from 119.

31. Cabanis, *Relations* 1:50–51.

32. Ibid., 51.

33. Ibid., 86.

34. Ibid., 235.

35. Cheryl B. Welch, *Liberty and Utility: The French Idéologues and the Transformation of Liberalism* (New York: Columbia University Press, 1984), 77–78.

36. Cited in Welch, *Liberty and Utility*, 157–58.

37. Jean-Baptiste Say, *A Treatise on Political Economy, or, The Production, Distribution, and Consumption of Wealth*, trans. C. R. Princep, new edition (Philadelphia: Lippincott, 1859), xxvii.

38. Ibid., xxvii.

39. Ibid., xlvii.

40. Ibid. Second italics and boldface emphases mine. Martin Staum argues that Say's methods were much more "rationalist" and less "empiricist" than he was aware of, citing Say's unwillingness to allow statistical evidence to bear against his generalizations. By our present standards Staum may be correct, but by the standards of early nineteenth-century political economy Say was on the empiricist end of the methodological spectrum. See Martin Staum, "French Lecturers in Political Economy, 1815–1848: Varieties of Economic Liberalism." *History of Political Economy* 30 (1998): 95–120.

41. Say, *Political Economy*, xxxv.

42. Ibid., 62.

43. Ibid., 285.

44. Ibid., 287.

45. Ibid., xl–xli.

46. Ibid., 86.

47. Ibid., 299–300, including note on 300.

48. Ibid., 82.
49. Ibid., 84–85.
50. Ibid., 83–84.
51. Ibid., 391.
52. Ibid., 397–98.
53. Ibid., 411.
54. Ibid., 408.
55. Ibid., 372.
56. Ibid., 98.
57. On de Tracy's first *Memoire,* see Emmet Kennedy, *A Philosophe in the Age of Revolution: Destutt de Tracy and the Origins of "Ideology"* (Philadelphia: American Philosophical Society, 1978), 44–48. Citation is from 47.
58. Cited in Kennedy, *Destutt de Tracy,* 48.
59. See Olson, *Emergence of the Social Sciences,* 123–24, on Boisguilbert and his qualified focus on free market exchanges.
60. De Tracy, *Treatise on Political Economy,* 83. (Note: there is a double pagination in the 1817 edition. In general, because the second pagination is the more extensive, page numbers refer to it. In this case, however, the reference is to the first pagination.)
61. Ibid., 85–86, first pagination.
62. Ibid., 15–17.
63. Ibid., 170–73.
64. Ibid., 143.
65. Ibid., 35, 47, first pagination.
66. Ibid., 47–48.
67. Ibid., 112.
68. Ibid., 113.
69. Ibid., 115.
70. Ibid., 129.

Chapter 2: Saint-Simon, Saint-Simonianism, and the Birth of Socialism

1. Claude Henri, Comte de Saint-Simon, *Lettres d'un habitant de Geneve,* Paris, 1803, 33, cited in *The Political Thought of Saint-Simon,* ed. Ghita Ionescu (Oxford University Press, 1976), 23.
2. See *Oeuvres de Saint-Simon et d'Enfantin,* 47 vols. (Paris, 1865–78).
3. See Robert Wokler, "Saint-Simon and the Passage from Political to Social Science," chap. 14 in Christopher Fox, Roy Porter, and Robert Wokler, eds., *Inventing Human Science: Eighteenth-Century Domains* (Berkeley: University of California Press, 1995), 333.
4. Ionescu, *Political Thought of Saint-Simon,* 78–79.
5. Ibid., 78.
6. Cited in Frank Manuel, *The Prophets of Paris* (Cambridge, Mass.: Harvard University Press, 1962), 130–31.
7. See Xavier Bichat, *Recherches physiologigues sur la vie et sur la mort* (Paris, 1799–1800).
8. Ionescu, *Political Thought of Saint-Simon,* 76.
9. Ibid., 77.

10. Ibid., 77–78.

11. Ibid., 72.

12. Ibid., 81.

13. On the functions of the Religion of Newton, see especially Frank Manuel, *The New World of Henri Saint-Simon* (Cambridge, Mass.: Harvard University Press, 1956), 69–70.

14. Ionescu, *Political Thought of Saint-Simon,* 73.

15. Mary Midgely, *Science as Salvation: A Modern Myth and Its Meaning* (London: Routledge, 1992), 141–42, 140.

16. Ionescu, *Political Thought of Saint-Simon,* 93–97.

17. Ibid., 99, 108.

18. Cited in Manuel, *New World of Henri Saint-Simon,* 311.

19. Cited in ibid., 192.

20. Cited in ibid., 203.

21. Cited in ibid., 350–51.

22. Ionescu, *Political Thought of Saint-Simon,* 142.

23. Ibid., 155.

24. Ibid., 156.

25. Ibid., 158.

26. Ibid., 201.

27. Ibid., 159.

28. Cited in Manuel, *New World of Henri Saint-Simon,* 242.

29. Cited in ibid., 302.

30. Cited from *Henry Saint-Simon a MM. les ouvriers* (n.d., n.p.), 1–2, in Manuel, *New World of Henri Saint-Simon,* 255.

31. Ionescu, *Political Thought of Saint-Simon,* 206, 210.

32. Gustave d'Eichthal to John Stuart Mill, December 1, 1829, cited in Manuel, *New World of Henri Saint-Simon,* 421, n. 10.

33. Manuel, *New World of Henri Saint-Simon,* 363.

34. On the linkages between science and religion in Scottish Common Sense Philosophy, see part 1 of Richard Olson, *Scottish Philosophy and British Physics: The Foundations of the Victorian Scientific Style* (Princeton: Princeton University Press, 1975).

35. See E. M. Butler, *The Saint-Simonian Religion in Germany* (Cambridge: Cambridge University Press, 1926), passim.

36. On the first Saint-Simonian use of the term "socialism" and its key role in defining later socialist movements, see F. A. Hayek, *The Counter-Revolution of Science: Studies on the Abuse of Reason* (New York: Free Press, 1952), 282, 291–320.

37. A. Transon, *De la religion Saint-Simonienne* (Paris, 1830), 48–49, cited in F. A. Hayek, *Counter-Revolution of Science,* 276.

38. *The Doctrine of Saint-Simon: An Exposition,* ed. George Iggers (New York: Schocken Books, 1972), 86.

39. Ibid., 59.

40. Ibid., 1–2.

41. Ibid., 12.

42. Ibid., 14.

43. Ibid., 14–15.

44. Ibid., 15.

45. Ibid., 95–96, 97–98.

46. Ibid., 105–7.
47. Ibid., 108–9.
48. Ibid., 109.
49. See especially F. A. Hayek, *Counter-Revolution of Science,* 314–18.
50. Cited in ibid., 315, n. 54.
51. See M. Wallon, *Les Saint-Simoniens et les chemins de fer* (Paris, 1908).
52. Iggers, *Doctrine of Saint-Simon,* 37.
53. Ibid., 154.
54. Ibid., 156.
55. Ibid., 222.
56. Ibid., 206.
57. Ibid., 219.
58. Ibid., 219.
59. Ibid., 236–37.
60. See ibid., 236, for a discussion of the hypothetical character of heliocentrism.
61. Iggers, *Doctrine of Saint-Simon,* 219.
62. Ibid., 234.
63. Ibid., 205.
64. Ibid., 237.
65. Ibid., 266–67.
66. K. Grun, *Die soziale Bewegung in Frankreich und Belgien* (1845), 182, cited in Hayek, *Counter-Revolution of Science,* 320.
67. Hayek, *Counter-Revolution of Science,* 24.

Chapter 3: Auguste Comte and Positivisms

1. See Walter M. Simon, *European Positivism in the Nineteenth Century* (Ithaca: Cornell University Press, 1963), 126–52. Hereafter cited as *European Positivism.*
2. Simon, *European Positivism,* 154–57.
3. On Positivism in India, see Geraldine Hancock Forbes, *Positivism in Bengal: A Case Study in the Transmission and Assimilation of an Ideology* (Calcutta: Minerva Associates, 1975), and Gayan Prakash, *Another Reason: Science and the Imagination of Modern India* (Princeton: Princeton University Press, 1999). On Positivism in Latin America, see especially Jao Cruz Costa, *A History of Ideas in Brazil,* (Berkeley: University of California Press, 1964); Leopoldo Zea, *Major Trends in Mexican Philosophy* (South Bend: University of Notre Dame Press, 1966); and Ralph Lee Woodward Jr., *Positivism in Latin America, 1850–1900,* (Lexington: D. C. Heath and Co., 1971).
4. See Simon, *European Positivism,* 68.
5. Auguste Comte, *System of Positive Polity,* in Gertrude Lenzer, ed., *Auguste Comte and Positivism: The Essential Writings* (New York: Harper and Row, 1975), 319. All further references to the *System of Positive Polity* will be listed as in Lenzer, *Essential Writings.*
6. On the self-consciousness of early Indian Positivists about this issue, see Forbes, *Positivism in Bengal,* especially chap. 4, "The Early Indian Positivists," 50–72.
7. In Lenzer, *Essential Writings,* 54.
8. Harriet Martineau did a translation and condensation of the *Cours de philosophie positive* under the title, *The Positive Philosophy* (New York: C. Blanchard, 1855), which Comte admired so much that he had it retranslated into French to replace his own version

in his collected works. All citations are to that edition, reprinted in 1974 by AMS Press in New York.

9. John Stuart Mill, *Auguste Comte and Positivism*, 3rd ed. (London: Tubner and Co., 1882), 5.

10. George Sarton, "Auguste Comte, Historian of Science: With a Short Digression on Clotilde de Vaux and Harriet Taylor," *Osiris* 10 (1952): 345, 352.

11. See Donald G. Charlton, *Positivist Thought in France during the Second Empire* (Oxford: Clarendon Press, 1959), passim, especially chaps. 2–5.

12. Comte to Clotilde de Vaux, March 11, 1846, cited in Frank Manuel, *The Prophets of Paris* (Cambridge, Mass.: Harvard University Press, 1962), 291–92.

13. See, for example, Simon's assessment in *European Positivism*, 6–8.

14. The range of persons drawn to some degree by the content of the Religion of Humanity is explored very nicely in T. R. Wright, *The Religion of Humanity: The Impact of Comtean Positivism on Victorian Britain* (Cambridge: Cambridge University Press, 1986).

15. Simon, *European Positivism*, 48–70.

16. Comte, *Positive Philosophy*, 25.

17. Ibid.

18. Ibid., 26.

19. Ibid., 32.

20. Ibid., 28.

21. Ibid., 37.

22. Ibid., 30.

23. Ibid., 39–40.

24. Ibid., 34.

25. Ibid., 29, 33.

26. Ibid., 33.

27. Ibid.

28. Simon, *European Positivism*, 35.

29. Comte, *Positive Philosophy*, 44.

30. Ibid.

31. Ibid., 45–46.

32. Ibid., 47.

33. Ibid., 49–50.

34. William Gillespie, *The Philosophy of Mathematics: Translated from the Cours de philosophie positive of Auguste Comte* (New York: Harper and Brothers, 1851).

35. Comte, *Positive Philosophy*, 200.

36. Ibid.

37. Ibid.

38. Ibid.

39. Ibid., 399–403.

40. Ibid., 404.

41. Ibid., 408, 412.

42. Ibid., 420.

43. Ibid., 410.

44. Ibid., 452.

45. Ibid., 453.

46. Ibid., 27.

47. Ibid., 454.
48. Ibid., 457.
49. Ibid., 461.
50. Ibid., 462.
51. Ibid., 477.
52. Ibid., 477–78.
53. Ibid., 459.
54. Ibid., 466.
55. Ibid., 481.
56. Ibid., 468.
57. Ibid., 831–32.
58. Lenzer, *Essential Writings,* 332.
59. Ibid., 311.
60. Ibid., 319.
61. Ibid., 327.
62. Ibid., 328.
63. Ibid., 318–19.
64. Ibid., 356–57.
65. Ibid., 358–59.
66. Ibid., 357.
67. Ibid., 361–62.
68. Ibid., 360, 386.
69. Ibid., 319.
70. Ibid., 373.
71. Ibid., 379.
72. Ibid.
73. Ibid., 373–74.
74. See Richard Olson, "Sex and Status in Scottish Enlightenment Social Science: John Millar and the Sociology of Gender Roles," *History of the Human Sciences* 11 (1998): 73–100.
75. Lenzer, *Essential Writings,* 378–79.
76. Ibid., 378.
77. Ibid., 382–83.
78. Ibid., 396.
79. Ibid.
80. Ibid., 400–401.
81. Ibid., 401.
82. Ibid., 401–2.
83. Ibid., 403.
84. Ibid., 407–8.
85. Ibid., 462.
86. Auguste Comte, *The Catechism of Positive Religion,* trans. Richard Congreve, 3rd edition (London, 1891), reprinted (Clifton, N.J.: Augustus M. Kelley, 1973), 68.
87. Comte, *Catechism,* 72.
88. Ibid., 90–96.
89. Ibid., 99.
90. Ibid., 109.

Chapter 4: Naturphilosophie, *Romanticism, and Nationalism*

1. As is the case with the term "Positivism," the term "*Naturphilosophie*" is open to a variety of interpretations. The problem is nicely discussed by Kenneth Caneva in "Physics and *Naturphilosophie:* A Reconnaissance," *History of Science* 35 (1997): 36–106. My own preference is to be somewhat looser than Caneva, incorporating more Kantian and Goethean perspectives, rather than trying to distinguish *Naturphilosophie* from perspectives associated directly with Kant or Goethe. More critical of the use of the term in connection with German biology is Tim Lenoir, whose *The Strategy of Life: Teleology and Mechanics in Nineteenth- Century German Biology* (Dordrecht: Reidel, 1982), constitutes a sustained argument against a substantial role for *Naturphilosophie* in favor of a more directly Kantian tradition. Lenoir has convinced me that one can pick out Kantian elements that have had long-term impacts on biological thinking and that have been dissociated by their users from the more imaginative and mystical elements of the followers of *Naturphilosophie;* but he has not convinced me that a number of these scholars were not initially drawn to their approaches by the selective appropriation and adaptation of these elements out of their incorporation into *naturphilosophisch* texts and training.

2. For an account of the temporal development of Kant's conception of the mathematical sciences, see Michael Friedman, *Kant and the Exact Sciences* (Cambridge, Mass.: Harvard University Press, 1992).

3. Friedman, *Kant and the Exact Sciences,* for example, shows quite conclusively that the widely held understanding that Kant considered arithmetic to be the science of pure time, as opposed to geometry, which was the science of pure space, was a misreading of Kant's arguments (see Friedman, 104–14). For our purposes, it is more important to understand that he was read in this way and that William Rowan Hamilton developed the mathematical system of quaternians to combine his understandings of Kantian space and time into a single fourfold "space."

4. Johann Herder, *Letters on the Advancement of Humanity* (1793–97), cited in Willibald Klinke, *Kant for Everyman: An Introduction for the General Reader* (New York: Collier Books, 1962), 35–36.

5. Immanuel Kant, *Prolegomena to Any Future Metaphysics,* Paul Carus translation, republished in T. V. Smith and Marjorie Grene, eds., *Berkeley, Hume, and Kant* (Chicago: University of Chicago Press, 1940), 268. Further references to the *Prolegomena* will be to this edition, but section numbers will be provided to aid in locating the passages in other editions.

6. This point is particularly clearly articulated in the preface to *The Metaphysical Foundations of Natural Science,* trans. James W. Ellington (Indianapolis: Hackett Publishing Company, 1985), 4. All future references will be to this edition.

7. Kant, *Prolegomena to Any Future Metaphysics,* section 36, 316–18.

8. Toward the end of his life, Kant attempted to show that other aspects of what we call science, including chemistry, could be established as real sciences, that is, as being certain; but this attempt was never completed. See part 2 of Friedman, *Kant and the Exact Sciences,* 213–341.

9. Kant, *Prolegomena,* "remarks" to section 13, 289–96.

10. Robert Perceval Graves, *Life of Sir William Rowan Hamilton,* 3 vols. (London: Long-

mans, Green and Co., 1882), 1:501, cited in Thomas Hankins, *Sir William Rowan Hamilton* (Baltimore: Johns Hopkins, 1980), 104.

11. Kant, *Prolegomena,* section 15, 297.

12. Kant, *Metaphysical Foundations of Natural Science,* 40.

13. Ibid., 41.

14. Ibid., 56.

15. On the relation of Kantian Philosophy to field theories, see L. Pearse Williams, *The Origins of Field Theory* (Lanham, Mo.: University Press of America, 1980).

16. Kant, *Metaphysical Foundations of Natural Science,* 43.

17. Kant, *Critique of Judgement,* trans. Werner S. Pluhar (Indianapolis: Hackett Publishing Co., 1987), 249.

18. Ibid., 253.

19. See, for example, Timothy Lenoir, "Kant, Blumenbach, and Vital Materialism in German Biology," *Isis* 71 (1980): 77–108. The so-called vitalism of the Montpellier school in France was a vastly different sort than that in Germany and was based primarily on the claim that since no previously known universal forces could account for organic development, there must be another kind of force. See on this issue, chapter 1 of Elizabeth A. Williams, *The Physical and the Moral: Anthropology, Physiology, and Philosophical Medicine in France, 1750–1850* (Cambridge: Cambridge University Press, 1994), 20–66. My preference is to call this organicism, rather than vitalism.

20. Alexander Gode-Von Aesch, *Natural Science in German Romanticism* (New York: Columbia University Press, 1941), 3–4, identified Goethe's science as the paradigm for Romantic science, and that view has predominated ever since.

21. See Karl J. Fink, *Goethe's History of Science* (Cambridge: Cambridge University Press, 1991), 16.

22. Fink, *Goethe's History of Science,* 17–19.

23. Ibid., 9.

24. Cited in John Theodore Merz, *A History of European Thought in the Nineteenth Century,* vol. 2 (New York: Dover Publications, 1965; from 1897 original), 252.

25. See especially Gode-Von Aesch, *Natural Science in German Romanticism,* 53–73.

26. I owe my understanding of Goethe's struggle with the notion of type to my student, Miriam Zeldich Walters, whose senior thesis at the University of Santa Cruz in 1970, "A History of the Conceptual Development of *Naturphilosophie,*" continues to inform my understanding of many of the topics discussed in this and the following sections. On Goethe's changing concept of type, see 54–65. Goethe also presented his own account of these changes in his "morphological notebooks" of 1817. This discussion is analyzed in Fink, *Goethe's History of Science,* 29–31.

27. Cited from Goethe's poem, "The Metamorphosis of Plants," in Paul Carus, *Goethe* (Chicago: Open Court Publishing Co., 1915), 254.

28. See Robert J. Richards, *The Romantic Conception of Life: Science and Philosophy in the Age of Goethe* (Chicago: University of Chicago Press, 2002), 526–33.

29. Cited in Rudolph Magnus, *Goethe As a Scientist,* trans. H. Norden (New York: Collier Books, 1961), 172.

30. Magnus, *Goethe as a Scientist,* 172.

31. Cited in Charles Coulston Gillispie, *The Edge of Objectivity* (Princeton: Princeton University Press, 1960), 196.

32. For an excellent brief introduction to Goethe's optical work, see Fink, *Goethe's History of Science,* 31–41.

33. Friedrich Wilhelm Joseph von Schelling, *Ideas for a Philosophy of Nature as an Introduction to the Study of This Science,* based on both the first, 1797, and the second, 1803, editions, trans. Errol E. Harris and Peter Heath (Cambridge: Cambridge University Press, 1988), 40.

34. Ibid., 42.

35. Ibid., 39.

36. Ibid., 165.

37. Ibid., 216.

38. Ibid., 224.

39. Ibid., 226.

40. Ibid., 263.

41. Ibid., 128.

42. Ibid.

43. Ibid., 117.

44. Ibid., 118.

45. Ibid., 67.

46. Ibid., 37.

47. See Walter D. Wetzels, "Aspects of Natural Science in German Romanticism," *Studies in Romanticism* 10 (1971): 44–59, esp. 45.

48. The *naturphilosophisch* perspective always had important opponents among experimental and mathematical physicists, but by the late nineteenth century, the all but universal German condemnation of *Naturphilosophie* was well expressed in Ferdinand Rosenberger's *Geschichte der Physik* (3 vols., Braunschweig, 1887–90; rept. Hildesheim: Georg Olms, 1965), vol. 3, p. 9: "Soon after [Kant], philosophy of nature became discredited, due to the badly grounded actions of individual successors and pseudo successors to Kant, who passed off castles in the air as scientific structures on the basis of critical philosophy."

49. Coulston Gillispie, *Edge of Objectivity,* 201.

50. On the short-lived triumph of *Naturphilosophie* at Berlin, see Frederick Gregory, "Kant, Schelling, and the Administration of Science in the Romantic Era," *Osiris* (new series) 5 (1989): 17–35.

51. See the discussion of early nineteenth-century physics texts in Christa Jungnickel and Russell McCormmach, *Intellectual Mastery of Nature: Theoretical Physics from Ohm to Einstein,* vol. 1. (Chicago: University of Chicago Press, 1986), 23–33.

52. See Jungnickel and McCormmach, *Intellectual Mastery of Nature,* 1, 17.

53. Cited in Caneva, "Physics and *Naturphilosophie,*" 50.

54. Cited in ibid., 51.

55. Cited in R. A. R. Tricker, *Early Electrodynamics: The First Law of Circulation* (Oxford: Pergammon Press, 1965), 12.

56. Hans Christian Oersted, "Experiments on the Effect of a Current of Electricity on the Magnetic Needle," *Annals of Philosophy* 16 (October 1820).

57. See Wetzels, "Aspects of Natural Science in German Romanticism," 53–54.

58. For a discussion of this work see Caneva, "Physics and *Naturphilosophie,*" 44–47.

59. See ibid., 54–62.

60. Lorenz Oken, *Die Zeugung* (1805), cited in William Coleman, *Biology in the Nine-*

teenth Century: Problems of Form, Function, and Transformation (New York: John Wiley and Sons, 1971), 25.

61. See Erik Nordenskjold, *The History of Biology: A Survey* (New York: Tudor Publishing Co., 1936), 393.

62. Döllinger is discussed in Lenoir, *Strategy of Life*, 65–73, and Burdach is discussed by Lenoir on 73–77.

63. See, for example, Coleman, *Biology in the Nineteenth Century*, 42.

64. See, for example, Lenoir, *Strategy of Life*, 87.

65. Cited in Coleman, *Biology in the Nineteenth Century*, 50.

66. Cited in Erik Nordenskjold, *History of Biology*, 383.

67. Cited in Lenoir, *Strategy of Life*, 105.

68. See ibid., 103–11.

69. My discussion of the pre-1790s use of the terms "Romantic" and "Romanticism" is dependent on Raymond Immerwahr, "The Word 'Romantisch' and Its History," in Siegbert Prawer, ed., *The Romantic Period in Germany* (New York: Schocken Books, 1970), 34–63.

70. See Frank Manuel, *The Eighteenth Century Confronts the Gods* (Cambridge, Mass.: Harvard University Press, 1959), passim.

71. Friedrich Schlegel, *Kritische Fragmente*, 115, cited in Paul Roubiczek, "Some Aspects of German Philosophy in the Romantic Period," 303, in Prawer, *Romantic Period in Germany*.

72. H. A. Korff, *Geist der Goethezeit* 1:28, cited in Alexander Gode-Von Aesch, *Natural Science in German Romanticism* (New York: Columbia University Press, 1941; rept. AMS Press, New York, 1966), 31.

73. See Immerwhar, "'Romantisch' and Its History," 52.

74. Cited in Roubiczek, "Some Aspects of German Philosophy in the Romantic Period," 315.

75. Discussed in Immerwahr, "'Romantisch' and Its History," 57.

76. Cited in Roubiczek, "Some Aspects of German Philosophy in the Romantic Period," 315.

77. Cited in Gode-Von Aesch, *Natural Science in German Romanticism*, 140.

78. See ibid., 150–54.

79. On the organicist character of German late-Enlightenment historical thought, see Peter Hanns Reill, *The German Enlightenment and the Rise of Historicism* (Berkeley: University of California Press, 1975).

80. On this anti-statist element in Herder, see especially Frank Manuel's "editor's introduction" to Johann Gottfried von Herder, *Reflections on the Philosophy of the History of Mankind* (Chicago: University of Chicago Press, 1968).

81. Johann Fichte, *Addresses to the German Nation*, trans. R. F. Jones and G. H. Turnbull (Chicago: Open Court Publishing Co., 1922), 52.

82. Ibid., 232.

83. Ibid., 4–6.

84. Ibid., 144–45.

85. Ibid., 264.

86. Ibid., 13.

87. Ibid., 185.

88. Ibid., 217.

89. Ibid., 197.

90. On this issue, see especially Joseph Ben-David, *The Scientist's Role in Society: A Comparative Study* (Chicago: University of Chicago Press, 1984), chap. 7, 108–38. Although it is certainly true that German natural scientists continued to view themselves as underfunded and underappreciated relative to humanists throughout the first half of the nineteenth century-see, for example, Kurt Bayertz, "Spreading the Spirit of Science: Social Determinants of the Popularization of Science in Nineteenth Century Germany," in Terry Shinn and Richard Whitley, eds., *Expository Science: Forms and Functions of Popularization* (Dordrecht: D. Reidel Publishing Co., 1985), 212–17–English, French, and American scientists looked on with envy from about 1825 to the end of the century. Arleen Tuchman has documented the growth of a research orientation and research support at Heidelberg in *Science, Medicine, and the State in Germany: The Case of Baden, 1815–1871* (New York: Oxford University Press, 1993).

91. Fichte, *Addresses to the German Nation,* 223–24.

92. Ibid., 227.

93. G. W. F. Hegel, *Reason in History: A General Introduction to the Philosophy of History,* trans. Robert S. Hartman (Indianapolis: Bobbs-Merrill, 1953; from the 1837 German original), 87.

94. Ibid., 53.

95. Ibid., 46–47.

96. Leopold von Ranke, "Analecta," *Historische Zeitschrift* 244 (1887), cited by Rudolph Vierhaus in "Historiography between Science and Art," in George Iggers and James M. Powell, eds., *Leoplod von Ranke and the Shaping of the Historical Discipline* (Syracuse: Syracuse University Press, 1990), 61.

97. Cited by Friedrich Jager and Jorn Rusen, *Geschichte des Historismus* (Munich: C. H. Beck, 1992), 83.

98. Roger Smith, *The Norton History of the Human Sciences* (New York: W. W. Norton Co., 1997), 386.

Chapter 5: The Rise of Materialisms and the Reshaping of Religion and Politics

1. See, for example, Donald Rohr, *The Origins of Social Liberalism in Germany* (Chicago: University of Chicago Press, 1963), passim. Karl Heinrich Brueggemann, who had studied law at Bonn and Heidelberg, was actually sentenced to death for his leadership at the Hambach Festival; but cooler heads prevailed and his sentence was first commuted to life in prison before he was released after eight years (see Rohr, *Origins of Social Liberalism in Germany,* 99).

2. See Karl Hill, ed., *The Management of Scientists* (Boston: Beacon Press, 1964), essay by Everett Mendelsohn.

3. See Rohr, *Origins of Social Liberalism in Germany,* 16, 38–40.

4. On the centrality of natural theology to seventeenth- and eighteenth-century religious discourse, see John Hedley Brooke, *Science and Religion, Some Historical Perspectives* (Cambridge: Cambridge University Press, 1991), esp. chaps. 4–6.

5. Cited from Norman Kemp Smith, ed., *Immanuel Kant's Critique of Pure Reason* (New York: St. Martin's Press, 1965), 528–29.

6. G. W. F. Hegel, "Preface," in H. F. W. Hinrich, *Religion in Its Internal Relation to Science,* cited in Frederick Gregory, *Nature Lost: Natural Science and the German Theo-*

logical Traditions of the Nineteenth Century (Cambridge, Mass.: Harvard University Press, 1992), 37.

7. My comments on Hegel and Strauss are based largely on Frederick Gregory, *Nature Lost,* chap. 3.

8. Cited in Gregory, *Nature Lost,* 80.

9. See Gregory, *Nature Lost,* 88–111, on Strauss's turn to Darwinism.

10. Gregory, *Nature Lost,* 81.

11. Cited in Eugene Kamenka, *The Philosophy of Ludwig Feuerbach* (New York: Praeger Publishers, 1969), 15.

12. See Ludwig Feuerbach, *Thoughts on Death and Immortality* (Berkeley: University of California Press, 1980), 89.

13. Ibid., 17.

14. Cited in Frederick Gregory, *Scientific Materialism in Nineteenth Century Germany* (Dordrecht: D. Reidel, 1977), 16.

15. Cited in Kamenka, *Philosophy of Ludwig Feuerbach,* 158, n. 26.

16. Cited in Gregory, *Scientific Materialism,* 19.

17. Ludwig Feuerbach, *The Essence of Christianity,* trans. Marian Evans (George Eliot) (London: Kegan Paul Trench and Tubner, 1854), viii.

18. Feuerbach, *Essence of Christianity,* 270.

19. Ibid., xi.

20. Cited in Kamenka, *Philosophy of Ludwig Feuerbach,* 17.

21. Cited in ibid., 16.

22. Cited in Gregory, *Scientific Materialism,* 24.

23. See ibid., 61, 64.

24. See ibid., 100–104.

25. See ibid., 213.

26. See ibid., 41, on the philosophical neutrality of most German scientists.

27. The central role played by Say's work in the Germanies is being explored currently by Keith Tribe, and was mentioned in his "Classicism before Neo-Classicism: Political Economy between Natural Law and Historicism," a paper presented at a conference on "Science and the Social Sciences in the Late Eighteenth and Early Nineteenth Centuries," May 30–31, 1997, at the Clark Library, UCLA, arranged by Theodore M. Porter and Dorothy Ross.

28. Cited in Arleen Tuchman, *Science, Medicine, and the State in Germany: The Case of Baden, 1815–1871* (New York: Oxford University Press, 1993), 8. Tuchman's introductory chapter summarizes beautifully the case for a utilitarian emphasis on science emerging from liberal civil servants, though she does not see the cameralist curriculum as particularly important.

29. Cited in Tuchman, *Science, Medicine, and the State in Germany,* 11.

30. Ibid.

31. Cited in Gregory, *Scientific Materialism,* 105.

32. Ludwig Büchner, *Force and Matter: or, Principles of the Natural Order of the Universe. With a System of Morality Based Thereon,* Translated from the Fifteenth German Edition, Reprinted from the Fourth English Edition (New York: Peter Eckler Publishing Co., 1920), vii–viii.

33. Ibid., vi.

34. Ibid., 51.
35. Ibid., 7.
36. Ibid., 8.
37. Ibid., 46–47. There is no good historical evidence to believe the truth of this claim. Gnostic religions were much more hostile to matter in the ancient world than was Christianity.
38. Ibid., 52
39. Ibid.
40. Ibid., 57.
41. Ibid., 76.
42. Ibid., 77–78.
43. Ibid., 77.
44. Ibid., 85–86.
45. Ibid., 101.
46. Ibid., 173–74.
47. Ibid., 175–76.
48. Ibid., 205.
49. Ibid., see especially 214, 222–23.
50. Ibid., see 219–20.
51. See Gregory, *Scientific Materialism,* 120.
52. Büchner, *Force and Matter,* 238.
53. Ibid., 253.
54. Ibid., 244.
55. Cited in Ernest Jones, *The Life and Work of Sigmund Freud* (Garden City, N.J.: Doubleday and Co., 1961), 30.
56. See William Coleman, *Biology in the Nineteenth Century: Problems of Form, Function, and Transformation* (New York: Wiley, 1971), 150–54.
57. Hermann von Helmholtz, "Address Delivered August 2, 1877, on the Anniversary of the Foundation of the Institute for Education of Army Surgeons," in *Science and Culture: Popular and Philosophical Essays,* edited and with an introduction by David Cahan (Chicago: University of Chicago Press, 1995), 321.
58. Hermann von Helmholtz, "Über das Verhaltnis der Naturwissenschaften zur Gesammtheit der Wissenschaften," in *Vortrage und Reden,* 3 vols. (Braunscheig: F. Vieweg and Sohn, 1896–1903), 1:180–81, cited in Timothy Lenoir, *Instituting Science: The Cultural Production of Scientific Disciplines* (Stanford: Stanford University Press, 1997), 89.
59. Selection reprinted in Roland Stromberg, ed., *Realism, Naturalism, and Symbolism: Modes of Thought and Expression in Europe, 1848–1914* (New York: Harper and Row, 1968), 29–30.
60. Lenoir, *Instituting Science,* 77, 79.
61. Hermann von Helmholtz, "Autobiographical Sketch," in Cahan, ed., *Science and Culture,* 387.
62. Ibid., 391.
63. Virtually the entire story of Du Bois-Reymond told below comes from Lenoir, *Instituting Science,* chap. 4, "Social Interests and the Organic Physics of 1847," 75–95.
64. Cited in Lenoir, *Instituting Science,* 81.
65. Cited in ibid., 90.

66. Cited in ibid., 92.

67. On the role of dialectical materialism in Soviet science, see Loren Graham, *Science and Philosophy in the Soviet Union* (New York: Alfred Knopf, 1972).

68. Friedrich Engels, *Anti-Duhring* (Moscow: Progress Press, 1959), 16–17.

69. Cited in Sir Isaiah Berlin, *Karl Marx,* 4th ed. (New York: Oxford University Press, 1996), 87.

70. See Berlin, *Karl Marx,* 80.

71. See V. I. Lenin, *Karl Marx: A Brief Biographical Sketch with an Exposition of Marxism,* in *Collected Works,* 4th ed., vol. 21 (Moscow: Foreign Languages Publishing House, 1964), 50.

72. Both Karl Marx's father and his first philosophy teacher in Berlin, Edward Jans, are identified as Saint-Simonian followers by Ionescu in *Political Thought of Saint-Simon,* 25.

73. See Frank Manuel, *A Requiem for Karl Marx* (Cambridge, Mass.: Harvard University Press, 1995), 10–11.

74. See ibid., 12–13.

75. Karl Marx and Friedrich Engels, *Marx-Engels Collected Works in English* (London: Lawrence and Wishart, 1970), 5:247.

76. Karl Marx, *Early Writings* (London: Penguin Books, 1974), appendix, 425–26.

77. Karl Marx, *The German Ideology, Collected Works* 5:77–78.

78. See Olson, *Science Deified and Science Defied* 2, chaps. 5 and 6, 191–283, passim.

79. Cited in Ionescu, *Political Thought of Saint-Simon,* 24.

80. Berlin, *Karl Marx,* 15.

81. Marx and Engels, *Collected Works* 1:88.

82. Karl Marx, *The Holy Family, Collected Works* 4:128.

83. Reprinted in Marx, Engels, Lenin, *On Dialectical Materialism* (Moscow: Progress Publishers, 1977), 59.

84. Ibid.

85. From Engels, *Dialectics of Nature,* reprinted in Marx, Engels, Lenin, *On Dialectical Materialism,* 117.

86. From *Dialectics of Nature,* reprinted in Marx, Engels, Lenin, *On Dialectical Materialism,* 122.

87. From *Anti-Duhring,* reprinted in Marx, Engels, Lenin, *On Dialectical Materialism,* 82–84.

88. Cited in *Anti-Duhring,* reprinted in Marx, Engels, Lenin, *On Dialectical Materialism,* 86.

89. Engels, *Anti-Duhring,* in Marx, Engels, Lenin, *On Dialectical Materialism,* 86.

90. Published as an appendix to Friedrich Engels, *Ludwig Feuerbach and the Outcome of Classical German Philosophy* (New York: International Publishers, 1941; from 1888 German original), 84.

91. In Engels, *Ludwig Feuerbach,* 83.

92. In Engels, *Dialectics of Nature,* reprinted in Marx, Engels, Lenin, *On Dialectical Materialism,* 108.

93. Cited in Michael E. DeGolyer, "Science and Society, Justice and Equality: An Historical Approach to Marx" (Ph.D. diss., Claremont Graduate School, 1985), 128.

94. See Olson, *Science Deified and Science Defied* 2:201–2.

95. See ibid., 216–17.

96. Karl Marx, *Capital,* vol. 1 (London: Penguin Books, 1976), 181–82.

97. From Karl Marx and Friedrich Engels, *The German Ideology,* selections reprinted in Eugene Kamenka, *The Portable Karl Marx* (New York: Penguin Books, 1983), 177–78.

98. Marx, *Capital* 1:799.

99. Ibid., 477–79.

100. Karl Marx, *The Communist Manifesto,* in Kamenka, *Portable Karl Marx,* 203–4.

101. Ibid., 225.

102. Ibid., 235.

103. See especially part 5 of Karl Mannheim, *Ideology and Utopia* (New York: Harcourt Brace and World, 1936).

104. See Helena Sheehan, *Marxism and the Philosophy of Science: A Critical History* (Atlantic Highlands, N.J.: Humanities Press, 1985), for a survey of the complex tradition of Marxist philosophy of science, beginning with the dialectical materialism of Engels.

Chapter 6: Early Victorian Culture: Public Science and Political Science

1. For the contemporary critiques of the Royal Society and the paucity of government support for science, See Charles Babbage, *Reflections on the Decline of Science in England* (London: B. Fellowes, 1830), esp. 50–200, and David Brewster, Gerard Moll et al., *Debates on the Decline of Science* (New York: Arno Press, 1975).

2. On the founding of the British Association for the Advancement of Science, see Susan Faye Cannon, *Science in Culture: The Early Victorian Period* (New York: Science History Publications, 1978), 167–224.

3. In addition to a plethora of specialized studies, I would recommend David Knight, *The Age of Science* (Oxford: Basil Blackwell, 1986), Susan Faye Cannon, *Science in Culture: The Early Victorian Period,* and above all, the venerable but amazingly valuable first two volumes of John Theodore Merz, *A History of European Thought in the Nineteenth Century* (London: Blackwood, 1904), esp. vol. 1:226–301.

4. See Olson, *Science Deified and Science Defied* 2, chap. 8, 316–44.

5. See, for example, Robert Fox, "Science, Industry, and the Social Order in Mulhouse, 1798–1871," *British Journal for the History of Science* 17 (1984): 127–68; Arnold Thackray, "Natural Knowledge in Cultural Context: The Manchester Model," *American Historical Review* 79 (1974): 672–709; and Jack Morrell, "*Wissenschaft* in Worstedopolis: Public Science in Bradford, 1800–1850," *British Journal for the History of Science* 18 (1985): 1–23. Morrell describes a case in which a Literary and Philosophical Society oriented toward manufacturing interests folded for lack of support after a very brief existence in 1808–10.

6. For a list of such societies founded between 1812 and 1842, see Thomas Kelly, *A History of Adult Education in Great Britain* (Liverpool: Liverpool University Press, 1962), 113.

7. See Richard Olson, *Science and Religion, 1450–1900: From Copernicus to Darwin* (Westport, Conn.: Greenwood Press, 2004), 83–110.

8. See Olson, *Science Deified and Science Defied* 2, chap. 3.

9. See Robert Kargon, *Science in Victorian Manchester: Enterprise and Expertise* (Baltimore: Johns Hopkins University Press, 1977), 5–14.

10. Installments of the Library of Useful Knowledge came out biweekly in thirty-two-page pamphlets with double columns in small print for six pence each. Brougham's

opening *Objects, Advantages, and Pleasures of Science,* sold forty-two thousand copies and Mark Roget's two essays on *Electricity* sold over twenty-five thousand each, though most offerings had substantially smaller sales. See Jonathan Topham, "Science and Popular Education in the 1830s: The Role of the *Bridgewater Treatises,*" *British Journal for the History of Science* 25 (1992): 413–14.

11. For a characterization of this literature, see Susan Sheets-Pyenson, "Popular Science Periodicals in Paris and London: The Emergence of a Low Scientific Culture, 1820–1875," *Annals of Science* 42 (1985): 549–72.

12. See especially T. W. Heyck, *The Transformation of Intellectual Life in Victorian England* (Chicago: Lyceum Books, 1982), especially 56–57, 64–76.

13. There are still relatively few studies of middle-class and artisan science that view it as anything other than watered-down elite science. Two important exceptions are Susan Sheets-Pyenson, "Popular Science Periodicals," and Anne Secord, "Science in the Pub: Artisan Botanists in Early Nineteenth Century Lancashire," *History of Science* 32 (1994): 269–315.

14. Henry Brougham to James Reddie, December 17, 1796; National Library of Scotland, Ms. 3704.

15. See Steven Shapin and Barry Barnes, "Nature, Science, and Control: Interpreting Mechanics Institutes," *Social Studies of Science* 7 (1977): 7, for this numerical estimate.

16. See Thomas H. Cook, "Science, Philosophy, and Culture in the Early Edinburgh Review, 1802–1829," unpublished Ph.D. dissertation, University of Pittsburgh, 1976, 94 (pagination approximate, since there is no original pagination).

17. See Jack Morrell, "Individualism and the Structure of British Science in 1830," *Historical Studies in the Physical Sciences* 3 (1971): 199.

18. On the centrality of scientific issues to the early *Edinburgh Review,* see Cook, "Science, Philosophy, and Culture in the Early *Edinburgh Review.*"

19. On the role model that Stewart offered to the founders of the *Edinburgh Review,* see Cook, "Science, Philosophy and Culture in the Early Edinburgh Review," 6ff. On their special respect for Black, see 89.

20. J. Robison to G. Black (August, 1, 1800), cited in A. E. Musson and Eric Robinson, *Science and Technology in the Industrial Revolution* (London: Curtis Brown Ltd., 1969), 6.

21. Thomas Cochrane's *Notes from Doctor Black's Lectures on Chemistry, 1767/8,* D. McKie, ed. (1966), 2–4, cited in Musson and Robinson, *Science and Technology in the Industrial Revolution,* 5–6. Emphasis mine.

22. Adam Smith, *An Inquiry into the Nature and Causes of the Wealth of Nations* (Oxford: Clarendon Press, 1976; from 1776 original), 782.

23. Ibid., 783–84.

24. Ibid., 782.

25. On this issue, see Olson, *Science Deified and Science Defied* 2:216–18.

26. Dugald Stewart, *The Collected Works of Dugald Stewart,* Sir William Hamilton, ed. (Edinburgh: Thomas Constable and Co., 1859), vol. 2:59.

27. Ibid., 61.

28. Cited in James Hole, *An Essay on the History and the Management of Literary, Scientific, and Mechanics' Institutions* (London: Longman, Brown, Green, and Longmans, 1983), 15.

29. Cited in ibid., 15–16.

30. See Shapin and Barnes, "Science, Nature, and Control," 49.

31. On the sensitivity of the founders to limits of time for workers, see the Third Report of the Directors of the Edinburgh School of Arts, cited in Michael Stephens and Gordon Roderick, "British Artisan and Technical Education in the Early Nineteenth Century," *Annals of Science* 29 (1972): 89.

32. At some institutions the number of women members became very substantial. At Leeds, for example, in the early 1850s, 392 out of 2,166 members (18 percent) were women. See Hole, *Essay on the History and Management,* 18.

33. One of the first to condemn the movement as a failure, in part because of the drift away from technical content in lectures and courses, was James Hole. See his *Essay on the History and Management,* especially 24–35.

34. Shapin and Barnes, "Science, Nature, and Control," thoroughly documents the interest of many Mechanics' Institute patrons in issues of "social control," but their claim that this was the overriding concern of virtually all patrons is unconvincing to me. Indeed, the "memorials" from various groups addressed to the Royal Commissioners for the Exhibition of the Industries of all Nations in 1851 (collected as appendices to Hole, *An Essay on the Management*) suggest that many industrialist patrons were most concerned with the development of a more technically capable workforce.

35. Cited in Kelly, *Adult Education in Great Britain,* 121.

36. See ibid.

37. See Stephens and Roderick, "British Artisan and Technical Education," 92–93.

38. Michael Stephens and Gordon Roderick, "Science, the Working Class, and Mechanics' Institutes," *Annals of Science* 29 (1972): 352.

39. See Morrell, "*Wissenschaft* in Worstedopolis," 9–11. The citation is from 11.

40. See Margaret Jacob, *Living the Enlightenment: Freemasonry and Politics in Eighteenth-Century Europe* (New York: Oxford University Press, 1991), passim.

41. See Hole, *Essay on the History and Management,* 30.

42. These figures are derived from Hole, *Essay on the History and Management,* 30–31.

43. See, for example the figures for the Manchester Mechanics' Institute classes reported in Kelly, *Adult Education in Britain,* 128, and those for the Sheffield Institute reported in Ian Inkster, "Science and the Mechanics' Institutes, 1820–1850: The Case of Sheffield," *Annals of Science* 32 (1975): 458.

44. See Stephens and Roderick, "Science, the Working Classes, and Mechanics Institutes," 355–57.

45. Thomas Chalmers, "On Mechanic Schools," *The Christian and Civic Economy of Large Towns, III* (Glasgow, 1826), 386–92, cited in Shapin and Barnes, "Science, Nature, and Control," 55.

46. See, for example, Secord, "Science in the Pub," 269–315.

47. See J. N. Hays, "Science and Brougham's Society," *Annals of Science* 20 (1964): 227–41.

48. Shapin and Barnes, "Science, Nature, and Control," 50.

49. Ibid., 48–65.

50. John Playfair, *Outlines of Natural Philosophy, Being Heads of Lectures Delivered in the University of Edinburgh* (Edinburgh: Archibald Constable and Company, 1814), 1–6. The citation is from 2.

51. See on this subject, part 1 of Richard Olson, *Scottish Philosophy and British Physics, 1750–1870: A Study in the Foundation of the Victorian Scientific Style* (Princeton: Princeton University Press, 1975), 11–156.

52. See Inkster, "Science and the Mechanics' Institutes," 453–65.

53. Cited in Sheets-Pyenson, "Popular Science Periodicals," 550.

54. See ibid., especially 552–54.

55. *Working Man's Friend and Family Instructor 1* (1850): 35, cited in Sheets-Pyenson, "Popular Scientific Periodicals," 554.

56. Henry Brougham, review of "*Illustrations of the Huttonian Theory of the Earth* by John Playfair," *Edinburgh Review* 1 (1802–3): 201.

57. David Brewster, review of "*Cours de Philosophie Positive* par Auguste Comte," *Edinburgh Review* 67 (1838): 305.

58. Shapin and Barnes, "Science, Nature, and Control," 50.

59. Frank Whiston Fetter, "Authorship of Economic Articles in the *Edinburgh Review*," *Journal of Political Economy* 41 (1953): 232.

60. See Jack Morrell, "The University of Edinburgh in the Late Eighteenth Century: Its Scientific Eminence and Academic Structure," *Isis* 62 (1971): 158–71.

61. Harvey W. Becher, "Voluntary Science in Nineteenth-Century Cambridge University to the 1850s," *British Journal of the History of Science* 19 (1986): 60.

62. Ibid., 80–81.

63. See Heyck, *Transformation of Intellectual Life in Victorian England*, especially 64–76.

64. This very crude estimate is based on the numbers reported in Heyck, *Transformation of Intellectual Life*, 71 and 56–57.

65. See Jack Morrell, "The Leslie Affair: Careers, Kirk and Politics in Edinburgh in 1805," *Scottish Historical Review* 54 (1975): 63–82.

66. On the fusion of scientific and religious concerns at the origins of Common Sense Philosophy, see Richard Olson, *Scottish Philosophy and British Physics*, 26–54.

67. See ibid., 94–153.

68. James Clerk Maxwell, appendix to R. V. Jones, "James Clerk Maxwell at Aberdeen, 1856–1860," *Notes and Records of the Royal Society of London* 28 (1973): 57–81.

69. See D. W. Gundry, "The Bridgewater Treatises and Their Authors," *History* 31 (1946): 140–52.

70. See Sir Alexander Grant, *Story of the University of Edinburgh*, vol. 2:285.

71. See Robert M. Young, *Darwin's Metaphor: Nature's Place in Victorian Culture* (Cambridge: Cambridge University Press, 1985), 127–28.

72. William Whewell, *Astronomy and General Physics Considered with Reference to Natural Theology* (London: William Pickering, 1833), vi.

73. Jonathan Topham, "Science and Popular Education in the 1830s: The Role of the *Bridgewater Treatises*," *British Journal for the History of Science* 25 (1992): 400.

74. See ibid., 397, n. 2, for the total sales figure.

75. See ibid., especially 419–30.

76. See, "The Religious Census, 1851," 91–92, in Edward Royale, *Radical Politics, 1790–1900: Religion and Unbelief* (London: Longmans, 1971).

77. Robert Cooper, *The Immortality of the Soul* (London: 1853), 100, cited in Edward Royle, *Radical Politics*, 93.

78. Howard R. Murphy, "The Ethical Revolt against Christian Orthodoxy in Early Victorian England," *American Historical Review* (July 1955): 800.

79. See Howard Murphy, "Ethical Revolt against Christian Orthodoxy," 817.

80. This pattern is established throughout Howard Murphy's "Ethical Revolt against Christian Orthodoxy," 800–817.

81. See Olson, *Science Deified and Science Defied* 1:146–60.

82. Cited in T. R. Wright, *The Religion of Humanity: The Impact of Comtean Positivism on Victorian Britain* (Cambridge: Cambridge University Press, 1986), 26.

83. See ibid., 5.

84. John Stuart Mill, *The Collected Works of John Stuart Mill,* Francis E. Mineka, ed., 20 vols. (Toronto, 1963, 1991), 13:738–39, cited in Wright, *Religion of Humanity,* 44.

85. John Stuart Mill, *The Letters of John Stuart Mill,* A. S. R. Eliot, ed., 2 vols. (Toronto, 1910), 2:361–63, cited in Wright, *Religion of Humanity,* 44–45.

86. See Wright, *Religion of Humanity,* 43, 45.

87. Cited in ibid., 73.

88. See ibid., 77.

89. See Christopher Kent, *Brains and Numbers: Elitism, Comtism, and Democracy in Mid-Victorian England* (Toronto: University of Toronto Press, 1978), esp. 69–83, and Wright, *Religion of Humanity,* 88–124.

90. On associationist psychology, see Olson, *Science Deified and Science Defied* 2, chap. 6, 236–83.

91. Cited in Margaret Cole, *Robert Owen of New Lanark* (London: Batchworth Press, 1953), 9.

92. See Cole, *Robert Owen,* 34.

93. Cited in ibid., 89–90.

94. Robert Owen, *A New View of Society and Other Writings,* G. D. H. Cole, ed. (London: Dent, 1927), 16.

95. Robert Owen, "Determinism," from *The Book of the New Moral World* (London, 1836), 94–95, cited in Edward Royle, *Radical Politics,* 111.

96. Owen, *New View of Society,* 17–18.

97. On Bentham's emphasis on punishment or sanctions, see Jeremy Bentham, *The Principles of Morals and Legislation* (1789).

98. Owen, "Observations on the Effect of the Manufacturing System," in *A New View of Society,* 124–25.

99. See Cole, *Robert Owen,* 94–103.

100. In Owen, *New View of Society,* 156–58.

101. Robert Owen, "Report to the Committee for the Relief of the Manufacturing Poor," in *New View of Society,* 158.

102. The most comprehensive version of Owen's plan, including all of those features mentioned here, is to be found in Robert Owen, "Report to the County of Lanark" (May 1, 1820) in *New View of Society,* 245–98.

103. Robert Owen, "Report to the Committee for the Relief of the Manufacturing Poor," 164–67.

104. See, for example, I. L. Goldsmith, letter to Robert Owen, December 22, 1823, cited in Ronald G. Garnett, *Co-operation and the Owenite Socialist Communities in Britain, 1825–45* (Manchester: Manchester University Press, 1972), 45–46.

105. See Garnett, *Co-operation and the Owenite Socialist Communities,* especially chaps. 3 (on Orbiston), 4 (on Ralahine), and 6 (on Queenwood).

106. See Cole, *Robert Owen,* 176.

107. See, for example, Johnston Birchall, *The International Co-operative Movement* (Manchester: Manchester University Press, 1997), 4–8.

108. Cited in Cole, *Robert Owen,* 198.

109. *The New Moral World* 5 (November 29, 1834): 38, cited in Garnett, *Co-operation and the Owenite Socialist Communities,* 143.

110. See Garnett, *Co-operation and the Owenite Socialist Communities,* 147–53.

111. See ibid., 150.

112. See Kent, *Brains and Numbers,* 6.

113. See Stefan Collini, Donald Winch, and John Burrow, *That Noble Science of Politics* (Cambridge: Cambridge University Press, 1983), especially 33, for Stewart's view on the reason for the necessity of political science.

114. On Martineau as a popularizer of political economy, see Claudia Klaver, "Moralizing the Economy: Contestations of Economic Authority in Mid-Victorian Literature and Culture," Ph.D. dissertation, Johns Hopkins University, 1994, 67–91.

115. Cited in Collini, Winch, and Burrow, *Noble Science of Politics,* 36.

116. On McCulloch's role in promoting interest and confidence in political economy, see Klaver, "Moralizing the Economy," 91–106.

117. Thomas Robert Malthus, *On Population,* Gertrude Himmelfarb, ed. (New York: Modern Library, 1960; from the first, 1798 ed.), 11, 13.

118. Ibid., 14–15.

119. Ibid., 17.

120. Frank Whitson Fetter, *The Economist in Parliament: 1780–1868* (Durham: Duke University Press, 1980), 148.

121. Raymond G. Cowherd, *Political Economists and the English Poor Laws* (Athens: Ohio University Pres, 1977), 9–11.

122. Fetter, *Economist in Parliament,* 148.

123. Malthus, *On Population,* 30–31.

124. Ibid., 33.

125. See Fetter, *Economist in Parliament,* 152–54.

126. See ibid., 156–59.

127. Cited in Fetter, *Economist in Parliament,* 158–59.

128. Cited in ibid., 40.

129. Cited in ibid., 40–41.

130. Cited in ibid., 42.

131. Fetter, *Economist in Parliament,* 269–70.

132. Ibid., 64–67.

133. Brundage, *England's Prussian Minister,* 23–29.

134. Cited in Fetter, *Economist in Parliament,* 73.

135. Ibid., 71.

136. Cited in ibid., 75.

Chapter 7: The Rise of Evolutionary Perspectives: 1820–59

1. John Stuart Mill, *The Logic of the Moral Sciences,* intro. A. J. Ayer (La Salle, Ill.: Open Court, 1988), 90. Originally published as "Book VI. On the Logic of the Moral Sciences," in Mill's *A System of Logic, Ratiocinative and Inductive,* 8th ed. (1872).

2. Charles Booth, quoted in Beatrice Webb, *My Apprenticeship* (1936), 192, cited in J. W. Burrow, *Evolution and Society: A Study in Victorian Social Theory* (Cambridge: Cambridge University Press, 1966), 88.

3. This claim is made often by many mid-Victorians, but one of the most powerful

expressions is in Herbert Spencer's *Social Statics: The Conditions Essential to Human Happiness Specified, and the First of Them Developed* (New York: Robert Schalkenbach Foundation, 1995 [reprint of London, 1877 ed.]), 32–33.

4. John Stuart Mill, *Autobiography* (Oxford: Oxford University Press, 1954 [1873]), 151, cited in Burrow, *Evolution and Society,* 65.

5. Matthew Arnold, *Culture and Anarchy,* 2nd ed. (New York: Cambridge University Press, 1993), 149.

6. Comte, *Positive Philosophy,* viii.

7. G. H. Lewes, "The State of Historical Sciences in France," 73, cited in Burrow, *Evolution and Society,* 94.

8. Spencer, *Social Statics,* 288–89.

9. Mill, *Logic of the Moral Sciences,* 83.

10. Ibid., 65.

11. Ibid., 101.

12. Ibid., 101–2.

13. Ibid., 103.

14. John Stuart Mill, *On Liberty* (Oxford: Basil Blackwell, 1946), 9, cited in J. W. Burrow, *Evolution and Society,* 158.

15. Burrow, *Evolution and Society,* 128.

16. See Peter Bowler, *Evolution: The History of an Idea,* rev. ed. (Berkeley: University of California Press, 1989), 73.

17. See ibid., 74.

18. Note that the possibility of the frequent spontaneous generation of life was not decisively challenged until Pasteur's experiments in the mid nineteenth century.

19. Cited in John Greene, *The Death of Adam: Evolution and Its Impact on Western Thought* (New York: New American Library, 1959), 63–64.

20. Cited in Charles Coulston Gillespie, *Genesis and Geology* (New York: Harper and Row, 1959), 99–100.

21. Cited in Phillip Appleman, ed., *Darwin,* 3rd ed. (New York: W. W. Norton and Company, 2001), 50.

22. On the appropriation of Lamarckian transformism by Geoffroy Saint-Hilaire and Serres, see Adrian Desmond, *The Politics of Evolution: Morphology, Medicine, and Reform in Radical London* (Chicago: University of Chicago Press, 1989), 52–54.

23. See Adrian Desmond, "Artisan Resistance and Evolution in Britain, 1819–1848," *Osiris* 3 (1987): 77–110.

24. Cited in Desmond, *Politics of Evolution,* 264.

25. Cited in ibid.

26. See Robert Chambers, *Vestiges of the Natural History of Creation and Other Evolutionary Writings,* James A. Secord, ed. (Chicago: University of Chicago Press, 1994 [London: John Churchill, 1844]), 205–11.

27. Ibid., 212–13.

28. Ibid., 346.

29. Ibid., 307.

30. Ibid., 215–317.

31. Ibid., xxvi–xxvii.

32. Ibid., 376.

33. On the reviews of *Vestiges,* see Secord's introduction, xxviii–xxxiii.

34. Adam Sedgwick, *Edinburgh Review* (1845), cited in Bowler, *Evolution,* 147.

35. D. Duncan, *Life and Letters of Herbert Spencer* (1908), 578, cited in Burrow, *Evolution and Society,* 186.

36. Ibid., 187.

37. Spencer, *Social Statics,* 31–32.

38. Ibid., 405, 408.

39. Ibid., 361–62.

40. Ibid., 289–90.

41. Robert C. Bannister explores the changing emphasis on sympathy and benevolence in Spencer in *Social Darwinism: Science and Myth in Anglo-American Social Thought* (Philadelphia: Temple University Press, 1979), 49.

42. Herbert Spencer, *First Principles* (New York: A. L. Burt, 1980 [1862]), 343.

43. Ibid., 247.

Chapter 8: Darwinian Evolution: Natural Selection, Group Selection, and Sexual Selection

1. See Bowler, *Evolution: The History of an Idea,* 163–64.

2. Cited by Charles Darwin in "Extract from an Unpublished Work on Species by C. Darwin, Esq., Consisting of a Portion of a Chapter Entitled, 'On the Variation of Organic Beings in a State of Nature; of the Natural Means of Selection; On the Comparison of Domestic Races and True Species,'" *Journal of the Linnaean Society, Zoology* (1858): 47.

3. Charles Darwin, *The Origin of Species by Means of Natural Selection or the Preservation of Favored Races in the Struggle for Life, and The Descent of Man and Selection in Relation to Sex* (New York: Modern Library, n.d., from the 6th ed. of *The Origin of Species* and the 1st ed. of *The Descent of Man),* 421–22, 436–37, 447–48 (hereafter, *Origin of Species* and *Descent of Man,* respectively).

4. George Bernard Shaw, *Back to Methuselah* (New York: Brentano's, 1921), xlv–xvi.

5. See Alvar Ellegård, *Darwin and the General Reader: The Reception of Darwin's Theory of Evolution in the British Periodical Press, 1859–1872* (Chicago: University of Chicago Press, 1990; republication of 1958 original).

6. On the philosophical debates regarding scientific method, especially as they relate to Darwin's work, see especially Michael Ruse, *The Darwinian Revolution: Science Red in Tooth and Claw* (Chicago: University of Chicago Press, 1979).

7. Darwin to Henry Fawcett, September 18, [1861], in Francis Darwin, ed., *More Letters of Charles Darwin: A Record of His Work in a Series of Hitherto Unpublished Letters* (New York: D. Appleton and Company, 1903), 195.

8. Darwin, *Origin of Species,* 373.

9. See Ellegård, *Darwin and the General Reader,* passim.

10. Darwin, *Descent of Man,* 442.

11. Darwin, *Origin of Species,* 15–35.

12. Ibid., 46–51.

13. Ibid., 41.

14. Ibid., 210.

15. Ibid., 47.

16. Ibid., 104–6.

17. Ibid., 153.
18. Ibid., 346–52.
19. Ibid., 128.
20. Ibid., 239–40.
21. Ibid., 236–38.
22. Ibid., 182–83.
23. Ibid., 133–35.
24. Ibid., 152.
25. Ibid., 160.
26. Darwin, *Descent of Man,* 441.
27. W. K. Clifford, "On the Scientific Basis of Morals," reprinted in *Lectures and Essays,* ed. Leslie Stephen and Frederick Pollock, 2nd ed. (London: Macmillan, 1879), 291.
28. Darwin, *Descent of Man,* 443.
29. Ibid., 498.
30. Ibid.
31. Ibid., 444.
32. Ibid., 472.
33. Ibid., 479.
34. Ibid., 486.
35. Ibid., 493.
36. Ibid., 501–2.
37. Ibid., 521.
38. Ibid., 493.
39. Henry Maudsley, *Body and Mind* (New York: D. Appleton, 1874), 48–49.
40. Henry Maudsley, *Pathology of Mind* (New York: D. Appleton, 1882; from London 1879 original), 107.
41. Henry Maudsley, *The Physiology and Pathology of Mind* (London: Macmillan, 1867), 202.
42. Maudsley, *Pathology of Mind,* 130.
43. Ibid., 174.
44. Ibid., 99.
45. Ibid., 174.
46. Ibid., 134–35.
47. Ibid., 115.
48. Maudsley, *Body and Mind,* 43.
49. Maudsley, *Pathology of Mind,* 88.
50. For an excellent discussion of the politics of psychiatric Darwinism see Elaine Showalter, *The Female Malady: Women, Madness, and English Culture: 1830–1980* (New York: Random House, 1985).
51. Darwin, *Descent of Man,* 556.
52. Ibid., 573.
53. Ibid., 872–73.
54. Ibid., 581–82.
55. Ibid., 873.
56. See Andrew Sinclair, *The Emancipation of the American Woman* (New York: Harper and Row, 1965).

57. Cited in Cynthia Eagle Russett, *Sexual Science: The Victorian Construction of Womanhood* (Cambridge, Mass.: Harvard University Press, 1989), 1.
58. Cited in ibid., 122.
59. Darwin, *Descent of Man,* 581.
60. On the extension of Darwin's energy argument in the writings of Spencer and Geddes, see Jill Conway, "Stereotypes of Femininity in a Theory of Sexual Evolution," *Victorian Studies* 14 (1970): 47–62.
61. Cited in Eagle Russett, *Sexual Science,* 122.
62. Ibid., 123.
63. Cited from P. Geddes and J. A. Thompson, *The Evolution of Sex* (1889) in Conway, "Stereotypes of Femininity," 53.
64. Darwin, *Descent of Man,* 874.
65. Ibid., 871.
66. Ibid., 901.
67. Ibid.
68. Ibid., 904–5.
69. Cited in Peter Raby, *Bright Paradise: Victorian Scientific Travellers* (Princeton, N.J.: Princeton University Press, 1997), 161.
70. Cited in ibid., 172.

Chapter 9: "Social Darwinisms" and Other Evolutionary Social Theories

1. See Robert C. Bannister, *Social Darwinism: Science and Myth in Anglo-American Social Thought* (Philadelphia: Temple University Press, 1979), for an extended argument that social Darwinism was largely a myth created by reformers to oppose extreme individualism. Initial uses of the term are identified on 4, 34–35.
2. Walter W. Rostow, *The Stages of Economic Growth: A Non-Communist Manifesto* (Cambridge: Cambridge University Press, 1961); see chart facing p. 1.
3. Taken from *Discours et Opinions de Jules Ferry,* ed. Paul Robiquet (Paris: Armand Colin and Cie, 1897), 199–201. Translated by Ruth Kleinman.
4. Charles Darwin, *Descent of Man,* 502.
5. See Paul Crook, *Darwinism, War, and History* (Cambridge: Cambridge University Press, 1994), passim.
6. Walter Bagehot, *Physics and Politics, or: Thoughts on the Application of the Principles of "Natural Selection" and "Inheritance" to Political Society,* Roger Kimball, ed. (Chicago: Ivan R. Dee, 1999), 182.
7. Ibid., 183–84.
8. See Michael Adas, *Machines as the Measure of Men: Science, Technology, and Ideologies of Western Dominance* (Ithaca: Cornell University Press: 1990), passim.
9. Bagehot, *Physics and Politics,* 23–24.
10. Ibid., 52.
11. Ibid., 71.
12. Ibid., 71, 191.
13. Ibid., 190–91.
14. Ibid., 70, 112.

15. Ibid., 57.
16. Ibid., 63–65.
17. Ibid., 143–45, 180.
18. Ibid., 73.
19. Ibid., 58, 57.
20. Ibid., 46.
21. Ibid., 48.
22. On Wallace's attitude toward the special situation of man, see Crook, *Darwinism, War, and History,* 54–58.
23. Cited in Crook, *Darwinism, War, and History,* 59.
24. Alfred Russel Wallace, *Contributions to the Theory of Natural Selection* (London: 1870), Essay 10, "The Limits of Natural Selection as Applied to Man." On Wallace's religion and spiritualism, see Michael Shermer, *In Darwin's Shadow: The Life and Science of Alfred Russel Wallace: A Biographical Study on the Psychology of History* (New York: Oxford University Press, 2002); and Ross A. Slotten, *The Heretic in Darwin's Court: The Life of Alfred Russel Wallace* (New York: Columbia University Press, 2004).
25. Wallace, "Human Selection," *Fortnightly Review* 48 (1890): 331ff.
26. Alfred Russel Wallace, *The Wonderful Century* (1898), 340, cited in Paul Crook, *Darwinism, War, and History,* 59.
27. See Crook, *Darwinism, War, and History,* 59–60.
28. Excerpt from T. H. Huxley, *Evolution and Ethics,* in Phillip Appleman, ed., *Darwin* (New York: W. W. Norton, 1970), 403–4.
29. In Appleman, ed., *Darwin,* 404.
30. Excerpted from *Mutual Aid,* in Appleman, ed., *Darwin,* 519.
31. In Appleman, ed., *Darwin,* 521.
32. See Crook, *Darwinism, War, and History,* 109.
33. See ibid., 110–11.
34. See ibid., 112–14.
35. See D. P. Crook, *Benjamin Kidd: Portrait of a Social Darwinist* (Cambridge: Cambridge University Press, 1984), 52.
36. Crook, *Benjamin Kidd,* 39.
37. Ibid.
38. Ibid., 65
39. Ibid., 55, 65–66.
40. Ibid., 55.
41. Cited in ibid., 101.
42. Crook, *Benjamin Kidd,* 118–19.
43. Ibid., 120–21.
44. Ibid., 123–24.
45. Cited in ibid., 124.
46. See John Hope Franklin, "Edward Bellamy and the Nationalist Movement," *New England Quarterly* 11 (1938): 739–72.
47. On Bellamy's popularity among German workers, see Alfred Kelly, *The Descent of Darwin: The Popularization of Darwinism in Germany, 1880–1914* (Chapel Hill: University of North Carolina Press, 1981), 129–37.
48. Edward Bellamy, *Looking Backward, 2000–1887* (New York: New American Library, 1960), "Postscript," 220.

49. Ibid., 45.
50. Ibid., 49.
51. Ibid., 52.
52. Ibid., 53–54.
53. Ibid., 179–80.
54. Ibid., 40.
55. Ibid., 194.
56. See Ernst Haeckel, *The Riddle of the Universe* (Buffalo: Prometheus Books, 1992 [reprint of 1900 New York ed., trans. Joseph McCabe]), 80.
57. Cited in Kelly, *Descent of Darwin*, 25.
58. See ibid.
59. See ibid., 10–35.
60. See ibid., 123–41.
61. I am inclined to take Daniel Gasman's assessment of Haeckel's cultural significance much more seriously than Alfred Kelly and many other German historians, while admitting that the case for direct ties to the Volkist elements of Nazi ideology may be overstated. See Daniel Gasman, *The Scientific Origins of National Socialism: Social Darwinism in Ernst Haeckel and the German Monist League* (London: Macdonald and American Elsevier, Inc., 1971).
62. Cited in Kelly, *Descent of Darwin*, 113.
63. See Haeckel, *Riddle of the Universe*, 81.
64. Ibid., 8.
65. Cited from Haeckel's *History of the Creation*, in Gasman, *Scientific Origins of National Socialism*, 91.
66. Haeckel, *Riddle of the Universe*, 8–9.
67. Ibid., 35.
68. Ibid., 268–69.
69. Ibid., 269–70.
70. Ibid., 224–25. Haeckel even argues that Goethe recognized this continuity of psychic life and exploited it in his novel, *Affinities*.
71. Haeckel, *Riddle of the Universe*, 20.
72. Ibid., 131.
73. Ibid., 89–90.
74. Cited from Haeckel's *History of Creation*, in Gasman, *Scientific Origins of National Socialism*, 36.
75. Haeckel, *Riddle of the Universe*, 350.
76. Ibid., 354.
77. Ibid., 357.
78. Ibid., 103.
79. Ibid., 125.
80. Ibid., 182.
81. See Gasman, *Scientific Origins of National Socialism*, xxv.
82. Cited from Ernst Haeckel, *The Wonders of Life* (1904) in Gasman, *Scientific Origins of National Socialism*, 40.
83. Cited in Gasman, *Scientific Origins of National Socialism*, 39.
84. Cited in ibid., 127.

85. Bert James Lowenberg, "Darwinism Comes to America, 1859–1900," *Mississippi Valley Historical Review* 28 (1941): 363.

Chapter 10: The Problem of Degeneration in the Late Nineteenth Century

1. Max Nordau, *Degeneration* (New York: D. Appleton, 1895), 537.
2. Ibid., 541.
3. Morel, *Degenerescences,* vii–viii, cited in Daniel Pick, *Faces of Degeneration: A European Disorder, c. 1848–1918* (Cambridge: Cambridge University Press, 1989), 54.
4. Cited in Pick, *Faces of Degeneration,* 54.
5. See ibid., 50–51.
6. Cited in ibid., 47.
7. Cited in ibid., 64.
8. Cited in ibid., 66.
9. Hippolyte Taine, *Notes on Paris,* John Austin Stevens, ed. (New York, 1879), 280, cited in Leo Weinstein, *Hippolyte Taine* (New York: Twayne, 1972), 48.
10. Hippolyte Taine, *Lectures on Art* 1 (1896): 104, cited in Weinstein, *Hippolyte Taine,* 81.
11. Hippolyte Taine, *On Intelligence* 2 (1889): 120, cited in Weinstein, *Hippolyte Taine,* 48. Emphasis mine.
12. Hippolyte Taine, *The French Revolution* 2:22–23, cited in Weinstein, *Hippolyte Taine,* 131.
13. Hippolyte Taine, *The French Revolution* 1:52–53, cited in Pick, *Faces of Degeneration,* 71.
14. Émile Zola, *Correspondence* 1, cited in Elliott M. Grant, *Émile Zola* (New York: Twayne, 1966), 23.
15. Émile Zola, *Claude,* cited in Philip Walker, *Zola* (London: Routledge and Kegan Paul, 1985), 61.
16. Walker, *Zola,* 70
17. Ibid., 71.
18. Cited in Grant, *Émile Zola,* 32.
19. See Walker, *Zola,* 81.
20. Émile Zola, *Therese Raquin, Roman* (Paris: American Edition., n.d.), 8–10.
21. Cited in Grant, *Émile Zola,* 45.
22. Cited in Walker, *Zola,* 89.
23. Nordau, *Degeneration,* 500, 497.
24. Émile Zola, *Nanna* (Harmondsworth, Middlesex, England: Penguin Books, 1972), 221.
25. Émile Zola, *La Bête Humaine* (London: Penguin Books, 1977), 66.
26. Émile Zola, *La Debacle,* 322, cited in Daniel Pick, *Faces of Degeneration,* 83.
27. Zola, *La Bête Humaine,* 364, 366.
28. Émile Zola, *Germinal* (London: Penguin Books, 1954), 5–6.
29. Ibid., 319, 320.
30. Ibid., 339–41.
31. Ibid., 352.
32. Nordau, *Degeneration,* 495.

33. Ibid., 495–96.

34. A good introduction to social statistics in the nineteenth century is Eileen Janes Yeo, "Social Surveys in the Eighteenth and Nineteenth Centuries," 83–99, in Theodore Porter and Dorothy Ross, eds., *The Cambridge History of Science, Volume 7: The Modern Social Sciences* (Cambridge: Cambridge University Press, 2003).

35. Nordau, *Degeneration,* 496.

36. William Schneider, "The Eugenics Movement in France, 1890–1940," in Mark B. Adams, ed., *The Wellborn Science: Eugenics in Germany, France, Brazil, and Russia* (New York: Oxford University Press, 1990), 72.

37. Francis Galton, *Inquiries into the Human Faculty* (London: Macmillan, 1883), 24–25, cited in Daniel Kevles, *In the Name of Eugenics: Genetics and the Uses of Human Heredity* (Berkeley: University of California Press, 1985), ix.

38. On Dugdale's work, see Mark H. Haller, *Eugenics: Hereditarian Attitudes in American Thought* (New Brunswick, N.J.: Rutgers University Press, 1963), 20–23.

39. Cited in Kevles, *In the Name of Eugenics,* 318, n. 6.

40. Victoria Woodhull Martin, *The Rapid Multiplication of the Unfit* (London, 1891), 38, in Kevles, *In the Name of Eugenics,* 85.

41. Thomas Crothers, *The Disease of Inebriety from Alcohol, Opium, and Other Narcotic Drugs* (New York: E. B. Treat and Co., 1904), 161, cited in Haller, *Eugenics,* 30–31.

42. See John S. Haller, *Outcasts from Evolution: Scientific Attitudes of Racial Inferiority, 1859–1900* (Urbana: University of Illinois Press, 1971), 19–39.

43. Ibid., 58.

44. Alfred Ploetz, *Die Tuchtigkeit unsrer Rasse und der Schutz der Swachen: Ein Versuch uber Rassenhygiene und ihr Verhaltnis zu den humanen Idealenm, besonders zum Socialismus* (Berlin: Gustav Fischer, 1895), 5, cited in Mark B. Adams, "The Race Hygiene Movement in Germany, 1904–1945," in Adams, ed., *Wellborn Science,* 17.

45. Adams, *Wellborn Science,* 19.

Conclusion

1. On the intellectual and political impact of Positivism in India, see especially Geraldine Forbes, *Positivism in Bengal: A Case Study in the Transmission and Assimilation of an Ideology* (Calcutta: Minerva, 1975), and Gayan Prakash, *Another Reason: Science and the Imagination of Modern India* (Princeton: Princeton University Press, 1999). Ralph Lee Woodward, ed., *Positivism in Latin America, 1850–1900* (Lexington: D. C. Heath and Co., 1971), provides a useful introduction to Positivism and politics in Latin America.

INDEX

RICHARD G. OLSON is a professor of history and chair of the faculty at Harvey Mudd College. He is the author of *The Emergence of the Social Sciences, 1642–1792; Science Deified and Science Defied: The Historical Significance of Science in Western Culture, Vol. 1: From the Bronze Age to 1620* and *Vol. 2: 1620–1820;* and *Scottish Philosophy and British Physics, 1750–1850: Foundations of the Victorian Scientific Style.*

The University of Illinois Press
is a founding member of the
Association of American University Presses.

Composed in 10.5/13 Adobe Minion
with Adobe Woodtype One display
by Celia Shapland
at the University of Illinois Press
Manufactured by Sheridan Books, Inc.

University of Illinois Press
1325 South Oak Street
Champaign, IL 61820-6903
www.press.uillinois.edu